ABHANDLUNGEN
DER AKADEMIE DER WISSENSCHAFTEN DER DDR

Abteilung Mathematik — Naturwissenschaften — Technik

Jahrgang 1983 · Nr. 1 N

Darwin today

The 8th Kühlungsborn Colloquium on Philosophical and Ethical Problems of Biosciences

organized by the
Society of Physical and Mathematical
Biology of the GDR
tagether with the
Research Centre of Molecular Biology and
Medicine and the Central Institute
of Philosophy of the Academy of
Sciences of the GDR
This conference was sponsored by
the UNESCO-Commission of the GDR
and supported by UNESCO.

Kühlungsborn 8th—12th November 1981

Edited by:
E. Geissler and W. Scheler

Akademie-Verlag · Berlin 1983

Herausgegeben im Auftrag des Präsidenten
der Akademie der Wissenschaften der DDR
von Vizepräsident Prof. Dr. Heinrich Scheel

Erschienen im Akademie-Verlag, DDR-1086 Berlin, Leipziger Straße 3—4

Lizenznummer: 202 · 100/486/83
Printed in the German Democratic Republic
Gesamtherstellung: VEB Druckerei „Thomas Müntzer", 5820 Bad Langensalza
Redaktionsschluß: 30. 3. 1982
LSV 1305, 1385
Bestellnummer: 763 097 4 (2001/83/1 N)
DDR 65,— M

Preface

The 100th anniversary of Charles DARWIN's death was taken as the occasion by the Society for Physical and Mathematical Biology of the German Democratic Republic to hold, together with the Research Centre for Molecular Biology and Medicine of the Academy of Sciences of the GDR and the Central Institute for Philosophy of the Academy of Sciences of the GDR, the VIII Kühlungsborn Colloquium on Philosophical and Ethical Problems of the Life Sciences from the 8th to the 12th of November 1981. The colloqium "DARWIN today" took place under the patronage of the UNESCO Commission of the GDR and was supported by UNESCO.

This volume contains the lectures presented at the VIII Kühlungsborn Colloquium and the contributions to the discussion. Here we wish to report in addition on some of the activities that took place but could not be mentioned directly in the minutes of the proceedings.

THE ROYAL SOCIETY OF LONDON
TO
THE ACADEMY OF SCIENCES
OF THE GERMAN DEMOCRATIC REPUBLIC

The Royal Society sends greetings to the Academy of Sciences on the occasion of the VIIIth Kühlungsborn Colloqium, held to mark the 100th anniversary of the death of Charles DARWIN.

I am particularly glad that on this auspicious occasion the Royal Society is represented by a great-grandson of Charles DARWIN whose own scientific work is so much in the distinguished family tradition.

November 1981

Andrew HUXLEY
President

Prof. KEYNES took this address as the occasion to present the representative of the President of the Academy of Sciences of the GDR with the facsimile of a photograph of his great-grandfather and a copy of the volume "The Beagle Record: Selections from the Original Pictorial Records and Written Accounts of the Voyage of H. M. S. Beagle" that he published in 1979 at the Cambridge University Press. The participants of the meeting were also addressed by the Chairman of the UNESCO Commission of the GDR, Bernhardt Neugebauer, Deputy Minister of Foreign Affairs of the GDR.

Prof. Richard Darwin KEYNES, President of the International Union of Pure and Applied Biophysics (IUPAB) attended the colloquium as the official representative of the British Royal Society and conveyed the following greeting of the Royal Society to the participants of the colloquium.

Apart from Prof. KEYNES the colloquium was attended by further representatives of national and international scientific organizations, including Prof. Dr. W. BRAUER, Rector of the Wilhelm-Pieck-University Rostock, Prof. Dr. I. T. FROLOV, President of the Scientific Council for Philosophical and Social Problems of Science and Technology of the Academy of Sciences of the USSR, Academician M. S. GHILAROV, Vice-President of the IUPAB and Chairman of the National Committee of Soviet Biologists, Prof. Dr. H. KLINKMANN, President of the International Society for Artificial Organs and Chairman of the Medical Research Council of the GDR, Prof. Dr. F. KLIX, President of the International Union of Psychological Societies, Prof. Dr. S. M. RAPOPORT, President of the Coordinating Council of the Medical Societies of the GDR, and further leading personalities.

After the opening session Prof. Dr. W. SCHELER, President of the Academy of Sciences of the GDR gave a reception for the participants at the venue of the colloquium. In addition, the participants were invited by W. Marlow, Chairman of Rostock County Council, to attend a reception at the Warnow Hotel in Rostock.

True to the tradition set by earlier Kühlungsborn Colloquia, the programme included various artistic and cultural events that played no small part in ensuring the success of the meeting.

In a slide lecture held late one evening Richard Darwin KEYNES, a great-grandson of Charles DARWIN, showed impressive pictures of the famous voyage of the "Beagle" that DARWIN took part in and during which he made the observations that were so decisive for the development of his theory. Most of the pictures were the work of Conrad MARTENS who accompanied the "Beagle" as the official ship's artist for nine months of her voyage. Some of the etchings and watercolours were not discovered by KEYNES until the summer of 1968, when he found them in the possession of a collector in Buenos Aires. The "Beagle Record" published by KEYNES contains reproductions of over 50 of these pictures, as well as others, together with diary entries and letters. The pictures together with the animated comments given by KEYNES left the participants of the colloquium with a vivid impression of what

Charles DARWIN saw in the ports, sailing along the coast, and on expeditions ashore during his "Beagle" voyage.

Thanks to the obliging cooperation of the National Film Archives of the GDR we were able to attend a special performance of the Soviet-German coproduction "Salamander", a film produced in 1928, at the „Strandbühne" cinema in Kühlungsborn. The theme of the film was the tragic fate of the Austrian zoologist Paul KAMMERER. In the nineteen-twenties KAMMERER, using salamanders and toads, had attempted to prove that acquired characters are heritable — a topic that also played a minor role during one of the scientific sessions (cf. GEISSLER, PASTERNAK, this volume). KAMMERER was finally proved to have falsified results (the background for which has never been satisfactorily clarified). This very dramatic silent film was extraordinarily interesting in many respects, not least because the scenario was written by Anatoli LUNATSCHARSKI then Commissar for Popular Education, who did not consider it beneath his dignity to take one of the roles in the film: that of himself ... Then scenes taken on location in the city centres of Berlin, Leipzig, and Erfurt also left a deep impression.

An exibition of postage stamps carefully and cleverly arranged by H. MEISSNER, Güstrow, in the rooms where the sessions were held also aroused considerable interest among the participants. The theme of the exhibiton was "From DARWIN's voyage with the 'Beagle' to the synthetic theory of evolution".

Prof. Hans-Anselm PERTEN, Director of the Volkstheater Rostock, has shown constant interest in enlivening the cultural programmes of the Kühlungsborn colloquies, and it was thanks to this interest that the participants were able to attend a performance of Rolf HOCHHUTH's play „Ärztinnen" on the stage of the „Großes Haus" in Rostock. The performance was followed by a debate, in which the principal actors, Christine VAN SANTEN and Michael CHRISTIAN together with Prof. Dr. sc. med. Horst KLINKMANN, who in addition to his several other duties also acts as medical adviser to the Volkstheater, and Prof. PERTEN took part, regarding the form, content, and production of the play. The discussion, during which the floor was taken by several colloquium participants, including GRANTHAM, Villeurbanne, FLEISCHHACKER, Berlin-Buch, MEISSNER, Güstrow, FROLOV, Moscow, MENDE, Berlin, PFEIFFER, Dresden, PLATZER, Berlin-Buch, RUSE, Guelph, H. A. ROSENTHAL, Berlin, S. ROSENTHAL, Berlin-Buch, SCHUSTER, Vienna, and VOGEL, Heidelberg, surprised even the travel-seasoned people from the Rostock theatre with its vehemence and broadness of range, and left a persistent impression on all concerned.

The staff of the "Ernst Grube" holiday hotel, which belongs to the hotel chain run by the Confederation of Free German Trade Unions (FDGB) and was generously placed at our disposal by the National Executive of the FDGB for the VIII Kühlungsborn Colloquium, played an outstanding part in ensuring that the whole colloquy ran smoothly. The staff of this hotel, which served simultaneously as venure and accomodation for the participants of the colloquium, broke all records for hospitality. One small detail, but a typical one, may serve as an illustration: Without being requested by the organizers of the colloquium the mangement of the hotel arranged for the dish washers to be switched off during the day and for the washing up to be done at night to avoid disturbing the sessions ...

Naturally, even boundary conditions as pleasant as these are not sufficient on their own to guarantee success. During the preparation of the wellbalanced scientific programme the conference committee composed of Herbert HÖRZ, Günter PASTERNAK, Werner SCHELER and the present writer received active support and imaginative advice from numerous colleagues both at home and abroad. We feel obliged to express our special thanks in this respect to Dr. Ilse JAHN, Berlin, H. BACH, Jena, H. A. FREYE, Halle, M. S. GHILAROV, Moscow, R. LÖTHER, Berlin, E. MAYR, Cambridge, MA, M. A. MIKULINSKI, Moscow, K. MOTHES, Halle, G. PETERS, Berlin, H. J. SCHARF, Halle, H. STUBBE, Gatersleben, and K. WENIG, Berlin. The programme was arranged in accordance with past traditions to leave about 50% of the available time for free discussion.

The colloquium, finally, could not even have taken place but for the painstaking and careful work of the head of the organization committee, Dr. M. KOCH, and his co-workers, Mrs. A. KOLLAT and Mrs. S. FISCHER, and the interpreter, Mrs. R. SCHNEIDER. And for her extraoridinarily valuable assistance both in the colloquium office and in conducting most correspondence involved in the preparation and evaluation of the meeting as well as the typing of the manuscripts I am indebted to my indefatigable secretary, Mrs. E.-M. Myler.

E. GEISSLER

Contents

Zentralinstitut für Molekularbiologie und Zentralinstitut für Philosophie der Akademie der Wissenschaften der DDR, Berlin

DARWIN Today — Introductory Lecture

E. GEISSLER and H. HÖRZ

Every scientist today who does not contradict his scientific insight on ideological grounds recognizes the magnificent historic achievement of Charles DARWIN. With his book "Origin of Species", published in 1859, DARWIN helped the idea of development to its breakthrough in biology. He revealed the natural mechanisms underlying the change of species and oriented the development of science towards investigation of the development laws governing natural evolution. This gave rise to a colourful spectrum of research programs dealing with problems ranging from the physical and chemical foundations of prebiotic evolution to the emergence of human society. We have gathered here to show our respect for DARWIN's achievements and to honour him as the initiator of research into natural evolution by discussing critically and constructively recent results in the history of science, molecular and developmental biology and anthroposociogenesis. We wish here to submit a few suggestions to this end.

1. DARWIN'S Achievement — a Scientific Revolution

DARWIN ranks among such pioneers of science as COPERNICUS, NEWTON, MARX and EINSTEIN. Eminent scientists soon recognized the significance of DARWINS work. Friedrich ENGELS emphasized its importance as proof that all living beings, including Man, have resulted from protracted development processes that started with unicellular germs formed by chemical reactions. To him, the DARWINian theory, the discovery of the cell and the law of conservation of energy were the three major discoveries of the 19th century (F. ENGELS, [20] 259). According to HELMHOLTZ, DARWIN raised the status of biotic evolution "from the state of an accumulation of enigmatic curiosities to the coherence of a great development" [28, p. 176]. BOLTZMANN stated that "all salvation for philosophy could be expected from the theory of DARWIN" [5, p. 251]. And in E. MAYRS opinion "the DARWINian revolution of 1859 is perhaps the most fundamental of all intellectual revolutions in the history of mankind" [51].

Scientific revolutions bring to light principles that unite the numerous insights obtained empirically into the laws that hold in a certain field and into associated deductive and hypothetical systems to form a consistent theory which can be verified practically. They account for previously discovered salient characteristics of the subject. Such revolutions are rare in science. They have repercussions on other sciences, establish a new relationship between science and practice, and lead to a profound change in world pictures, which are the unity composed of illustrated theories, insight into weltanschauung and orientation for action. DARWINS achievement was a scientific revolution. It was a phenomenological systems theory which explained the emergence of the qualitatively new in terms of variation and selection from common origins although the elementary mechanisms involved were then unknown. This was a radical break with the traditional view of creation and the immutability of species.

The elaboration of the DARWINian theory infers the existence of previous and contemporary work in developmental theory. Without the work of predecessors, DARWINS theory could not have brought about a break-through. Since LEIBNITZ' time, there had been discussions about preordained harmony in fauna and flora. The preformation theory, which regarded development as the revelation of the preformed, emerged. This theory opposed occasionalism, which stated that natural variations in plants and animals had to be continually created anew. Certainly, continuous creation could explain the diversity of phenotypes, but it made natural scientific research impossible. Preformism, on the other hand, allowed discovery of the laws governing the formation of organisms. Preformism implies immutability of species. Proceeding from preformism, KANT had introduced the idea of development into cosmology. The step-ladder concept advocated by BONNET could indeed account for the multiplicity of species, but it eternalized their number at the same time. DARWINS achievement is only understandable if one comprehends the theoretical solution of the problems dealt with by his predecessors: Evolution and the accidental range of variation had to be explained meaningfully in terms of natural mechanisms. The tools available for doing this included a speculative developmental theory expressed in terms of natural philosophy, empirical evidence of the change of species and initial approaches to developmental thinking in the works of St. HILAIRE, GOETHE, and Erasmus DARWIN (HÖRZ and WENIG, [29]).

Charles DARWINS work provided not only an empirical basis for developmental thought but also a theoretical explanation of the empirically established transformation of existing, and the emergence of new, species. It was he who made descendance from common ancestors plausible, who showed that adaptation and reproduction of the species are part of the struggle for survival, and who recognized that selection by natural selective breeding is an evolutionary mechanism. He confronted theories involving inexplicable evolutionary leaps with the idea of continuous evolution. DARWIN discovered the developmental principle governing biotic evolution: species change by variation and selection. With this magnificent achievement he disproved the myths of creation and the assumed immutability of species that were so contradictory to empirical fact. Occasionalism suffered a defeat, and preformism was guided further towards the developmental approach.

With regard to the elementary mechanism systems changes DARWIN's theory was phenomenological. The elementary mechanisms responsible for the relative immutability of species and for their changing could not be recognized until DNA and viral RNA had been identified as genetic material and the occurrence of identical reduplication, mutation and recombination had become known. These further substantiated and sharpened the DARWINian evolutionary theory: selection of individuals is based on the variability of the gene pool, which provides the potential for the trial-and-error method by which nature brings forth organisms which are better able to cope with their surroundings.

The repercussions of Darwinism on other sciences were enormous. MARX called the DARWINian theory a natural scientific basis for his view of the class struggle. SPENCER generalized the idea of development in the notion of the integration of matter by the scattering of motion. Biological disciplines and other sciences accumulated material for or against Darwinism, which had become the object of a raging battle in terms of world-outlock, because it established in terms of natural science, the material unity of the world in its genesis.

Friedrich ENGELS had advocated the thesis that Nature makes no jumps because it consists of nothing else. In this way he drew attention to the unity between continuity and discontinuity in development. Development is a dialectical process in which the trend towards a higher state is accompanied by stagnation, regression and the formation of all elements associated with a phase of development. There is no automatism of higher development. This is evident from random variations, the stagnation of possible lines of development and the extinction of species. Laws of development are therefore of a statistical character. They are trend laws. They encompass both the trend towards higher development in the course of time by the qualitative transformation of forms which provide a whole range of possibilities and the conditionally accidental realization of variants with certain probabilities. The complexity of development processes has often been hidden by one-sided views that emphasized either fitness or chance.

At the beginning of our century, MAUTHNER [49, p. 283] wrote that those who had pricked their ears to scientific catchwords in the same way as to everyday words could hear the struggle between purpose and chance even from the all-explaning terms of Darwinism: from the word development, which for not quite fifty years had given rise to a wealth of literature on development or realization of potential. DARWIN solved the task which had been set by the development of the sciences themselves: the natural explanation of biotic evolution. The natural system of the species had long been sought, but under the one-sided assumption that species remain completely immutable. DARWIN saw in the common origin the secret connection that explorers of nature had been unconsciously seeking as the natural system. The natural system unites immutability with evolution and the variability of species by common origin and selection. Every structure is recognized as a coagulated development which forms the basis for further development. The time for the empirical substantiation and theoretical foundation of the principle of development had come. DARWIN's advantages were his knowledge of the empirical material he had collected and his theoretical interpretation of the facts which was based on the idea of development. This led to a novel view which accounted for the dialectics of accidental variability and purposeful adaptation in selective evolution and had repercussions on the development of science, practice, and world-outlook.

Hence, we are justified in stating that DARWIN's achievement was a scientific revolution.

2. DARWIN's Work and Modern Biology

DARWIN's work embraced several basic insights which have the character of theories in their own right. Thus, for him, descendence theory as the doctrine of descent from common ancestors is connected with the theory of evolution which gives a theoretical explanation of biotic evolution and therefore covers various aspects (S. TSCHULOK, [64]), E. MAYR went further in differentiating the theoretical structure. According to him, DARWIN's work contains five distinct theories (MAYR, [51]):

"First, evolution in general.

Second, descent from common ancestors.

Third, the gradual character of evolution.
Fourth, the emergence of species as a source of variety.
Fifth, selection as the basic mechanism of change."
These theories have different degrees of generality. Thus, the first, third and fifth theories can be subsumed to the theory of evolution, and the second, fourth and fifth theories to the descendence theory. However, analysis of the theoretical aspects mentioned by MAYR is interesting in contemporary discussion.

So many facts have been accumulated to underpin the theories of evolution and descendence that they can be regarded as scientifically substantiated fundamental theories whose validity can no longer be doubted by anyone wishing to be taken seriously. In recent times, their universal character has been underlined by the universality of the genomic amino acid code and the validity of the so-called central dogma of molecular biology in man, animals, plants and microbes.

During the past decade in particular, genetic engineering has provided further overwhelming evidence for the correctness of the theoretical structure which has meanwhile been fused with the chromosome theory of heredity, population genetics, and other biological and paleontological concepts to form a "synthetic theory" which is sometimes called "Neodarwinism". It not only answers many of the questions raised in recent years, but also poses new ones. In this context we regard the elaboration of methods for sequence analysis of polynucleotides and for the elucidation of the structures of prokaryotic and eukaryotic genes as especially important. These allow us to investigate evolution and descent even beyond the geological fossil record: the absence of "missing links" can be compensated for by molecular biological analysis. Moreover, they have extended past estimates of the duration of evolution hitherto, which since DARWIN's time had remained fairly constant at 500 million years, by one order of magnitude [37, 67]. Thus, life can be considered to have emerged almost as far back as the formation of the Earth.

After comparative analysis of amino acid sequences — not only of cytochrome c but also of haemoglobins, immunoglobulin, etc. — during the early seventies had made it possible to construct phylogenetic trees in terms of molecular biology, analysis of small RNA molecules — for instance of t-RNA by EIGEN and associates and other teams and of 16S-ribosomal-RNA by WOESE and associates — not only confirmed to a large extent the classical phylogenetic trees but also detected a new third kingdom of organisms, the archaebacteria. Archaebacteria differ in several respects from the prokaryotes and eukaryotes. The ancestors of at least the methanogenic archaebacteria were probably able to live in the kind of atmosphere thought to have existed on the primeval earth due to their special methanogenic metabolism.

Furthermore, it seems possible that both archaeobacteria and eubacteria are descendents of a common ancestor, a proposed "progenote" defined as "an entity having simple properties, imprecise and general functions and a rather small complement of genes" (WOESE, [67]). Progenotes are assumed to consist of "urgenes" whose presumed structure has also been deduced by analysis of rRNAs (EIGEN, [18]). New insights into the evolution of the eukaryotic cell have also been gained (FOX et al. [22]). But sequence analysis is also contributing our finally understanding the descendence of Man. In this way modern molecular biology is proving to be of major importance in solving some of the open fundamental questions of the synthetic theory.

Although the theories of evolution and descent have been accepted almost without exception, gradualism, an integral part of DARWIN's evolution theory, has still not found universal acceptance. This is due, perhaps, to an incomplete understanding of the transition to other, new and higher basic qualities. New basic qualities which differ from their predecessors in structure and function result from qualitative and quantitative changes within the framework of the old basic quality. There thus exists a constant unity of continuity, i.e., the existence in time of an uninterrupted sequence of structures with certain functions, and discontinuity, i. e., structural and functional change. Where the temporal and spatial dimensions of evolution are great, the qualitative jump appears to be a discontinuity. As G. MAYNARD SMITH [62] points out, such a debate is at least partly caused by a semantic difficulty: "A new species arising in 500,000 years is sudden to a palaeontologist, but gradual to a geneticist."

The addition of the spatial and temporal conditions involved often leads to the qualitative jump being neglected. Let us consider the animal-man transition that brought forth primeval Man. Gradualism takes into account the dialectical unity of the relative immutability of species and of the change of species by variation and selection.

Moreover, contemporary results of molecular biology show that development has been both gradual and saltatory: gradual as a result of point and minor chromosome mutations and saltatory due to more or less profound rearrangements of the genetic material (cf. GEISSLER, this volume).

Studies on genetic variation at 16 protein and enzyme loci in *Cercopithecus aethiops* and other primate species indicate that the variation of extracellular (plasma) proteins is much more pronounced than that of most intracellular (red cell enzymes) (PALMOUR et al. [55] p. 807). "These rapidly evolving loci show a pattern of continuous change as taxonomic distance increases from the populational level through to congeneric species and to generic comparisons. This observation . . . is compatible with a model of phyletic gradualism at the structural gene level during speciation, rather than with a punctuated or rectangular model of evolution."

Other arguments for our dialectical point of view stem from the results described below regarding the sources of genetic variation.

However, in this context, it must be remembered that the synthetic theory of evolution must be further extended by the principles of self-organization and the actions of feedback control systems as proposed by EIGEN and SCHUSTER. This principle entails saltatory consequences.

DARWIN's fourth theory concerned the emergence of diversity. DARWIN was the first to recognize that the formation of species is the key to the problem of how biotic diversity emerged, and that this again is a process which involves not individuals but populations. Here, too, the general principle has been accepted, but opinions differ profoundly concerning the "how". There differences again involve the question of whether the formation of new species is gradual or saltatory. An international conference in 1981 showed that the disunity of standpoints regarding this question persists [50].

DARWIN's fifth theory was about the mechanism of evolution. Here DARWIN recognized that a two-phase process consisting of genetic variability and of natural selection was involved. Thanks to developments in genetics and molecular biology we now know, in contrast to DARWIN who was obviously unaware of MENDEL's famous work, that in the case of accidental genetic variability, which provides the range of possibilities permitting natural selection to work, the structures affected are DNA and the mechanisms responsible are point, chromosome and genome mutations and interchromosomal recombinations. The results of base and amino acid sequence analysis show not only the correctness of descendence theory but also the significance of base substitutions and duplications for the evolution of new information-carrying structures.

In this context the discussion about molecular evolution mechanisms has been enlivened especially by the discovery of split eukaryotic genes and of pseudogenes and "jumping genes" in both prokaryotes and eukaryotes. Since the suspected role of the latter will be the subject of a separate lecture (GEISSLER, this volume) we shall not discuss these interesting structures in greater detail here.

GILBERT [23, p. 10] regards "the gene not as a coding structure interrupted by introns, but as an intron background into which the exons have been either inserted by recombination or called forth by mutations". He wonders, whether "the genes (were) once whole, but became separated by the insertion of elements whose purpose is to keep the exons apart? Or were the parts of the genes always separated and then put together by the introns to create the evolving genetic element?" This could proceed either by alteration of the splicing pattern ("multiple-choice splicing". CHAMBON [7]) as observed, for instance, in the early region of adenoviruses or by increasing the frequency of legitimate and illegitimate recombination by expanding the gene in size. Illegitimate recombination could bring together exons from different genes, while legitimate recombination could join different mutated exons from homologous genes to produce a better gene product.

Furthermore, CHAMBON [7] proposes, "that new genes coding for longer proteins could arise through fusion of duplicated transcription units rather than by a much rarer event: the precise deletion of the spacer DNA separating the duplicated copies of the original gene".

In fact, first studies in this field reveal that the mechanisms proposed by GILBERT and CHAMBON actually did happen. CHAMBON [7] has evidence from his laboratory that the chicken gene for ovotransferrin, which is split into 17 exons, evolved by fused duplication of an ancestral gene that was already split into seven or eight exons. The chicken type Iα2 collagen gene which occupies 38 kb DNA and is subdivided into more than 50 exons even seems to be evolved by multiple duplication of a single genetic unit of 54 bp (YAMADA et al. [68]).

ARTYMIUK et al. [2] demonstrated very close agreement between the exons of the hen lysozyme and the structural homology between the hen lysozyme and the phage T 4 lysozyme.

Moreover, GŌ [24] and JENSEN [31] have shown that defined regions of the β-haemoglobin chain, known as "structural units", "sub-domains", or "compact structures" and corresponding to the exons of the haemoglobin gene, are related. The central exonic region seems to code for two compact structures and is regarded therefore to have evolved by "exon fusion".

Furthermore, the studies of CRAIK et al. [8,9] on the function of the central exonic region of human haemoglobin indicate relationships to b-type cytochromes. BLAKE [4] wonders therefore whether "the globins and the cytochromes may have evolved from a common haem-binding domain encoded by one or more exons".

Regarding the occurrence of base substitutions and other minor changes and gross aberrations, P. LEDER et al. [44] studied the mouse globin gene system in detail. They learned that strongly selected sequences of β-globin minor gene and of closely related β-major gene change slowly, if at all, and then by point mutation. On the other hand, unselected (flanking and intervening) sequences of both genes change quickly, altering large pieces of DNA at a single stroke by insertions or deletions. The coding sequences themselves are naturally also not immune to gross aberrations, as can be observed when studying pseudogenes, but selection prevents them from appearing in essential genes. Large segments, at least of the second intron, of the β-globin genes are obviously not subject to the same selective pressures that operate on its coding sequences, otherwise their homology would not be so tightly conserved.

Similar conclusions stem from the study of pseudogenes, which were probably functional genes until frameshift mutations, insertions and deletions prevented their expression (DE JONG and RYDEN [13]). Mouse globin pseudogene α-3, for instance, obviously diverged from the alpha-gene lineage after the alpha-β divergence but has lost both its introns [44]. It does not encode globin and it seems not to be active in the production of large amounts of stable mRNA. In comparison to α- 1 it contains about 10 deletions/insertions and appears to change in the same way like flanking and intervening sequences (NISHIOKA et al. [53]).

In contrast to DARWIN, we now also know that variability does not begin in a genetically more or less uniform population containing only a few variants, but rather that marked polymorphisms exist in natural populations. In particular it has been established by gel electrophoresis that the degree of heterozygosity in invertebrates is 13,4%, in vertebrates 6.6%, in humans 6.7%, and in plants 17% (AYALA [3]). This means that the major part of variability is not due to the appearance of new mutations in each generation but rather to a redistribution of the mutations that have already accumulated. This is quite compatible with the recent discovery, made by physical mapping, that the majority of sickle-cell anemia patients i.e. carriers of an HbS-allele, acquired it not by new mutation but by inheritance from an ancestor who had lived in a certain region of the Upper Volta or Ghana and whose descendant(s) had been deported to North America as slave(s) (KAN and DOZY [36]).

From the genetic variations formed in one way or another, but always randomly, the factors which confer advantages are selected in a second step. MAYR [51] points out that in this respect chance in fact plays no role: in selection a bonus is awarded in an almost teleological manner for being better. In reality, however, MAYR argues, it has nothing to do with teleology because it is not programmed in advance but, even better, is determined in each generation by the prevailing circumstances.

Selection is therefore not an unambiguously predetermined process but is subject to statistical laws with fields of possibilities and leads to the random realization of variants which have certain probabilities. This introduces the dialectics of the necessary tendency towards higher development and of an accidental range of variation in natural selection to the structure of development laws (HÖRZ and WESSEL [30]).

After the principle of selection had been largely accepted, discussions arose as to the target of selection: Does selection attack the gene, as suggested, for instance, by DAWKINS [12], is the genome the unit of selection, as will be proposed by GRANTHAM at our conference, is the individual with its phenes subject to selection, or is it rather the group or even the species?

As RUSE [59, p. 625] has shown DARWIN himself was "convinced that in the non-human world selection acts at, and only at, the level of the individual", whereas for Man he assumed that it acted at group level with the tribe being the critical unit.

We share the view of E. MAYR [51]: "Neither genes nor populations nor species (are) the target of selection, but individuals, solely and uniquely. It is individuals that survive or do not, that propagate or do not."

In this connection the current discussions as to whether repetitive DNA, transposons and similar DNA sequences can be called "selfish" because their "only 'function' is survival within genomes" while "it makes no specific contribution to the phenotype" (DOOLITTLE and SAPIENZA [14] ORGEL and CRICK [54], SAPIENZA and DOOLITTLE [60]) are irrelevant to our argument.

However, in all these discussions we cannot avoid one dilemma: Despite all the detailed knowledge we have acquired due to advances in molecular biology, we have a rough idea only of how the explicit genetic information is expressed, and no idea at all of the relationships between DNA and phenes, which are controlled mostly by multiple factors. Meanwhile, more than 50 loci are known to control pigmentation alone in the mouse, and we also know that one or several alleles of a large proportion of these genes have also morphogenetic effects (SILVERS [61]): Most variability, at least of higher organism, therefore is not caused by genes that can be individually identified.

And how are we to imagine the evolution of such complex organs as, for instance, the eye of vertebrates by variation and selection? Even DARWIN wrote: "To suppose that the eye with all its inimitable contrivances for adjusting the focus to different distances, for admitting different amounts of light, and for the correction of spherical and chromatic aberration, could have been formed by natural selection, seems, I freely confess, absurd in the highest degree" [11, p. 160]. RIEDEL's associate, KASPAR [38] has estimated that for the development of the vertebrate eye in the simplext case at least 10^{30} mutations would be necessary.

However, according to RIEDL [58] it is possible to avoid this dilemma by accepting the principle of the feedback of functional regulatory circuits not only for the prebiotic phase — for which it has been explicitly and convincingly developed by EIGEN und SCHUSTER [17, 19], but also for the organismic level. It can then be assumed that the organization of a genetic system of a species is causally reciprocally related to the organization of the morphological features of the individuals of that species. On the one hand, the genome generates a certain structure and, on the other, this structure is decisive for the probability of success of further mutants (KASPAR [38]). It is necessary to point out in this respect, however, that such harmonization in the evolution of morphological structures by connection of the appropriate genes must take place without hereditary transmission of acquired proporties and despite the validity of the central dogma of molecular biology — which was also not questioned by RIEDL [58].

With regard to the target of selection, the discussion on the significance of neutral mutations should not be ignored: KIMURA [39], as well as KING and JUKES [41] have postulated that many of the alleles whose products can be identified electrophoretically as polymorphisms in the organisms of a population are selectively **neutral** since they induce amino acid substitutions in parts of the enzyme molecules which have no fundamental significance. This also applies to the "silent exchanges" which, as same-sense mutations, bring about the exchange of synonymous codons.

In contrast to "selectionists" who assume that a mutant allele must have some selective advantage in order to spread through a species, KIMURA [39a] and other "neutralists" suggest that some mutants can spread through a population on their own **without** having any selective advantage. Some of them are lost by chance, others eventually become fixed in the population.

Indeed, much of what we know can be explained in neutralistic terms: amino acid and codon frequencies, polymorphisms, evolution of pseudogenes, enhanced frequency of nucleotide substitutions in introns and intervening regions, the different speeds of evolutionary clocks, both for different proteins and for the same protein in different species, the existence of overlapping genes in DNA and RNA viruses, etc. (JUKES [32, 33] KIMURA [39 a]).

But, firstly, it is wrong to say that the neutralistic hypothesis implies "non-DARWINian evolution", since, as we have already mentioned, DARWIN's theoretical structure consists not only of the principle of selection and because DARWIN himself thought such changes possible:
"Variations neither useful nor injurious would not be affected by natural selection, and would be left either a fluctuating element, as perhaps we see in certain polymorphic species, or would ultimately become fixed, owing to the nature of the conditions." [11, p. 70].

To explain the distribution of silent and neutral changes PERIER et al. [56] and others have developed a "hitch-hiking theory"; The driving force for fixation is positive neutral selection operating on some fraction of amino acid-altering changes, and such selected fixation carries along with it neutral alterations and silent changes. In the view of KIMURA [40] this theory cannot explain the actual results of DNA sequencing. On the other hand, "hitch-hiking" of a changed HpaI restrictase recognition site has in fact been observed on the 3′ side of the sickle cell anemia gene (KAN and DOZY [36]). Thus, most of the arguments forwarded by neutralists can be countered by arguments forwarded by selectionists (CROW [10]). We wonder therefore, whether there is an all-or-none answer at all. Furthermore, we believe that such discussions are fruitless as long as the relationship between multi-factorial coded phenes and their genetic determinants is unclear. How can one definitely distinguish between selection and no selection when only base or amino acid sequences are known?

Evolutionary forces are thought to have been optimally adjusted mutation rates that strike a balance between minimum and preservation of evolutionary plasticity. Phenomena of this sort were already suggested several years ago in connection with studies on mutation rates versus an increase in genome size (cf. DRAKE [15]).

Sequence data now allow such studies at the molecular level. MODIANO et al. [52] studied human α- and β-globin genes to ascertain whether codon usage or patterns of nucleotide substitutions or both are biased to reduce the possibility of "forbidden" missense mutations or of nonsense mutations. What they observed was a considerable reduction of mutational events causing drastic phenotypic effects. The question of how selection has succeeded in reducing the rate of deleterious substitutions and avoids pretermination codons which can mutate to nonsense codons is a matter for speculation.

In other words there are still plenty of questions to be answered regarding the molecular mechanisms of selection. We sincerely hope that our discussions will contribute to further clarification in this field. Furthermore, rapid progress in molecular and cell genetics will also have an impact on this very field.

Anyway, the existence of open problems does not make the synthetic theory any the less convincing. This would still be so even if a seemingly LAMARCKistic new "somatic selection hypothesis" forwarded by STEELE [63] could be supported by reproducible experiments. This hypothesis postulates a transfer of genes from (mutated) somatic cells to the germ line cells by retroviruses or similar vectors and is therefore still compatible with the central dogma (cf. GEISSLER, this volume, PASTERNAK, this volume).

3. The Influence of Darwinism on the Philosophical Concept of Development and a Few Discussions on Darwinism in Terms of World-Outlook

Darwinism, i.e. the entire biological theory of descendence and evolution derived from DARWIN's work, contributed decisively to the philosophical theory of development by introducing the notion of development as a heuristic principle for interpreting empirical results. E. HAECKEL characterized DARWIN's theory as an "insight modifying the whole world-outlook" [27, p. 23]. The significance of Darwinism in terms of world-outlook was firstly its explanation in natural scientific terms of the mergence of species by natural selection which refuted teleology with its theological consequences and, at the same time occasionalism. Secondly, it explained evolution in all its aspects theoretically as an irreversible process of maturation, differentiation and change of quality: the unity composed of reproducibility and novelty in biotic evolution became understandable in theoretical terms. Thirdly, the principle of development was applied to other fields. Thus, MARX wrote to ENGELS on 19 December 1860 that DARWIN's book "contains the natural historical basis for our view" [48, p. 13]. At the same time, the emergence of social Darwinism brought with it the danger of ignoring the specific character of social processes with their own system laws.

Darwinism encouraged the dialectical and deterministic understanding of development processes as accidental realization of possibilities and the trend towards higher development. Historism comprehends structures as coagulated development, as a proceeding structure and, thus, as a basis for qualitative changes. In this context, the development of biology itself has substantially contributed to the scientific underpinning of Darwinism as a natural scientific basis for the philosophical theory of development. For instance, the relative immutability of species by invariant reproduction on the basis of a genetic program can be explained only by the development of the genetic code. Errors in the genetic program create a natural basis for variability and, thus, for realizeable variants incorporating qualitative changes.

The discussion on Darwinism has various aspects from the philosophical viewpoint. They include the philosophical dispute with the proponents of creationists and mechanicistic views, the distinction between the heuristic process of development and statements that can be proved or refuted in evolutionary theories, and the necessary closer definition of general philosophical statements regarding development by means of new scientific knowledge.

When we started organizing the colloquium "DARWIN today", we thought it would be stimulating to discuss the various consequences of DARWIN's work on the occasion of the hundredth anniversary of his death. We did not realize two years ago that we might have another reason to summarize once again the importance and validity of DARWIN's work. In fact, the reappearance of any kind of creationism more than 100 years after the first publication of the "Origin of species" seemed inconceivable.

But the change of administration in Washington was accompanied by a new wave of creationism. In 1980, Ronald REAGAN stated during the presidential campaign that he had "a great many questions" about the theory of evolution. Whenever Darwinism is presented in the public schools, REAGAN suggested, "the biblical story of creation schould also be taught" (cf. GURIN [26] p. 16). Now that REAGAN is President, the creationists have launched a large-scale attack against the theory of evolution — more than half a century after SCOPES illfamed "monkey trial" in Dayton, Tenn. in 1925.

"Creationists" and "fundamentalists" maintain that the earth is roughly six to ten thousand years old, that the planets, stars and all living things were literally created in six days by a "Designer", that the different species of plants and animals were created separately, and that they did not evolve from any other species. According to the creationists, human beings have nothing to do with other primates, Adam and Eve were created specially, and we are all their descendants (GURIN [26]).

Interestingly enough they are convinced, however, that the fruit that brought the downfall of Adam and Eve was not an apple but more likely an apricot, plum, fig or quince [25].

In the meantime the fundamentalists have persuaded legislators in at least 14 states to introduce laws requiring creationist views to be presented in science classes (PIERCE, [57]). In Florida, the Hillsborough County School Board has already decided to require that scientific creationism be taught in the county's school system besides the theory of evolution. This decision "was supported by those who insist that belief in the power of a supreme being to create the Earth helps to maintain social order" [43]. Similar laws have been signed in Arkansas [25a] and in Georgia [26].

Georgia Judge Braswell DEEN argued, "This monkey mythology of DARWIN is the cause of permissiveness, promiscuity, pills, prophylactics, perversions, pregnancies, abortions, pornotherapy, pollution, poisoning and proliferation of crimes of all types" [57, p. 82]. And the head of the Institute for Creation Research, San Diego, Henry MORRIS said, "The evolutionary theory of origins . . . is, at its ultimate roots, the anti-God conspiracy of Satan himself" [34].

The textbook publishers have already been under pressure by creationists for many years. Today in 27 US states, textbook selection committees are forced to accept books that teach divine creation as biology [35]. PIERCE [57] found that "Between 1974 and 1977, the section on DARWINS life in **Biology**, a text published by Silver BURDETT, was cut from 1,373 words to 45. Discussions on the origins of life went from 2,023 words to to 322. The text devoted to DARWIN's view to evolution shrank from 2,750 words to 296. Sections on fossil formation and geologic eras were deleted entirely . . ." The 1973 edition of **Modern Biology**, published by HOLT, REINHART and WINSTON, stated, "Scientists do not doubt that organisms living today descended from species of previous ages". That sentence was omitted in the 1977 revision. The 1969 text said, 'Modern man has probably evolved from primitive, more generalized ancestors'. The 1977 version: 'DARWIN was suggesting thet humans may also have evolved from less specialized ancestors.' And PIERCE quotes Frank SPICA, a biology teacher in Evanston, Ill. who said, "I think the creationists have won. They've not passed any legislation, but they've got the textbooks changed."

Creationism is dogmatically fundamentalist and profoundly antiscientific. It ignores all results of modern science, of astronomy, astrophysics, physics, geology, paleontology and biology. Ever more scientists in the States are recognizing that creationism is "not a scientific or even a religious issue; it's a political issue" (N. ELDRIDGE, 1981, cf. [65]).

Even religious scientists are disturbed by the efforts of creationists. William S. POLLITZER, professor of anatomy at the University of North Carolina and immediate post president of the American Association of Physical Anthropologists said, "My main opposition to what the creationists are doing stems from the intimidating effect it has on the educational process. The function of public education is not to indoctrinate children in any particular way, but rather to expose them to the prevailing ideas of their time and let them make up their own minds" [65].

4. On Problems of Anthropo(socio)genesis and of the Relationships Between Man's Biotic and Social Evolution

Another consequence of DARWIN's descendance theory "met then with the utmost opposition, the inclusion of Man into the overall phylogenesis of the animal kingdom" (MAYR, 51). But particularly in recent years and again due mainly to molecular and cell biological studies but also to the results of comparative behavioural research and numerous hominid finds, such overpowering evidence has been accumalted that this problematic today almost culminates in the question: "How does man really differ from the animal?" It is difficult to analyze the kinship of Man with his presumably closest relatives, the gorilla and the chimpanzee by biochemical and genetic methods, because adequate fossils are lacking: anatomists, biochemists and cytogeneticists disagree on this subject, and even the extensive homology of the DNAs of their genomes allows no elucidation [42]. The investigations carried out by DUTRILLAUX [16] into chromosome rearrangements and the work done by A. C. WILSON on the physical mapping of mitochondrial DNA, which evolves more rapidly than the nuclear DNA [6], so that the differences between the various species become more distinct, will perhaps provide an answer. FERRIS et al. [21] compared about 50 restrictase recognition sites by the parsimony method for man, the chimpanzee (*Pan troglodytes*), the pygmean chimpanzee (*Pan paniscus*), the gorilla, the orangutan and the lar gibbon (*Hylobates lar*). As a result they were able to establish a common "family tree" showing that the two chimpanzee species were the most closely related, with the gorilla as their next closest relative, followed by man.

In contrast, the assertion of the sociobiologist E. O. WILSON [66] that the differences between our species and the primeval ape can only be explained as a result of specific human genes is completely unfounded.

Reason dictates that one-sided biologistic and sociologistic views be rejected. Man is neither a naked ape dominated by the urge of aggression, nor a social being who is no longer dominated by biological laws. He shares with his nearest relatives, the subhuman primates, not only predominantly identical DNA sequences in a largely identical set of chromosomes, but also numerous patterns of behaviour: Chimpanzees, macaques and other anthropoid apes also use tools and have the prerequisites necessary for their use, such as upright posture, spatial intelligence, binocular stereoscopic vision, prehensile hand and so on. In primates we also find the formation of social frames of reference, learning, handing down by tradition at least in primitive form, personal identity concepts, and the partial learning and mastery of artificial languages. In fact, these observations constitute a confirmation of Friedrich ENGELS' brillant consideration, which show that in the final analysis man differs fundamentally from the animal only by work, by the production and utilization of artificial instruments to produce the means of his subsistence. Something that still requires elucidation, especially from the standpoint of Marxist philosophy, is the mutual influence of work, conscience and language, the latter, of course, being enabled by specific morphological features resulting from equally specific evolutionary processes.

Is it, however, necessary to share E. O. WILSON's [66] assumption that there was also a "genetic evolution of human social behaviour" which occurred predominantly during the five million years before civilization began, when mankind consisted of scattered, relatively immobile populations of gatherers and hunters? And is WILSON right in saying that, on the contrary, the overwhelming part of cultural evolution has taken place since the emergence of agriculture and towns about ten thousand years ago? During this, most recent historical spurt genetic evolution has continued to a certain extent, but it cannot have formed more than a fraction of the features of human nature (p. 38—39). We wonder whether we can deal with this problem at least during the last part of our colloquium. Moreover we wonder whether we can subject to critical analysis the most recent speculation put forward by E. O. WILSON together with C. J. LUMSDEN: the theory of gene-culture coevolution. This theory is based on two surprising postulates: Fifty generations, 1000 years are said to be sufficient for important genetic shifts (cf. LEWIN [45]). Furthermore culture is viewed as a particular phenomenon, the units of which are termed 'culturgens' (LUMSDEN and WILSON [46, 47]). Culturgens are assumed to be traits that are possessed and transmitted by individuals. These traits can exist in one of only two possible discrete forms, e.g. outbreeding and brother-sister incest or preference or lack of preference for sugar. The probability of changing from one alternative form of the trait to another is determined by epigentic rules controlled by some genes. The hypothesis is described in length in a new book "Genes, Mind, and Culture" described by biologists as "marvelons . . . a once in a century occurrence" and "seriously flawed and dangerous". (LEWIN [45]) ALPER and LANGE [1] demonstrate that their theory is essentially reductionist. Furthermore they mention that there is absolutely no evidence that genes governing transitions from one cultural trait to another do exist.

We will probably also be unable to ignore the question of where mankind is going. Apart from the brand-new highly topical problem of preventing the arms race from escalating and culminating in the suicide of Mankind, this problem is still irrelevant. It will, however, be possible to tackle it thanks to the rapid development of genetic engineering and cell genetics, at least on a laboratory scale with the isolated cell, and it is also a world-wide topic of discussion not only in the sensation press.

We think that the positions adopted to date in discussions on such questions have been excessively biologistic and sociologistic and have not, furthermore, taken into account the dialectics of the forms of motion of matter. Similarly, they have ignored the fact that "directed biological evolution of Man" could today take place within a relatively short period by the horizontal transfer of suitable genes while also neglecting the fact that, at least for us Darwinists, Man is not the "crown of creation" and, hence at least in the **biological** sense, is capable of improvement. This is why we believe that nothing in this connection should be put under taboo. This is why we are opposed to corresponding antiscientist views which dictated by an unfounded fear of intervention on the part of geneticists in the human psyche. On the other hand, we likewise oppose unrealistic optimism regarding a future of "eternal life free of disease".

We condemn all antihumanist utilization of scientific insights. As humanists we respect the integrity of the personality. This demands that experiments performed on Man be based on strictly humanist considerations and principles. They must involve minimum risk, ensure freedom of decision and be conducted with a deep sense of responsibility to ensure that progress in science and technology is to the benefit of mankind. DARWIN's work is the point of departure for many new scientific insights into the behaviour of Man himself. Therefore we are called upon to make our responsible contribution to the development of science, and thus to the benefit of mankind.

References

[1] ALPER, J. S. and R. V. LANGE: LUMSDEN-WILSON theory of gene-culture coevolution, Proc. Nat. Acad. Sci. USA **78** (1981), 3976—3979

[2] ARTYMIUK, P. J., C. C. F. BLAKE, and A. E. SIPPEL: Genes pieced together — exons delineate homologous structures of diverged lysozymes, Nature **290** (1981), 287—288

[3] AYALA, F. J.: The mechanisms of evolution, Sci. Am. **239** (1978), 48—61

[4] BLAKE, C. C. F.: Exons and the structure, function and evolution of haemoglobin. Nature **291** (1981), 616

[5] BOLTZMANN, L.: Populäre Schriften, Hrsg.: E. BRODA, Braunschweig 1979

[6] BROWN, W. M., M. GEORGE, jr., and A. C. WILSON: Rapid evolution of animal mitochondrial DNA, Proc. Nat. Acad. Sci. USA **76** (1979), 1967—1971

[7] CHAMBON, P.: Split genes, Sci. Am. **244** (1981), 48—59

[8] CRAIK, C. S., S. R. BUCHMAN, and S. BEYCHOK: Characterization of globin domains: Heme binding to the central exon product, Proc. Nat. Acad. Sci. USA **77** (1980), 1384—1388

[9] CRAIK, C. S., S. R. BUCHMAN, and S. BEYCHOK: O_2 binding properties of the product of the central exon of β-globin gene, Nature **291** (1981), 87—90

[10] CROW, J. F.: The dilemma of nearly neutral mutations: How important are they for evolution and and human welfare?, J. Hered. **63** (1972), 306—316

[11] DARWIN, C.: Origin of Species, H. M. Caldwell Company, New York 1896

[12] DAWKINS, R.: The Selfish Gene, Oxford University Press, Oxford 1976

[13] DE JONG, W. W. and L. RYDEN: Causes of more frequent deletions than insertions in mutations and protein evolution, Nature **290** (1981), 157—159

[14] DOOLITTLE, W. F. and C. SAPIENZA: Selfish genes, the phenotype paradigm and genome evolution, Nature **284** (1980), 601—603

[15] DRAKE, J. W.: The genetic control of spontaneous and induced mutation rates in bacteriophage T4, Genetics, Suppl. **73** (1973), 45—64

[16] DUTRILLAUX, B.: Chromosomal evolution in primates: Tentative phylogeny from Microcebus murinus (prosimian) to man, Human Genetics **48** (1979), 251—314

[17] EIGEN, M.: Self-organization of matter and the evolution of biological macromolecules, Naturwissenschaften **58** (1971), 465—526

[18] EIGEN, M.: Das Urgen, Nova Acta Leopoldina NF **52** (1980), 5—40

[19] EIGEN, M. and P. SCHUSTER: The Hypercycle, Springer Verlag Berlin, Heidelberg, New York 1979

[20] ENGELS, F.: Ludwig Feuerbach und der Ausgang der klassischen deutschen Philosophie, In: MARX/ENGELS, Werke, Bd. **21**, Berlin 1962, p. 259—307

[21] FERRIS, S. D., A. C. WILSON, and W. M. BROWN: Evolutionary tree for apes and humans based on cleavage maps of mitochondrial DNA, Proc. Nat. Acad. Sci. USA **78** (1981), 2432—2436

[22] FOX, G. E., E. STACKEBRANDT, R. B. HESPELL, J. GIBSON, J. MANILOFF, T. A. DYER, R. S. WOLFE, W. E. BALCH, R. S. TANNER, L. J. MAGRUM, L. B. ZABLEN, R. BLAKEMORE, R. GUPTA, L. BONEN, B. J. LEWIS, D. A. STAHL, K. R. LUEHRSEN, K. N. CHEN, and C. R. WOESE: The phylogeny of prokaryotes, Science **209** (198), 457—463

[23] GILBERT, W.: Introns and exons: Playgrounds of evolution, In: Eucaryotic Gene Regulation. Academic Press, 1979, p. 1—12

[24] GŌ, M.: Correlation of DNA exonic regions with protein structural units in haemoglobin, Nature **291** (1981), 90—92

[25] GOLDMAN, A. L.: The Bible is in bloom at St. John the Divine, New York Times 31. 7. 1981

[25a] GUEST, W. and F. E. CLAYTON: Evolution-science versus creation-science, Evolution **35** (1981), 822

[26] GURIN, J.: The creationist revival, The Sciences, April 1981, p. 16—19, 34

[27] HAECKEL, E.: Der Kampf um den Entwicklungsgedanken, Leipzig 1953

[28] v. HELMHOLTZ, H.: Philosophische Vorträge und Aufsätze, Hrsg.: H. HÖRZ/S. WOLLGAST, Berlin 1971

[29] HÖRZ, H. und K. WENIG: Zur Genese des Entwicklungsdenkens in der Biologie bis zur Mitte des 19. Jahrhunderts, In: Materialistische Wissenschaftsgeschichte, Argument-Sonderband AS 54, Berlin 1981, p. 107—131
[30] HÖRZ, H. und K.-F. WESSEL: Philosophische Entwicklungstheorie — weltanschauliche, erkenntnistheoretische und methodologische Probleme der Naturwissenschaften, Berlin 1982
[31] JENSEN, E. Ø., K. PALUDAN, J. J. HYLDIG-NIELSEN, P. JØRGENSEN, and K. A. MARCKER: The structure of a chromosomal leghaemoglobin gene from soybean, Nature **291** (1981) 677—679
[32] JUKES, T. H.: Neutral changes revisited, In: D. S. SIGMAN and M. A. B. BRAZIER (eds.): The Evolution of Protein Structure and Function, Academic Press (1980), p. 203—219
[33] JUKES, T. H.: Silent nucleotide substitutions during evolution, Naturwiss. **67** (1980), 534—539
[34] JUKES, T. H.: Creationists persist, Nature **291**, (1981), 186
[35] JUKES, T. H.: Creationism and biochemical backfiring, Trends in Bioch. Sci. **6** (1981), No. 7, I—II
[36] KAN, Y. M. and A. M. DOZY: Evolution of the hemoglobin S and C genes in world populations, Science **209** (1980), 388—391
[37] KANDLER, O.: Archaebakterien und Phylogenie der Organismen, Naturwiss. **68** (1981), 183—192
[38] KASPAR, R.: Die Evolution des Lebendigen als Erkenntnisvorgang, Umschau **80** (1980), 493—498
[39] KIMURA, M.: Evolutionary rate at the molecular level, Nature **217** (1968), 624—626
[39a] KIMURA, M.: The natural theory of molecular evolution, Sci. Am. **241**, No. 5, (1979), 94—104
[40] KIMURA, M.: Estimation of evolutionary distances between homologous nucleotide sequences, Proc. Nat. Acad. Sci. USA **78** (1981), 454—458
[41] KING, J. L. and T. H. JUKES: Non-Darwinian evolution, Science **164** (1969), 788—798
[42] KING, M.-C. and A. C. WILSON: Evolution at two levels in humans and chimpanzees, Science **188** (1975), 107—116
[43] KLEIN, G.: Putting "creation" in the curriculum: Tampa agrees to do it — but now?, Christian Science Monitor, 30. 3. 1981, p. 1
[44] LEDER, P., J. N. HANSEN, D. KONKEL, A. LEDER, Y. NISHIOKA, and C. TALKINGTON: Mouse globin system: A functional evolutionary analysis, Science **209** (1980), 1336—1342
[45] LEWIN, R.: Cultural diversity tied to genetic differences, Science **212** (1981), 908—912
[46] LUMSDEN, C. J. and E. O. WILSON: Translation of epigenetic rules of individual behaviour into ethnographic patterns, Proc. Nat. Acad. Sci. USA **77** (1980), 4382—4386
[47] LUMSDEN, C. J. and E. O. WILSON: Gene-culture translation in the avoidance of sibling incest, Proc. Nat. Acad. Sci. USA **77** (1980), 6248—6250
[48] MARX, K.: Brief an F. ENGELS vom 19. 12. 1860, In: MARX/ENGELS, Werke **30**, Berlin 1964, S. 130—131
[49] MAUTHNER, F.: Wörterbuch der Philosophie, München, Leipzig 1914
[50] MAYR, E.: Das Problem der Artbildung, Vortrag Leopoldina Halle, 12. 5. 1981
[51] MAYR, E.: Der gegenwärtige Stand des Evolutionsproblems, 2. Bernhard-RENSCH-Vorlesung, Münster 20. 5. 1981
[52] MODIANO, G., G. BATTISTUZZI, and A. G. MOTULSKY: Nonrandom patterns of codon usage and of nucleotide substitutions in human α- and β-globin genes: An evolutionary strategy reducing the rate of mutations with drastic effects, Proc. Nat. Acad. Sci. USA **78** (1981), 1110—1114
[53] NISHIOKA, Y. A. LEDER, and P. LEDER: Unusual α-globin-like gene that has cleanly lost both globin intervening sequences, Proc. Nat. Acad. Sci. USA **77** (1980), 2806—2809
[54] ORGEL, L. E. and F. H. C. CRICK: Selfish DNA: the ultimate parasite, Nature **284** (1980), 604—607
[55] PALMOUR, R. M., J. E. CRONIN, A. CHILDS, and B. W. GRUNBAUM: Studies of primate protein variation and evolution: Microelectrophoretic detection, Biochem. Genet. **18** (1980), 793—808
[56] PERIER, F., A. EFSTRATIADIS, P. LOMEDICO, W. GILBERT, R. KOLODNER, and J. DODGSON: The evolution of genes: the chicken preproinsulin gene, Cell **20** (1980), 555—556
[57] PIERCE, K. M.: Putting Darwin back in the dock, Time 117, No. **11**, (1981), 80—82
[58] RIEDL, R.: Die Ordnung des Lebendigen, Hamburg, Berlin: Parey 1975

[59] Ruse, M.: Charles Darwin and group selection, Ann. Science **37** (1980), 615—630
[60] Sapienza, C. and W. F. Doolittle: Genes are things you have whether you want them or not, Cold Spring Harbor Symp. Quant. Biol. **45** (1981), 177—182
[61] Silvers, W. K.: The Coat Colors of Mice. A Model for Mammalian Gene Action and Interaction, Springer Verlag, Berlin 1979
[62] Smith, J. Maynard: Macroevolution, Nature **289** (1981), 13—14
[63] Steele, E.: Somatic Selection and Adaptive Evolution Williams-Wallace International and Croom Helm, Toronto, and London 1979
[64] Tschulok, S.: Deszendenzlehre, Jena 1922
[65] Williamson, D.: Fundamentalist theory troubles scientists at UNC, The Chapel Hill Newspaper, 21. 6. 1981
[66] Wilson, E. O.: On Human Nature, Harvard College Press, Cambridge, MA 1978, (Biologie als Schicksal. Ullstein. Frankfurt/M., Berlin, Wien 1980)
[67] Woese, C. R.: Archaebacteria, Sci, Am. **244** (1981), 6, 94—106
[68] Yamada, Y., V. E. Arredimento, M. Mudryj, H. Ohkubo, G. Vogeli, M. Irani, I. Pastan, and B. De Crombrugghe: The collagen gene: Evidence for its evolutionary assembly by amplification of a DNA segment containing an exon of 54 bp, Cell **22** (1980), 887—892

Department of Physiology, University of Cambridge, England

The Origins of the Origin — DARWIN's Early Scientific Development

R. D. KEYNES

I must begin this talk with a disclaimer. My presence here today owes more to my ancestry than to any profound knowledge of the topics to be discussed at this Symposium, for although I am a biologist by profession, my field is the biophysics of the nerve membrane, and is thus nearly as remote from evolutionary biology as it could be. However, thanks to my accidental discovery in 1968 of a forgotten cache of drawings made by Conrad MARTENS on board H. M. S. "Beagle" in 1833 and 1834, a selection from which was published in "The Beagle Record" a couple of years ago (KEYNES, 1979), I have perforce had to learn something about the opening stages of my great-grandfather's scientific career. Since this part of his story may be less familiar to some of you than the later years, I offer you on this occasion an account of DARWIN's formative years as a scientist.

As the son of a prosperous physician, Dr. Robert DARWIN, he was sent to the nearest 'public' school at Shrewsbury, where he received the education typical for the middle and upper classes of 19th century England. This cannot be given much credit for his subsequent intellectual development, for it consisted mainly of Latin and Greek, occasionally enlivened by lectures on the geometry of Euclid. According to DARWIN, the best part of his education at school was the time that he spent with his elder brother Erasmus working, often until late at night, in the chemical laboratory that they built in the garden tool-shed at home. This earned him the nickname 'Gas' among his schoolfellows, and a public reprimand from the Headmaster for wasting his time on such useless pursuits. Dr. BUTLER of Shrewsbury was not the only person to take a rather harsh view of his scholastic talents, for his father also castigated him with the words 'You care for nothing but shooting, dogs and rat-catching, and you will be a disgrace to yourself and all your family.' The earliest mention of DARWIN's activities as a naturalist is his comment in his comment in his "Autobiography" (BARLOW, 1958) that already as a schoolboy he was enthusiastically collecting beetles and butterflies, and watching birds. 'In my simplicity I remember wondering why every gentleman did not become an ornithologist.'

The next phase of his upbringing was the year and a half that he spent at Edinburgh University studying medicine. His medical studies never really flourished, but he attended Professor HOPE's chemistry lectures, and enrolled in Professor JAMESON's geology class. Although JAMESON's lectures attracted large audiences, 'The sole effect they produced on me was the determination never as long as I lived to read a book on Geology or in any way to study the science.' Fortunately, he did not stick to this resolution. He made friends with Robert GRANT (1793—1874), whose reputation as a zoologist earned him the Professorship of Comparative Anatomy and Zoology at University College London the following year. GRANT introduced him to the Plinian Society, a student society for scientific discussion, and indoctrinated him into the mysteries of marine biology. His first scientific papers were written for presentation to the Plinian Society in 1827. One of them was the demonstration that the globular black objects which the local fishermen called 'great peppercorns', often found sticking to shells and associated with the seaweed *Focus loreus*, were not the seeds of the seaweed, as everyone including Grant had assumed, but were actually the egg-cases of the vermiform leech *Pontobdella muricata*. Another showed that the small yellow bodies found with the marine invertebrate *Flustra carborea* were not after all its ova, but had the power of independent movement, and were really its larvae. In neither respect was DARWIN quite correct (FREEMAN, 1978), but 5 years before he embarked on the "Beagle" he was nevertheless diplaying one of the most important characteristics of a good field naturalist — the ability to make direct observations, and to check for himself the correctness of the standard explanation for what he saw, taking nothing for granted. Of GRANT he wrote: 'He did nothing more in science, a fact which has always been inexplicable to me'. T. H. HUXLEY was less charitable: 'I met nobody, except Dr. GRANT of University College, who had a word to say for Evolution — and his advocacy was not calculated to advance the cause.'

It was while he was in Edinburgh that DARWIN heard AUDUBON lecture; and he took lessons in taxidermy from a coloured man who had travelled with WATERTON in South America.

DARWIN felt that he had no vocation to practise medicine, and it was therefore agreed that he should proceed to Cambridge University to study theology. He entered Christ's College in October 1827. The formal instruction that he received in Cambridge included even less science than it had in Edinburgh, and apart from just sufficient attention to his theology to obtain a somewhat undistinguished degree, most of his energies were devoted to field sports such as riding, hunting, shooting and fishing, and to entomology. Of this period he wrote (Autobiography, p. 62): 'But no pursuit at Cambridge was followed with nearly so much eagerness or gave me so much pleasure as collecting beetles. It was the mere passion for collecting, for I did not dissect them and rarely compared their external characters with published descriptions, but got them named anyhow. I will give a proof of my zeal: one day, on tearing off some old bark, I saw two rare beetles and seized one in each hand; then I saw a third and new kind, which I could not bear to lose, so that I popped the one which I held in my right hand into my mouth. Alas it ejected some intensely acrid fluid, which burnt my tongue so that I was forced to spit the beetle out, which was lost, as well as the third one.' An amateur he may have been, but hardly a dilettante!

Early in 1830, he became a close friend of the Rev. John Stevens HENSLOW, Professor of Mineralogy and then Botany in Cambridge, one of the founders of the Cambridge Philosophical Society, whose weekly gatherings for students interested in natural history were an important feature of the University's scientific life. Soon he was botanizing almost daily with the Professor, and was known by some of the dons as 'the man who walks with HENSLOW'. To HENSLOW he owed, of course, his selection as the „Beagle's" naturalist, and it was to HENSLOW that he wrote the famous letters reporting his first impressions of the geology and natural history of South America. The great indebtedness to HENSLOW that DARWIN always felt did not, however, rest on his scientific distinction, for 'I do not suppose that anyone would say that he possessed much original genius', and his deeply felt religious views prevented him from ever fully accepting the revolutionary theories of his pupil. But he was a man of sterling character, and since as DARWIN said 'His strongest taste was to draw conclusions from long-continued minute observations', the example of his scientific approach was a very valuable one to his protegé.

The following year, HENSLOW persuaded DARWIN to overcome the distaste for geology that he had acquired in Edinburgh, and to accept some instruction from the Rev. Adam SEDGWICK, Woodwardian Professor of Geology from 1818 to 1873. In August 1831, DARWIN accompanied SEDGWICK on a 3-week geological tour of North Wales, an experience for which he was always intensely grateful. In his „Autobiography" DARWIN wrote (p. 10): 'On this tour I had a striking instance how easy it is to overlook phenomena, however conspicuous, before they have been observed by anyone. We spent many hours in Cwm Idwal, examining all the rocks with extreme care, as SEDGWICK was anxious to find fossils in them; but neither of us saw a trace of the wonderful glacial phenomena all around us; we did not notice the plainly scored rocks, the perched boulders, the lateral and terminal moraines. Yet these phenomena are so conspicuous that, as I declared in a paper published many years afterwards in the „Philosophical Magazine", a house burnt down by fire did not tell its story more plainly than did this valley. If it had still been filled by a glacier, the phenomena would have been less distinct than they are now.' Through SEDGWICK, DARWIN was brought into contact with the controversies over the past history of the Earth that had convulsed geology for the previous half century. SEDGWICK was not always the most flexible of men, his eventual reaction to a copy of "The Origin of Species" sent him by DARWIN in 1859 being to write: 'I have read your book with more pain than pleasure. You have deserted — after a start in that tram-road of all solid physical truth — the true method of induction, and started us in machinery as wild, I think, as Bishop WILKIN's locomotive that was to sail with us to the moon. 'But he was an honest scientist, not afraid to shift his viewpoint in the face of the evidence, and this too was a useful scientific lesson for DARWIN.

Teacher need not always be encountered in the flesh, and there were 3 books that particularly influenced the young DARWIN. One of them was Sir John HERSCHEL's "Preliminary Discourse on the Study of Natural Philosophy", in which he extolled the virtues of scientific enquiry as a way of life. His conclusion that studies of natural history could be undertaken without an advanced knowledge of mathematics must have encouraged DARWIN, who confessed himself unable at that time 'to see any meaning in the early in algebra', although in later life he deeply regretted his ignorance of mathematics. Another work that filled him with a burning zeal for science was HUMBOLDT's "Personal Narrative of Travels to the Equinoctial Regions of America during the Years 1799—1804". I like to think that it may have been the insistence with which he frequently inflicted lengthy readings from HUMBOLDT on his Cambridge friends whether they liked it or not that led directly to the most fruitful of HENSLOW's interventions in his career — the proposal that he should sail on the "Beagle" with FITZROY.

The third and ultimately the most important of the books that formed his scientific reading was Charles LYELL's "Principles of Geology, being an Attempt to explain the former Changes of the Earth's Surface, by reference to Causes now in operation". He did not meet LYELL in person until soon after returning to England in 1836, when they quickly became close friends, LYELL and HOOKER being the only people to whom he confided his innermost scientific secrets during the long years of study that preceeded the joint communication with WALLACE to the Linnean Society in 1858, and the publication of "The Origin of Species" a year later. The influence of LYELL's "Principles" in helping to form DARWIN's scientific views was crucial, as he acknowledged in his "Autobiography"

(p. 101): 'The science of Geology is enormously indebted to LYELL — more so, I believe, than to any other man who ever lived. When I was starting on the voyage of the "Beagle", the sagacious HENSLOW, who like all other geologists believed at that time in successive cataclysms, advised me to get and study the first volume of the "Principles", which had then just been published, but on no account to accept the views therein advocated. How differently would anyone now speak of the "Principles"! I am proud to remember that the first place, namely St. Jago, in the Cape Verde Archipelago, which I geologised, convinced me of the infinite superiority of LYELL's views over those advocated in any other work known to me.' The actual copy of Volume 1 of the "Principles" that DARWIN took to South America has on its title page the pencilled inscription 'Given me by Capt. F. R., C. DARWIN', for having recruited a geologist, FITZROY presumably wished him to be fully informed about current thinking on the subject, somewhat ironically as it turned out. In Volume 2, published early in 1832, DARWIN noted in ink that the book reached him in Montevideo that November; he had probably ordered it before leaving England.

This, then, sets the stage for DARWIN's departure from Plymouth on board the "Beagle" on 27th December 1831. By the standards of today, he could hardly be considered to have had sufficient training to equip him for the tasks that lay ahead, but it should not be forgotten that the days of Ph. D. courses in the minutiae of geology, or taxonomy, or ecology, or recombinant DNA, were still in the distant future, and FITZROY would have been hard put to it to find anyone with much greater experience in the field. DARWIN had, after all, impressed several of the leading scientists of the day with his promise, and had already displayed an ability to think for himself, a single-mindedness in pursuing his objectives, and a strength of motivation, that would have been difficult to match.

Within two weeks of sailing, DARWIN was displaying the hall-marks of a practised professional. In his diary (BARLOW, 1933) for 10th January 1832 he noted: 'I proved today the utility of a contrivance which will afford me many hours of amusement & work, it is a bag four feet deep, made of bunting, & attached to a semicircular bow: this by lines is kept upright, & dragged behind the vessel. This evening it brought up a mass of small animals & tomorrow I look forward to a greater harvest.' On 11th January he continued: 'I am quite tired having worked all day at the produce of my net. The number of animals that the net collects is very great & fully explains the manner so many animals of a large size live so far from land. Many of these creatures, so low in the scale of nature, are most exquisite in their forms & rich colours. It creates a feeling of wonder that so much beauty should be apparently created for such little purpuse.' This was only the second recorded use of a plankton net, preceeding by some years the invention often credited to Johannes MÜLLER. The first was that of J. V. THOMPSON a few years earlier, of which GRANT had told him. Although in the course of the voyage, DARWIN devoted many hours to the study of specimens collected in his net, the results were never published. The labour was not wholly wasted, however, despite T. H. HUXLEY's dismissal of the notes on surface fauna made on the "Beagle" as 'a vast accumulation of useless manuscript', when he wrote DARWIN's obituary for the Royal Society. For in DARWIN's own view, the knowledge that he gained helped him later with his barnacles, and in any case like the trouble that he took to keep his journal up to date, it was 'good practice'.

DARWIN's comments on the harvest of plankton are typical in two respects of many later entries in his diary. One is the manner in which he invariably commented or speculated with critical ingenuity on the why and wherefore of all his zoological and ornithological observations. Who else at that period, having discovered the relative abundance of these tiny organisms in the ocean, would have immediately related his observation to the rest of the food chain? The other characteristic feature is that he invariably retained a sense of wonder at the beauty of nature, whether from the strictly scientific or from the purely aesthetic viewpoint.

Shortly afterwards he had his first taste of geology, which he described in his "Autobiography" (p. 11) as 'far more important' than natural history because it involved reasoning as well as observation. This hardly seems a fair comment on his actual method of approach, because as I have just pointed out, he never failed to exercise his powers of reasoning to the full in considering the logical implications of everything that he came across. He continued (p. 81): 'The geology of St. Jago is very striking yet simple: a stream of lava formerly flowed over the bed of the sea, formed of triturated recent shells and corals, which has baked into a hard white rock. Since then the whole island has been upheaved. But the line of white rock revealed to me a new and important fact, namely that there had been afterwards subsidence round the craters, which had since been in action, and had poured forth lava. It then first dawned on me that I might perhaps write a book on the geology of the various countries visited, and this made me thrill with delight. That was a memorable hour to me, and how distinctly I can call to mind the low cliff of lava beneath which I rested, with the sun glaring hot, a few strange desert plants growing near, and with living corals in the tidal pools at my feet. Later in the voyage FITZROY asked to read some of my Journal, and declared it would be worth publishing; so here was a second book in prospect!'

His enthusiasm fully kindled, he wrote to his father after little more than six weeks on the "Beagle": 'I think, if I can so soon judge, I shall be able to do some original work in Natural History — I find there is little known about many of the tropical animals.' In the course of the voyage, there were some periods when the geology claimed most of his attention, and as he wrote to one of his sisters: 'There is nothing like geology; the pleasure of the first day's partridge shooting or first day's hunting cannot be compared to finding a fine group of fossil bones, which tell their story of former times with almost a living tongue.' At other times, especially towards the end, he concentrated more on natural history. But wherever his interests were directed, he made his observations with the same meticulous attention to detail and his preliminary interpretations of their meaning with the same critical judgement.

The place where the seeds of "The Origin of Species" were sown, and where DARWIN discovered the first of the lines of evidence that led him to question and finally to reject the doctrine of the fixity of species, was Punta Alta, at the northern extremity of Patagonia. On 22nd September 1832, he recorded in his pocket note-book (BARLOW, 1945): 'Entrance of creek, dark blue sandy clay much stratified dipping to NNW or N by W at about 6°. On the beach a succession of thin strata dipping at 15° to W by S — conglomerate quartz and jasper pebbles — with shells — vide specimens. Proceeding to NW there is a horizontal bed of **earth** containing much fewer shells — but armadillo — this is horizontal but widens gradually, hence I think conglomerate with broken shells was deposited by the action of tides — **earth** quietly. Megatherium like Armadillo case, teeth.' In the journal entry for the same day he wrote: 'We staid some time on Punta Alta about ten miles from the ship; here I found some rocks. These are the first I have seen, and are very interesting from containing numerous shells and the bones of large animals.' His identification of the large armadillo-like animals as *Megatherium*, both in his notes and in letters to his sister and Professor HENSLOW, was incorrect. It was actually a glyptodont called *Hoplophorus*. But he may be forgiven, for as he wrote in his Journal on 8th October: 'I obtained a jaw bone, which contained a tooth: by this I found that it belongs to the great antediluvial animal the Megatherium. This is particularly interesting as the only specimens in Europe are in the King's collection at Madrid, where for all purposes of science they are nearly as much hidden as if in their primaeval rock.' The mistake was corrected when Professor OWEN examined the specimens in January 1837.

Although in the first edition of the "Journal of Researches", published in 1839, DARWIN pointed out the contrast in size between the fossil animals that he had found and the similar species now living in South America, and speculated on the reasons for the extinction of the former, he did not at this stage mention the possible implications of the close resemblance in everything but size between the extinct and living species. In the privacy of his notebooks, he had, nevertheless, begun to think about them, and a journal entry for the year 1837 (de BEER, 1959) runs as follows: 'In July opened first note book on "transmutation of species" — Had been greatly struck from about month of previous March on character of S. American fossils — & species on Galapagos Archipelago. These facts origin (especially latter) of all my views.' DARWIN's earliest enquiries into the question of the distribution of species through time and through space are to be found in his 'Red Notebook', and have been dated by Dr. Sandra HERBERT (1980) as having been made in March 1837.

One of the crucial issues to which it is clear from his field notes that his attention constantly returned was the significance of the distribution of similar species of plants and animals as between isolated oceanic islands and the nearest mainland, and on either side of a physical barrier such as an extensive mountain range. As early as 2nd March 1833, he noted in his pocket-book on landing in the Falkland Islands for the first time (BARLOW, 1945): 'Is not the closer connection of insects and plants [with the mainland equivalents] as well as this fact point out closer connection than migration? To what animals did the dung beetles in S. America belong? Scarcity of Aphidians? Vide Annales des Sciences for Rio Plata. The peat not forming at present, and but little of the Bog Plants of Tierra del F; no moss; perhaps decaying vegetables may slowly increase it. Beds ranging from 10 to one foot thick. Great scarcity in Tierra del of Corallines, supplanted by Fuci: Clytra prevailing genus.' Three weeks later he wrote: 'It will be interesting to observe difference of species and proportionate numbers; what also appear characters of different habitations. Migrations of Geese in Falkland Islands, as connected with Rio Negro?' When the "Beagle" returned to the Falklands again the following year, the differentiation of species associated with geographical isolation was again much in his mind, and the sentence in his notes 'Rats and mice and Foxes on small islands of Georgia?' was heavily underlined.

A related issue was the question of representative species. Why in neighbouring areas of South America did different yet closely similar species replace one another, sometimes with overlapping ranges? DARWIN's unpublished zoology notes in the Cambridge University Library show how from 1833 onwards he was constantly trying to relate the distribution of what seemed to be allied species to possible changes in habitat and the existence of physical barriers like the Cordilleras of the Andes. One example of the principle of representative species was the presence on the pampas of two kinds of rhea, differing only a little in colouring and size. By the end of 1833, DARWIN was familiar with the common *Rhea americana*, but had still not seen the smaller bird known by the gauchos as the 'Avestruz Petise' When the "Beagle" was at Port Desire in Patagonia, the artist Conrad MARTENS (KEYNES, 1979) wrote to his brother Henry that 'While here, as there was but little to be done in the way of sketching, I used generally to take my gun, and was fortunate enough one day to bring home an ostrich, the only one indeed which as yet we had been able to kill, altho great numbers had been seen. It was a young one, and excellent eating.' But its importance was more than gastronomic, for let DARWIN take up the tale: 'Mr. MARTENS shot an ostrich; I looked at it, and from most unfortunate forgetting at the moment the whole subject of the Petises, thought it was a two-third grown one of the common sort. The bird was skinned and cooked before my memory returned. But the head, neck, legs, wings, many of the larger feathers, and a large part of the skin, had been preserved. From these a very nearly perfect specimen has been put together, and is now exhibited in the museum of the Zoological Society.' The type specimen of *Rhea darwinii* thus had a somethat chequered history!

Despite the fact that in his pocket-book entries, and in the zoological notes written up towards the end of the voyage, he returned to this problem again and again, there is little hint of his interest in it in the diary about his general activities that he kept for the for 5th August 1834, when the "Beagle" had just arrived at Valparaiso (BARLOW, 1933), runs as follows: 'I have taken several long walks in the country. The vegetation here has a peculiar aspect; this is owing to the number and variety of bushes which seem to supply the place of plants; many of them bear very pretty flowers and very commonly the whole shrub has a strong resinous or aromatic smell. In climbing amongst the hills one's hands, and even clothes, become strongly scented. With this sort of vegetation I am surprised to find that insects are far from common; indeed this scarity holds good to some of the higher orders of animals — there are very few quadrupeds, and birds are not very plentiful. I have already found beds of recent shell yet retaining their colours, at an elevation of 1300 feet; and beneath this level the country is strewed with them. It seems not a very improbable conjecture that the want of animals may be owing to none having been created since this country was raised from the sea.' This passage was duly transferred to the "Journal of Researches", but with the significant omission of the final sentence.

To some extent, DARWIN was ready for what he found on landing in the Galapagos Islands, 700 miles off the coast of Ecuador. Two days later he noted in his pocket-book: 'The Thenca very tame and curious in these Islands. I certainly recognise S. America in Ornithology — would a botanist? $\frac{3}{4}$ of plants in flower.' In his diary for 26th and 27th September 1835 he wrote: 'I industriously collected all the animals, plants, insects and reptiles from this Island. It will be very interesting to find from future comparison to what district or "centre of creation" the organized beings of this archipelago must be attached.'

Full realisation of the significance of what he had seen did not, however, dawn upon him until he was completing his ornithological notes towards the end of the voyage in the following year (BARLOW, 1963). He wrote: 'These birds are closely allied in appearance to the Thenca of Chile or Callandra of La Plata. In their habits I cannot point out a single difference; they are lively, inquisitive, active, run fast, frequent houses to pick up the meat of the Tortoise which is hung up, sing tolerably well; are said to build a simple open nest; are very tame, a character in common with the other birds. I imagined however its note or cry was rather different from the Thenca of Chile? Are very abundant over the whole Island; are chiefly tempted up into the high and damp parts by the houses and cleared ground.

I have specimens from four of the larger Islands: the two above enumerated, a female from Albermarle Island, and a male from James Island. The specimens from Chatham and Albermarle Island appear to be the same; but the other two are different. In each Island each kind is exclusively found; habits of all are indistinguishable. When I recollect the fact that from the form of the body, shape of scales, and general size, the Spaniards can at once pronounce from which Island any Tortoise may have been brougth. When I see these Islands in sight of each other, and possessed of but a scanty stock of animals, tenated by these birds but slightly differing in structure and filling the same place in Nature, I must suspect they are only varieties. The only fact of a similar kind of which I am aware is the constant asserted difference between the wolf-like Fox of East and West Falkland Islands. If there is the slightest foundation for these remarks, the zoology of Archipelagoes will be well worth examining, for such facts would undermine the stability of Species.'

Thus it was that the accumulation of fresh evidence to which he had been exposed, starting with the fossils at Punta Alta and culminating in the Galapagos, coupled with the penetrating honesty with which he analyzed its true implications, had forced him by the end of the voyage to begin to admit to himself that it was necessary to abandon the orthodox viewpoint and seek an altogether new framework to accommodate his observations. However, he was still far from ready to expose his doubts in public, and when writing the first edition of the "Journal of Researches" in 1837 he was cautious in the extreme when dealing with the subject of the finches that came to bear his name: 'I have stated that in the thirteen species of ground-finches, a nearly perfect graduation may be traced, from a beak extraordinarily thick, to one so fine, that it may be compared to that of a warbler. I very much suspect, that certain members of the series are confined to different islands; therefore, if the collection had been made on any one island, it would not have presented so perfect a graduation. It is clear, that if several islands have each their peculiar species of the same genera, when these are placed together, they will have a wide range of character. But there is not space in this work, to enter on this curious subject.'

In the second and final edition of the "Journal of Researches", published in 1845, the chapter on the Galapagos was considerably expanded, and after describing 'the perfect gradation in the size of the beaks in the different species' of ground-finches, DARWIN concluded: 'Seeing this gradation and diversity of structure in one small, intimately related group of birds, one might really fancy that from an original paucity of birds in this archipelago, one species had been taken and modified for different ends.' Dealing with the same subject in Volume II of the "Narrative of the Surveying Voyages of His Majesty's Ships Adventure and Beagle", of which the 'Journal of Researches' was Volume III, Captain FITZROY, who had become an ardent fundamentalist soon after returning to England, had no doubts as to what those ends were, for he wrote: 'This appears to be one of those admirable provisions of Infinite Wisdom by which each created thing is adapted to the place for which it was intended.' But even those readers who would not have gone along with this extreme view cannot have appreciated the true inwardness of DARWIN's scrupulously guarded and yet highly significant sentence.

It should be appreciated, nevertheless, that although at this time DARWIN had started to have doubts about the fixity of species, he had not yet deserted, as he had done by the end of his life, his belief in God the Creator. As evidence on this point, there is a well known passage in his diary for 19th January 1836, when he was in Australia: 'I had been lying on a sunny bank & was reflecting on the strange character of the animals of this country as compared to the rest of the World. An unbeliever in everything beyond his own reason might exclaim, "Surely two distinct Creators must have been at work; their object, however, has been the same & certainly the end in each case is complete." Whilst thus thinking, I observed the conical pitfall of a Lion-Ant — a fly fell in & immediately disappeared; then came a large but unwary Ant. His struggles to escape being very violent, the little jets of sand described by Kirby were promptly directed against him. His fate, however, was better than that of the fly's. Without doubt the predacious larva belongs to the same genus but to a different species from the European kind. Now what would the Disbeliever say to this? Would any two workmen ever hit on so beautiful, so simple, & yet so artificial a contrivance? It cannot be thought so. The one hand has surely worked throughout the universe. A Geologist perhaps would suggest that the periods of Creation have been distinct and remote the one from the other; that the Creator rested in his labor.'

"The Beagle" docked at Falmouth on 2nd October 1836, and DARWIN was soon hard at work sorting out his specimens and turning them over to specialists for identification — OWEN for the fossils, WATERHOUSE for the mammals, GOULD for the birds, JENYNS for the fish, and BELL for the reptiles. Within a few weeks he had established the friendship with LYELL that led to his becoming, with HOOKER, one of the only two men to whom he entrusted the secret of his evolutionary theories during the long years that preceeded the publication of "The Origin of Species" in 1859. Even they were kept pretty much in the dark about the way in which his thoughts were developing until they were shown the Essay of 1844. Yet his first tentative enquiries about the questions of the geographical distribution of animals, of the relation between the spatial and temporal distribution of species, and of their reproduction, were set on paper in the 'Red Notebook' in March 1837, and the first of the series of four notebooks on "Transmutation of Species" was opened in July of the same year (de BEER, 1960). There were two main reasons for this extreme reticence on his part. One of them was undoubtedly his acute awareness of the opposition that his ideas would be bound to encounter from religiously minded people. The other was that he had set himself from the beginning to achieve impeccable standards of scientific reasoning, backed by as much hard evidence as he could muster, and the labour of accumulating sufficient facts was immense.

On 28th September 1838, 'fifteen months after I had begun my systematic enquiry', wrote DARWIN in his "Autobiography" (p. 120), 'I happened to read for amusement MALTHUS on "Population", and being well prepared to appreciate the struggle for existence which everywhere goes on from long-continued observation of the habits of animals and plants, it at once struck me that under these circumstances favourable variations would tend to be preserved, and unfavourable ones to be destroyed. The result of this would be the formation of new species. Here, then, I had at last got a theory by which to work; but I was so anxious to avoid prejudice, that I determined not for some time to write even the briefest sketch of it. In June 1842 I first allowed myself the satisfaction of writing a very brief abstract of my theory in pencil in 35 pages; and this was enlarged during the summer of 1844 into one of 230 pages, which I had fairly copied out and still possess.' This was the Essay that he showed to LYELL and HOOKER.

It has always puzzled historians of science to explain why a reading of MALTHUS was necessary in order to set DARWIN thinking about natural selection. The concept of the "struggle for existence" was familiar to him from LYELL, who in Volume 2 of the "Principles" had written: 'If we consider the vegetable kingdom generally, it must be reflected, that even of the seeds which are well ripended, a great part are either eaten by insects, birds, and other animals, or decay for want of room and opportunity to germinate. Unhealthy plants are the first which are cut off by causes prejudicial to the species, being usually stifled by more vigorous individuals of their own kinds. If, therefore, the relative fecundity or hardiness of hybrids be in the least degree inferior, they cannot maintain their footing for many generations, even if they were ever produced beyond one generation in a wild state. In the universal struggle for existence, the right of the strongest eventually prevails; and the strength and durability of a race depends mainly on its prolificness, in which hybrids are acknowledged to be deficient.' DARWIN himself, writing of the Fuegians in his diary for 24th February 1834, had remarked: 'They are surrounded by hostile tribes speaking different dialects, and the cause of their warfare would appear to be the means of subsistence.' And 'survival of the fittest', a term originally introduced by Herbert SPENCER, seems clearly to have been foreshadowed when he wrote in the first "Transmutation of Species" notebook (de BEER, 1960), a year before reading MALTHUS: 'With respect to extinction we can easily see that [the] variety of ostrich Petise may not be well adapted, and thus perish out, or on the other hand like Orpheus [a bird common in South America] being favourable, many might be produced. This requires [the] principle that the permanent varieties, produced by confined breeding and changing circumstances are continued and produce according to the adaptation [to] such circumstances, and therefore that death of species is a consequence (contrary to what would appear from America) of non-adaptation [to] circumstances.'

I incline to the view that rather than seeking for complex explanations in terms of differences between DARWIN's, LYELL's and MALTHUS's definitions of a struggle, what happened on 28th September 1838 was simply that the reading of MALTHUS triggered the surfacing of an idea already dormant in DARWIN's mind, but not hitherto consciously expressed. Alfred Russel WALLACE's account of the way in which he too was influenced by MALTHUS, contains a similar element: 'During one of these fits [of malaria], while again considering the problem of the origin of species, something led me to think of MALTHUS's "Essay on Population" (which I had read about ten years before), and the "positive checks" — war, diesease, famine, accident, etc. — which he adduced as keeping all savage populations nearly stationary. It then occurred to me that these checks must also act upon animals, and keep down their numbers; and as they increase so much faster than man does, while their numbers are always very nearly or quite stationary, it was clear that these checks in their case must be far more powerful, since a number equal to the whole increase must be cut off by them every year. While vaguely thinking how this would effect any species, there suddenly flashed upon me the idea of 'the survival of the fittest' — that the individuals removed by these checks must be, on the whole, inferior to those that survided. Then, considering the variations continually occurring in every fresh generation of animals or plants, and the changes of climate, of food, of enemies always in progress, the whole method of specific modification became clear to me, and in the two hours of my fit I had thought [out] the main points of the theory.'

Be that as it may, the young scientist whom FITZROY had engaged on HENSLOW's advice in 1831, had deplored on the penultimate page of his diary of the voyage 'a constant tendency to fill up the wide gaps of knowledge by inaccurate and superficial hypotheses.' Only in 1859, sure at last of his ground, would he write in the opening paragraph of his great work: 'When on board H. M. S. "Beagle" as naturalist, I was much struck with certain facts in the distribution of the inhabitants of South America, and in the geological relations of the present to the past inhabitants of that continent. These facts, as will be seen in the latter chapters of this volume, seemed to throw some light on the origin of species — that mystery of mysteries, as it has been called by one of our greatest philosophers.'

References

[1] BARLOW, N. (1933): Charles DARWIN's Diary of the Voyage of H. M. S. Beagle. Cambridge University Press.

[2] BARLOW, N. (1945): Charles DARWIN and the voyage of H. M. S. Beagle. Pilot Press, London.

[3] BARLOW, N. (1958): The Autobiography of Charles DARWIN. Collins, London.

[4] BARLOW, N. (1963): DARWIN's Ornithological Notes. Bulletin of the British Museum (Natural History), Historical Series. 2, 201—278.

[5] de BEER, G. (1959): DARWIN's Journal. Bulletin of the British Museum (Natural History), Historical Series. 2, 1—21.

[6] de BEER, G. (1960): DARWIN's Notebooks on Transmutation of Species Parts I—IV. Bulletin of the British Museum (Natural History) Historical Series. 2, 23—200.

[7] DARWIN, C. (1839): Journal of researches into the geology and natural history of the various countries visited by H. M. S. Beagle. Henry Colburn, London.

[8] DARWIN, C. (1845): Journal of researches into the natural history and geology of the countries visited during the voyage of H. M. S. Beagle round the world, under the Command of Capt. Fitz ROY, R. N. 2nd edition. John MURRAY, London.

[9] DARWIN, C. (1859): On the origin of species by means of natural selection, or the preservation of favoured races in the struggle for life. John MURRAY, London.

[10] FREEMAN, R. B. (1978): Charles DARWIN — a Companion. Dawson, London.

[11] HERBERT, Sandra (1980): The Red Notebook of Charles DARWIN. British Museum (Natural History), London.

[12] KEYNES, R. D. (1979): The Beagle Record. Selections from the original pictorial records and written accounts of the voyage of H. M. S. Beagle. Cambridge University Press.

Discussion

NOVAK: I have one question to Prof. KEYNES: Are data on the attitude of the very young Charles DARWIN to the writings of his grandfather, Erasmus, available? I think they might have produced a strong emotional stimulus in the boy and I would not be surprised, on the basis of what we know about his character, if he avoided speaking about it openly.

KEYNES: It is said that Charles DARWIN first read his grandfather's book "Zöonomia" when he was at Edinburgh. But in his "Autobiography" C. D. said: 'on reading it a second time after an interval of ten or fifteen years, I was much disappointed; the proportion of speculation being so large to the facts given'.

GRANTHAM: When did DARWIN first read LAMARCK and what was his initial reaction? (I know that he later rejected „LAMARCKISM"). I have seen it suggested that he learned of LAMARCK through LYELL.

KEYNES: I believe that there is evidence that DARWIN was using LAMARCK's classification of invertebrates as a student in Edinburgh. Robert GRANT was a follower of LAMARCK, so it was probably from him rather than LYELL that DARWIN learnt about LAMARCK.

PASTERNAK: It is an interesting fact that exactly 150 years ago Charles DARWIN went on board the HMS "Beagle" to start a round-the-world voyage. Was this the only journey he made during his life?

KEYNES: DARWIN had once very briefly visited the Continent before the voyage of the "Beagle", but he never made any other journies outside Great Britain.

URBANEK: Dr. KEYNES, what is your opinion of Gertrude HIMMELFARB's view that DARWIN could properly evaluate South American fossils and Galapagos Finches only after OWEN and GOULD identified and described specimens he had collected that is after he was back from his voyage on the "Beagle". HIMMELFARB rejects the widespread opinion that DARWIN's belief in the transmutation of species grew gradually due to observations made during the voyage. She expressed doubts regarding the purely empirical origin of DARWINian theory.

KEYNES: I think that HIMMELFARB has a point that DARWIN may not have fully appreciated all the implications of the speciation that he found on the Galapagos until GOULD had identified the specimens he had brought back; and the same may be true about the fossils until OWEN had reported on them. Nevertheless, the passage that I quoted about the Galapagos and the stability of species was definitely written on board the "Beagle" before the end of the voyage, which argues against HIMMELFARB. Moreover, his ideas about geographical distribution certainly began to develop as early as 1833, as proved by numerous entries in his "Beagle" pocket-book, and in his zoological and ornithological notes. I therefore conclude that his doubts about the fixity of species did begin to arise, no doubt quite gradually, while he was still in South America.

URBANEK: It is striking, however, that there are only 3 or 4 remarks on the original copy of "Generelle Morphologie der Organismen" which was sent to DARWIN by Ernst HAECKEL. They are all on rudiments and reduction or organs. It seems that DARWIN was little impressed by HAECKEL's great work — both represented quite different ways of thinking.

993 Park Avenue, New York 10028, U.S.A.

"I Never Wrote so Much About Myself": Charles DARWIN's 1861—1870 Autobiographical Notes

R. COLP, Jr.,

In 1859, when Charles DARWIN was 50 years old, he published his book "The Origin of Species". Over the next 13 years his life reached a climax. "The Origin" provoked a furious controversy, and then effected a revolution in almost every area of human thought. During these years DARWIN devoted most of his working time to closely following the reactions to his book, and then to publishing works which further supported his ideas. He also wrote — in response to the requests of three authors — a succession of autobiographical notes. What follows will describe how these notes came to be written, and what they reveal about how Darwin — who was always conscious of his increasing fame — viewed the events of this life personal attributes.

In the years prior to the publication of "The Origin" DARWIN had read, with interest, at least 30 literary biographies of diverse individuals.[1] He had carefully retained different papers which pertained to his biography: his "Beagle" diaries and notebooks; his evolutionary notebooks; and a variety of other notes, and annotations to books, which chronicled his scientific thoughts. Since 1838 he had kept a little "Journal" in which he recorded, year by year, the most important events of his personal and professional life. In 1838 he had written two short recollections of his childhood and youth.[2] These writings were shown to no one, and were evidently intended only for his private perusal.

1 From 1838—1859 DARWIN read biographies, autobiographies, diaries, and letters of the following individuals: John DRYDEN, Thomas GRAY, Robert BURNS, Percy SHELLEY, and Lord BYRON; Walter SCOTT, Edward GIBBIN, and Jonathan SWIFT; Wolfgang Amadeus MOZART and Joseph HAYDN; David HUME and Jean Paul RICHTER; James HUTTON, Isaac RAY, and Isaac NEWTON (DARWIN read David BREWSTER's biography of NEWTON 3 months after the publication of "The Origin of Species"); Thomas MORE; the Duke of St. Simon; Michel de MONTAIGNE; John WESLEY and the Roman Catholic Popes; the Bronte sisters and Mary Wollstoncraft; Constantine the Great, Oliver Cromwell, Frederick the Great, Napoleon, and the Duke of Wellington; Robert CLIVE and Thomas Stamford RAFFLES. When be was 16 years old he read BOSWELL's "Life of Samuel JOHNSON", a volume at a time; when he was 30 he reread the "Life". From 1854—1856 he read all of the volumes of Samuel PEPYS' "Diary", commenting that the early volumes were "excellent," but that the particular edition was "poor." In 1848 he read Alexander SOMERVILLE's just published, "The Autobiography of a Working Man", which described how a working-class Englishman was affected by the Industrial Revolution and 1832 Reform Bill, and he rated it "excellent." For a complete list of DARWIN's readings see: Peter J. VORZIMMER, "The DARWIN reading notebooks (1838—1860)," "Journal of the History of Biology", 1977, 10, 107—153. For DARWIN's first reading of BOSWELL's "Life of Samuel JOHNSON", see W. E. SWINTON, "Charles DARWIN in Scotland, "The Advancement of Science", 1958, 15, 201—204, p. 201.

2 Ralph COLP, Jr., "I was born a naturalist': Charles DARWIN's 1838 Notes about Himself," "Journal of the History of Medicine and Allied Sciences", 1980, 35, 8—39.

In the spring of 1861, about $1^1/_2$ years after the publication of "The Origin", Edward WALFORD — an English author and biographer[3] — sent Darwin a note requesting a sketch of the latter's life,[4] so that WALFORD could then publish a short biography. DARWIN complied with this request, and along with his biography sent WALFORD the following note:

June 13 [1861] Down Bromley Kent
Sir

I enclose the dates of the principle events of my Life, according to your request. — I presume that I have given them much too fully, but you can use such as you think fit. —

I have the honour to remain
Sir
your obedt. servt.
Charles DARWIN[5]

DARWIN's Life was then published in 1862 in WALFORD's book, "Men of the Time"[6].

3 Edward WALFORD (1823—1897). English author and editor of a series of biographical and genaelogical works of reference.

4 WALFORD may have first asked DARWIN to meet him in London. This is suggested by the following letter of DARWIN.

[? 1860—1861] 22d Down Bromley Kent
My dear Sir
I should of course be proud to be one of your Series; but I cannot spare a day to go to London, & I am not likely to be there soon: I may, however in course of summer, & will call on Mr. Edwards, if it be then not too late; & he can inform me. —

my dear Sir
yours very faithfully
Ch. DARWIN

The identity of "Mr. Edwards" is unkown, and the identity of Darwin's correspondent is uncertain. DARWIN's reference to "one of your Series" could refer to WALFORD's biographical subjects, or to the subjects of a London photographer (there is, however, no record that DARWIN went to a London photographer in 1860—1861). DARWIN's letter is in the Library of the American Philosophical Society, and is noted in: P. Thomas CARROLL, editor, "An annotated calendar of the letters of Charles DARWIN in the library of the American Philosophical Society" (Scholarly Resources, Wilmington, Delaware, USA, 1976), letter no. 237, p. 90.

5 This letter of DARWIN to WALFORD is in my private possession. It is written not on DARWIN's regular stationery but on a sheet of pale blue note paper, which DARWIN used to make notes for his scientific work, and sometimes to write medical prescriptions on.

6 Edward WALFORD, "Men of the Time: A Biographical Dictionary of Eminent Living Characters, (Including Women)", a new edition, thoroughly revised, and brought down to the present time (London: Routledge, Warne, & Routledge, 1862). This is a book of 842 pages, with an average of 2 biographies per page, for a total of 1664 biographies.

The 1879 edition of "Men of the Time", edited by Thompson COOPER, printed a longer biography of DARWIN. DARWIN was evidently not consulted for this biography, which contained several factual errors.

It was the first published biography of Darwin, and it may have been printed essentially as DARWIN wrote it. The biography reads as follows:

DARWIN, Charles R., born at Shrewsbury, February 12, 1809 is a son of Dr. R. W. DARWIN, F. R. S., of the same place, and grandson of Dr. Erasmus DARWIN, F. R. S., author of the "Botanic Garden", and "Zoonomia", &c. He was educated at the Grammar School at Shrewsbury. In 1825 he went to Edinburgh, and attended the Lectures at the University for two years. At Christmas, 1827, he entered Christ's College, Cambridge, and took his degree in 1831. In the autumn of that year, Captain FITZROY, R. N., offered to give up part of his own cabin to any one who would volunteer to accompany H. M. S. "Beagle" as naturalist. Mr. DARWIN offered his services, and sailed in the "Beagle", for the survey of South America and the circumnavigation of the globe, on December 27, 1831, and returned to England October 2, 1836. Mr. DARWIN published an account of the voyage under the title of "Journal of Researches into the Geology and Natural History of the Various Countries, &c.," which first appeared together with a general account of the voyage by Captain FITZROY; it has subsequently been published separately, and has had a large sale. Mr. DARWIN resided in London from 1831[7] down to 1842, when he removed to his country-house near Bromley, in Kent. In 1839 Mr. DARWIN married the grand-daughter of Josiah WEDGWOOD, F. R. S., the well-known improver and manufacturer of earthenware. Besides numerous papers on various scientific subjects, Mr. DARWIN edited the "Zoology of the Voyage of the "Beagle," and published three separate volumes on geology, under the titles of "The Structure and Distribution of Coral Reefs," 1842; "Geological Observations on Volcanic Islands," 1844; and "Geological Observations on South America", 1846. The most important of Mr. DARWIN's subsequent works are his Monography, in two large volumes, published in 1851 and 1853 by the Ray Society, on recent pedunculated and sessile Cirripedia; and two volumes on fossil Cirripedia, published by the Paleontographical Society. At the close of 1859, Mr. DARWIN published his "Origin of Species by means of Natural Selection," which has now (1861) gone through three editions. Mr. DARWIN has been elected a member of various Foreign and English scientific bodies. The Royal Society awarded, in 1853, the Gold medal to Mr. DARWIN for his various scientific works; and in 1859 the Geological Society awarded him the Wollaston Palladian medal.[8]

In the first part of 1868, Robert HUNT[9] — an English scientific writer, who had corres-

7 "1831" is a misprint. From 1831—1836 DARWIN was on the "Beagle" cruise. He then lived in London, in two different residences, from 1837—1842.

8 WALFORD, "Men of the Time" (n.6), pp. 211—212.

9 Robert HUNT (1807—1887). In the 1840s HUNT studied how the rays of the sun — when passed through different kinds of glass — sensibly accelerated the germination of seeds, and he then did research on photography. In the 1850s he read several papers, on the "Influence of Light on the Growth of Plants", at meetings of the British Association for the Advancement of Science, and he worked on the science parts of the Great Exhibition of 1851. In his later years HUNT gave up research and devoted himself to writing popular books on science, contributing to the science columns of the "Athenaeum", and writing biographies of men of science for the "Dictionary of National Biography".

ponded with DARWIN 13 years previously[10] — requested that DARWIN supply him information for a short biography. Darwin complied, and wrote HUNT as follows:

Down Bromley Kent
May 3 [1868]

My dear Sir
I am much obliged for your note. I enclose a copy of a sketch of the principal events in my life which will give you all the facts & more that you require.
Believe me my dear Sir
yours very faithfully
Ch. DARWIN[11]

HUNT, after receiving DARWIN's sketch, wrote a short biography of DARWIN. It was to be included in Hunt's book, "Biographical memoirs of men of eminence". The book, however, was never published.[12]

10 There are 5 extant DARWIN-HUNT letters, and they are held in four different libraries in England and America. DARWIN and HUNT met in the early 1850s, perhaps at the Great Exhibition of 1851 or at a meeting of the British Association for the Advancement of Science. At this time, as part of his evolutionary theory, DARWIN was studying conditions which affect the migration of plants, and one aspect of this was conditions affecting the germination of seeds. After meeting with HUNT DARWIN asked further information on the work HUNT had done on the influence of light on seeds. On July 19, 1855 HUNT sent DARWIN some of the glasses he had used in his light experiments, along with a letter in which he summarised some of his work. At the end of his letter HUNT wrote: "It is not probable that I shall ever be able to resume my investigations on these matters, and it would rejoice me much to find so able an observer as yourself fairly working over this fertile field of research." DARWIN then wrote HUNT, thanking him for the glasses, and asking where he could procure other kinds of glasses that HUNT had also used. HUNT promptly replied, giving DARWIN the adress of a glass company in Birmingham which would supply the glass. At the end of his letter HUNT, again, told DARWIN: "I shall have the greatest pleasure in aiding you to the utmost in your experiments — since I am not likely to renew them for many years, if ever. I am much pleased to find them taken up by one so able to investigate this interesting subject." In the summer and fall of 1855 DARWIN experimented with crossing different seeds, and the effects of long burial and salt water on seeds, but there does not appear to be any record of his experiments on the effects of light on seeds. (Robert HUNT to DARWIN, 19, July 1855, Jermyn Street, London; Robert HUNT to DARWIN, 24th July 1855, Museum Pract[ical] Geology, Jermyn Street, [London] both letters are in DARWIN Papers, Cambridge University Library; DARWIN to Robert HUNT, July 22 1855] Down Farnborough Kent, Library of the American Philosophical Society. Since this letter was recently acquired it is not mentioned in CARROLL's "Calendar" (n.4). DARWIN's 1855 work on plants is discussed in Mea ALLAN, "DARWIN and his flowers: the key to natural selection" (London: Faber and Faber, 1977), pp. 150—155).

11 The manuscript of the letter and DARWIN's biography of himself are held in the Library of the British Museum (Natural History). Both documents are in the handwriting of Emma DARWIN, except for the signature of the letter which is in DARWIN's handwriting. Both documents have been transcribed and published by: Sir Gavin de BEER, "Some Unpublished Letters of Charles DARWIN", "Notes and Records of the Royal Society of London". June 1959, Vol. 14. No. 1, pp. 12—66, pp. 36—37.

12 In the Library of the British Museum (Natural History) the manuscripts of DARWIN's letter to HUNT and biography of himself are kept bound up with the proof sheets of Robert HUNT's book "Biographical memoirs of men of eminence". HUNT's book, after being printed in proof sheets, was for some undetermined reason not published. HUNT's proof sheets contain his biography of "Charles Robert DARWIN, M. A., F. R. S., F. L. S. &c", which fill 4 printed pages, and which concludes as follows:

The biographies that DARWIN sent to WALFORD and HUNT were extremely short, and made no mention of DARWIN's intellectual and emotional traits, and his early scientific interests. The biographies, however, do suggest how DARWIN regarded some of his family relations, and some of the events in his life. In both biographies DARWIN first mentions that he was the son of Dr. Robert DARWIN of Shrewsbury, and then that he was: "grandson of Dr. Erasmus DARWIN, F. R. S., author of the 'Botanic Garden,' and 'Zoonomis,' &c." "Zoonomia" was Dr. Erasmus DARWIN's best known book, and the book in which he had stated some of his evolutionary ideas. Very soon after the publication of "The Origin" DARWIN's evolutionary ideas were compared with those of "Zoonomia"[13], and DARWIN then saw the need to disassociate his book from the work of his grandfather.[14] Yet as he was growing up he had heard his father talk of his grandfather, and he had read grandfather's books, and probably first heard about evolution from reading "Zoonomia".[15] He knew that, in any account of his life, he should mention his grandfather and "Zoonomia".

DARWIN's mother, Susannah WEDGWOOD, was the daughter of Josiah WEDGWOOD, an eminent and succesful pottery maker. Susannah had died when DARWIN was $8\frac{1}{2}$ years old, and throughout his life DARWIN would say that he had no memories of his mother, and he would rarely refer to her. For undetermined reasons he would, also rarely mention

"From the earliest work published by Mr. DARWIN to his latest, there will be observed by every careful student a constant desire to search out the secret springs of nature. This is not shown, as is too often the case, by any imperfectly considered hypothesis, but everywhere we discover the same painstaking experimental investigation, the same close and long-continued observation; and also everywhere that high power of drawing with clearness and simplicity his deductions from his well established facts, which distinguishes the true 'Philosopher'."

In June 1880 Darwin wrote HUNT acknowledging receipt of the latter's book "Panthea, the Spirit of Nature", and saying: ..I will read your work as soon as I have finished a book in hand . . ." (DARWIN to Robert HUNT, June 9, 1880, Down Beckenham Kent, in the Howard Dittrick Museum of Historical Medicine, Cleveland, Ohio).

13 Reviewing "The Origin of Species" in the July, 1860 issue of "The Quaterly Review", the Reverend Samuel WILBERFORCE wrote: "If we go back two generations we find the ingeious grandsire of the author of the 'Origin of Species' speculating on the same subject, and almost in the same manner with his more daring descendent." WILBERFORCE quoted some of Dr. Erasmus DARWIN's evolutionary ideas, and then he quoted an anti-evolutionary parody of these ideas by Dr. DARWIN's "Anti-Jacobin" critics. WOLBERFORCE then commented that this parody "admirably" applied to the ideas of "The Origin". ("The Quarterly Review", vol. 108, July-October, 1860, 225—264, pp. 254—255) "The 'Quarterly'", DARWIN wrote to his friend and defender Thomas HUXLEY, "is uncommonly clever; and I chuckled much at the way my grandfather and self are quizzed." He sent "The Quarterly Review" to his old friend Charles LYELL with the comment: "it may amuse you. The Anti Jacobin part amused me." (Francis DARWIN and A. C. SEWARD, eds., "More Letters of Charles DARWIN", 2 vols., New York, 1903, 1, p. 156; Francis DARWIN, ed. "The Life and Letters of Charles DARWIN", 2 vols., New York, 1896, 2, p. 124).

14 In the "Historical Sketch", which was first published in the 1860 German translation of "The Origin", and then in the third, 1861, English edition, and all subsequent editions of "The Origin". DARWIN described his grandfather's evolutionary ideas as "erroneous."

15 In his 1876 "Autobiography" DARWIN wrote that reading the evolutionary views of his grandfather in "Zoonomia" may have influenced him in "upholding" these views "under a different form in my "Origin of Species." (Nora BARLOW, ed. "The Autobiography of Charles DARWIN", London, 1958, p. 49).

Josiah WEDGWOOD; or he would refer to his maternal grandfather in a peculairly detached and distant manner.[16] His wife Emma was his maternal cousin, and a granddaughter of Josiah WEDGWOOD.

In his WALFORD biography, which he had (presumably) written himself, he was silent about his mother, and his relations to Josiah WEDGWOOD. He only mentioned the latter when he wrote that he had "married the granddaughter of Josiah WEDGWOOD, F. R. S., the well-known improver and manufacturer of earthenware." In his HUNT biography, which he had dictated to his wife,[17] he specified his maternal relations. First, after mentioning his father and paternal grandfather, he wrote that he was "grandson" by his "mother's side of Josiah WEDGWOOD F. R. S., the celebrated potter." He then wrote that he had "married his cousin Emma WEDGWOOD," and by her "had a large family."

In both biographies DARWIN gave a similar, sharp and succinct, account of the main outward events of his life. First the educational institutions he had attended; then the "Beagle" cruise, the only outward adventure of his life. He carefully indicates that he was able to obtain the "Beagle" position because FITZROY, the "Beagle's" captain, "offered to give up part of his own cabin" to any one who would volunteer as a ship's naturalist. This was a point that he would stress — repeatedly and at different times and different occasions — throughout his life.[18] Although he writes that FITZROY made his "Beagle" offer in "the autumn" of 1831, this is not quite correct. The offer was made in the summer of 1831, and first two naturalists — HENSLOW and JENYNS — who both turned it down.[19] HENSLOW then passed on the offer to DARWIN. It was in the fall of 1831 that DARWIN and FITZROY first met, and that FITZROY definitely decided that DARWIN was acceptable as his "Beagle" companion. DARWIN's recollecting that FITZROY's offer was made in "the autumn" indicates, perhaps, a desire on DARWIN's part to ignore other applicants, and to only consider his fall interview with FITZROY.

In both biographies DARWIN was happy, and very proud, to record that his "Beagle" "Journal of Researches" had achieved a "larger sale" when it was "published separately," than when it was published as the third volume in the "Narrative" history of the "Beagle's" voyages[20]. He then listed his other, main, scientific publications (he would say

16 Once, referring to a poem about Australia, which was written by Dr. Erasmus DARWIN, and then inscribed on a clay medallion manufactured by Josiah WEDGWOOD, DARWIN wrote to his sister Susan: "When my Grandfather wrote the lines of 'HOPE's visit to Sydney COVE' on Mr. WEDGWOOD's medallion he prophecyed most truly." (Charles DARWIN to his sister Susan DARWIN, January 28, 1836, published in Nora BARLOW, ed., "Charles DARWIN and the voyage of the Beagle", New York, 1946, p. 132). Here Dr. DARWIN is identified as "Grandfather", whereas grandfather Josiah WEDGWOOD is called " Mr. WEDGWOOD." It is not clear why DARWIN felt so distant from his maternal grandfather.

17 See note 11.

18 DARWIN stressed this in the 2 editions of his "Beagle, Journal of Researches", and in his 1876 "Autobiography". In 1864, when he was being considered for the Copley Medal, DARWIN wrote to his brother Erasmus: "The origin of the Voyage in the "Beagle" was that FITZ-ROY generously offered to give up half his cabin to any one who would volunteer to go as naturalist". (DARWIN, June 30, 1864, Down, in DARWIN and SEWARD, ed. "More Letters" (n. 13), 1, p. 247).

19 ALLAN, "DARWIN and his flowers" (n. 10), p.53.

20 "In 1845, I took much pains in correcting a new edition of my "Journal of Researches", which was originally published in 1839 as part of FITZ-ROY's work. The success of this my first literary child always my vanity more than that of any of my other books. Even to this day it sells steadily in England and the United States, and has been translated for the second time into German, and into French and other languages. This success of a book of travels, especially of a scientific one, so many years after its first publication

that his books were "the milestones" in his life, and that after mentioning them little else remained to be said about him[21]).

In his 1861 biography he had only been able to say of "The Origin" that, after it's publication, it had gone through three editions. In his 1868 biography he had much more to report: "Towards the close of 1859 Mr. DARWIN published his 'Origin of Species,' of which four English editions have appeared, & nine foreign editions in French, German, Dutch, Italian, & Russian. Above one hundred reviews, pamphlets & separate books have been published on this work." At this time DARWIN collected all "The Origin" reviews and pasted these in a scrap-book[22]. As he watched the scrap-book grow he, probably, appreciated how his work was changing human thought.

At the end of March 1869 DARWIN received a letter from a young and rising German scientist, named Wilhelm PREYER, then 28 years old, PREYER was a naturalist, a chemist and physiologist, and a doctor[23]. In 1861 he had read the just published third edition of "The Origin", and at once come to believe in DARWIN's ideas, and to feel that these ideas altered his view of the world.[24] PREYER published a popular lecture, in German, on "The Struggle for Existence"; and he gave a course on this subject, in the winter of 1868—1869, for the German students at the University of Bonn. In his letter to DARWIN (which was written in fluent English[25]), PREYER reported that 200—250 students had attended his course. "Never before", wrote PREYER, had so numerous an assembly followed the lectures of a 'Privatdocent' in this place, whatever the subject might have been. This single fact shows the enormous progress your doctrine is making in Germany." With his letter PREYER enclosed a copy of his lecture, and he asked for a bibliography of DARWIN's publications.[26] At this time PREYER was appointed Professor of Physiology at the University of Jena, and he moved from Bonn to Jena.

Delighted at the German response to his work, yet always striving to be modest and self effacing, DARWIN replied: "It is highly gratifying to me to hear of the great success of your lectures (on Darwinismus) though I must attribute this to your powers as a

is surprising. Ten thousand copies have now been sold in England of the second edition." (BARLOW, ed., "Autobiography" (n. 15), p.116).

21 Ibid., p.136.

22 Ibid., p.123.

23 PREYER had been born, July 4, 1841 in England, the son of a German industrialist. He was first educated at the Clapham Grammer School near London. He immigrated to German in 1857, and attended gymnasiums in Duisburg and then Bonn. In 1859 he begann to study science at Heidelberg concentrating on physiology and chemistry. He was also interested in natural history and observing the habits of animals, and in 1860 (while still a student at Heidelberg) he made a trip to Iceland with the geologst Ferdinand ZIRKEL, and then published a geological-natural history account of his trip. In 1862 he received his doctorate from Heidelberg. He then studied physiology and medicine in Paris. Berlin, Vienna, and Bonn. In 1866 he received his medical degree from Bonn, after completing a dissertation on the chemical properties of Hemoglobin. In 1865 he had qualified at Bonn as a lecturer in zoophysics and zoochemistry in the Faculty of Philosophy.

24 Wilhelm PREYER, „DARWIN. Sein Leben und Wirken" (Berlin: Ernst Hofmann & Co., 1896), p.131.

25 PREYER had learned English during his early education at the Clapham Grammar School (see note 23).

26 PREYER to DARWIN, Bonn, University, March 21, 1869, DARWIN Papers, Cambridge University Library. Original letter in PREYER's writing.

lecturer."[27] He discussed some aspects of "The Struggle for Existence" with PREYER, and sent the latter a bibliography of his writings.[28]

Soon after receiving the bibliography, in the spring of 1869, Preyer began to prepare a biographical sketch of DARWIN; and he sent his subject a long list of questions about relatives, schools and teachers, events of the "Beagle" cruise, and events after the cruise. The questions were answered both by DARWIN and his wife Emma, and were promptly returned.[29]

The answers were especially interesting in describing the impact of the universities of Edinburgh and Cambridge (previously, in his notes to WALFORD and HUNT. DARWIN had merely noted that he attended these universities). When PREYER asked about "Professors of influence" DARWIN replied: "The Revd. HENSLOW;[30] the Professor of Botany had great influence in deciding my taste for natural history." When PREYER asked about "Favourite University courses" Emma replied: "His education only really began when he went on board the Beagle." At the end, when PREYER asked for "Additional biographical notes", DARWIN wrote about himself as follows:

> "From my earliest youth I was fond of collecting all sorts of specimens and observing habits of wild animals. I was an ardent sportsman and this made me very idle. At Edinburgh though I had the opportunity I neglected from foolish disgust to study anatomy, and I have regretted this all my life. I observed at Edinburgh the lower marine animals. At Cambridge I collected Coleoptera energetically and dabbled a little in geology; but I studied no subject scientifically; only for amusement. I never worked till I joined the Beagle and then I worked with all my heart."

As an afterthought DARWIN then added: "At Edinburgh Dr. GRANT, now Professor at the London University, led me to collect and observe the lower marine animals."[31] In this passage DARWIN succinctly recollects (perhaps for the first time) some of his youthful scientific strengths and weaknesses.

After receiving the answers to his questions PREYER wrote his biography of DARWIN, in 1869. He first sent the manuscript to Preussische Jahrbücher, which returned it saying that it would be more appropriate for a scientific and scholarly journal. Nine months later

27 DARWIN to PREYER, March 29, 1869, DARWIN Papers, Cambridge University Library, Box 147. This letter, and all the other DARWIN to PREYER letters in this Box, exist in the form of handwritten copies. German translations of this and other DARWIN to PREYER letters are published in PREYER's biography of DARWIN (n. 24).

28 Ibid. The bibliography that DARWIN enclosed was incomplete, and PREYER was able to give a more complete bibliography by consulting the Catalog of Scientific Treatises of the Royal Society. PREYER, DARWIN (n. 24), p. 142.

29 PREYER, DARWIN (n. 24), pp. 142—146, contains a German translation of PREYER's questions, and the answers of DARWIN and his wife. An English copy of these questions and answers is in DARWIN Papers, Cambridge University Library, Box 147, pp. 9—13 (one page, page 11, is missing). My quotation from these questions and answers is from the Cambridge manuscript.

30 DARWIN associated "intimately" with HENSLOW when he was at Cambridge from 1828—1831. He would recollect that his association with HENSLOW would influence his "whole career" more than any other association. BARLOW, ed., Autobiography (n. 15), pp. 64—66.

31 Robert Edmont GRANT (1793—1874), zoologist and physician. Professor of Zoology and Comparative Anatomy at University College, London, 1827—1874. DARWIN and GRANT collected animals on the sea-shore, in 1826, and DARWIN then dissected some of these, BARLOW, ed., Autobiography (n. 15), pp. 49—50.

PREYER sent his work to Oskar PESCHEL, the pro-DARWINian editor of „Das Ausland", and it was accepted for publication.[32] Sometime in late 1869 or early 1870 PREYER wrote DARWIN a letter about several subjects. He described his physiological work on the chemistry of blood, and on the effects of the powerful poison, prussic acid, on blood and on different animals (he probably enclosed his publications on these subjects).[33] He then asked about a paper that DARWIN's father had written on vision,[34] and he urged DARWIN to write more autobiographical recollections.

In February—March 1870 PREYER left Jena for a 2 month trip to Italy. When he returned home, early in May, he found awaiting him the April 2nd issue of „Das Ausland" containing his DARWIN biography, and a February 17th letter from DARWIN, which replied to his letter.[35]

PREYER's „Das Ausland" article, entitled "Charles DARWIN. A biographical sketch."[36] summarised in about 6,000 words, DARWIN's work and the main events of his life. It was the longest, and most carefully researched, DARWIN biography yet published. It was illustrated with a large, half-page, wood-cut portrait of DARWIN (copied from a recent photo). In his "sketch" PREYER had commented on this portrait as follows: "The physiognomy of this eminent man is most impressive. His long, full white beard, the unusual curvature of his head, the strong forehead, the bushy white eyebrows, showing extraordinarily deepset eyes, which extolled a nobleness comparable to the Vatican busts of Greek wisemen. One barely knows what to admire about the head, the self-possession, the intelligence, the look of discovery, or the confidence of an awakening openess."[37] PREYER sent his publication to DARWIN; and, after reading it, the latter wrote to PREYER:

> "I received two days ago your article which has gratified me in a very high degree. Every word which you say is dictated by the kindest possible spirit; but your estimation of the value of my work would be considered, I fear by most persons as very much too high. Nothing so full or, as far as facts are concerned, so true has hitherto been published about me. There are one or two very trifling inaccuracies, i.e., about my shooting and

32 PREYER, DARWIN (n.24), pp. 148—149.

33 PREYER's letter to DARWIN is missing. The contents of this letter have been inferred from DARWIN's February 17, 1870 letter to PREYER, and from PREYER, DARWIN" (.24), pp. 155—156.

34 PREYER was interested in Dr. Robert DARWIN's essay on vision because of his study of GOETHE's essay on light. PREYER, DARWIN (n.24), p. 155.

35 On returning home PREYER then wrote DARWIN as follows: "I beg you to accept my very best thanks for the most interesting letter you had the kindness to address to me. I would long since have sent you the biographical sketch: Charles DARWIN, „eine biographische Skizze" with your portrait published in the „Ausland", had I not been absent from here for two months. I only arrived here, coming from Naples, a few days ago. I hope ("the enclosed" inserted, words are barely legible) german article may not displease you." (PREYER to DARWIN, Jena, University, May 9, 1870, DARWIN Papers, Cambridge University Library, Original letter in PREYER's writing.) PREYER's reference to „the most interesting letter you had the kindness to address to me" refers to DARWIN's letter of February 17, 1870. None of the biographical facts that DARWIN mentions in this letter are included in PREYER's „Ausland" article.

36 W. PREYER, Charles DARWIN. Eine biographische Skizze „Das Ausland" (Augsburg) April 2, 1870, pp. 314—320.

37 Ibid., p. 316.

hunting in the Highlands.[38] You give also too bad a character to our roads or rather lanes;[39] they are very narrow and rather bad; but when you come to England, I hope that you will prove by visiting me here that they are not impassable."[40]

Subject and (no doubt) author were both pleased by the „Ausland" biography; neither commented that DARWIN's February 17th letter to PREYER had arrived too late to be used in the biography. DARWIN's letter was especially intestering because he expressed a range of emotions, and recollected many events of his past.[41] What follows is a transcription of this letter, along with specific biographical commentaries, and then a general commentary.

Feb. (17, 1870)

Down.
"Beckenham" (inserted) ("Bromley" crossed out)
Kent. S. C.

My dear Sir

I am very much obliged for your extremly kind letter & for your several presents. Although your appreciation of my work is ("certainly" inserted) too high, yet it is very encouraging to me, especially as yesterday I read two pamphlets, just published in England, in which every form of abuse is heaped on me. I am called, for instance, a "filthy dreamer."[42] — You seem to be doing splendid work in physiology, the noblest of sciences, as I have long thought it. What you say about the differences of the blood-crystals is

38 „From early childhood on, DARWIN derived the greatest pleasure from collecting fruit and observing the habits of animals in the wild. He was a passionate huntsman and so devoted to hunting in the Scottish highlands that during his student days scientific events made little impression on him." Ibid., 0 p. 14. During his youth in England DARWIN never hunted in Scotland. He hunted at Maer, the estate of his uncle Josiah WEDGWOOD 11, about 20 miles from his Shrewsbury home; and he also hunted at Woodhouse, the estate of his friend Squire OWEN, which was near his Shrewbury home (BARLOW, ed., "Autobiography" (n. 15), p. 54).

39 "It seems that the meager provisions on board (the "Beagle"), the stress of his (DARWIN's) visits ashore, his declining health, and the need to complete his studies in peace caused him, shortly after his marriage, to avoid company almost completely, and so he moved to Down in 1842; a small village of about 500 people that is connected by a poor, oftentimes almost impassable road to the town of Bromley in Kent, southeast of London." PREYER, „Charles DARWIN. Eine biographische Skizze" (n. 36), p. 316. PREYER later stated that he made his error about DARWIN's hunting in the Scottish highlands, and about the impassable state of the Down road, from talking to DARWIN's daughter, Henrietta DARWIN, who traveled in Europe in 1869—1870 (PREYER, DARWIN (n. 24), pp. 150—151). DARWIN remenbered PREYER's statement about the Down road when he came to write his "Autobiography". After mentioning his move to Down he commented that he was pleased "with the extreme quietness and rusticity of the place." And then he added: "It is not, however, quite so retired a place as a writer in a German periodical makes it, who says that my house can be approached only by a muletrack!" (BARLOW, ed., "Autobiography" (n. 15), p. 114).

40 DARWIN to PREYER, May 15, 1870, DARWIN Papers, Cambridge University Library, Box 147.

41 This letter of DARWIN to PREYER is in my private possession. It is written entirely in DARWIN's handwriting, and it is written on his printed stationary which read "Down. / Bromley. / Kent. S.C." The heading on the stationery changed with changes in the railroad station. From 1855 until late in 1869 it was "Down./Bromley./Kent. S.C.". From September 1869 onwards the address was "Down./Beckenham./Kent." As is evident in this letter the change from "Bromley" to "Beckenham" had just occurred. DARWIN only dated his letter "February", and the day and year were inserted by PREYER (presumably) from the post mark on the envelope.

42 These two pamphlets have not been identified.

truly astonishing. I am also much interested by what you say of the different effects of Prussic acid on different individuals of the same species: I remember some years ago wishing in vain for information on this head. I think it arose from observing how differently in quickness [whether due to ("rate of" inserted) respiration or the direct action ("of the poison" inserted) I could not tell] the vapour acted on different insects. I remember Bees died instantly, but (I think) it was a Longicorn beetle which resisted all effects for a astonishing time.[43] —

R. W. DARWIN is my Father, but I believe he was greatly aided ("in his paper on optics" inserted) by his Father Erasmus DARWIN.[44]

I have really nothing of interest about myself, but as you desire it, I will scribble whatever occurs to me. — I derived no advantage from the Lectures of Edinburgh, for they were infinitely dull & cured me of ("any taste for" inserted) Geology for 3 years. Dr. GRANT was not a Professor, but worked at zoology out of pure love, & his society was a great encouragement. I used to amuse myself with examining marine animals, but I did so solely for amusement. I believe I was the first ("person" inserted) who ever saw the earliest locomotive egg-like state of a Bryozoon: I showed it to GRANT who stated so at the meeting of the Wernerian Nat. Hist. Soc. & this little discovery was an immense encouragement.[45] I was disgusted at anatomy & attended only 2 or 3 Lectures & this has been ("ever since" inserted) an irreparable loss to me. — When I went to Cambridge, I became a most enthusiastic collector of Cleoptera; but again only for amusement. If any one told me the name of a Beetle, I thought I knew all that anyone could desire; & I believe I never ever looked even at the oral organs of an insect! Yet I worked like a slave at collecting.[46] HENSLOW's society was a great charm & benefit to me, & I liked much his Lectures on Botany. All my early life I was mad for

43 John Maurice HERBERT (1808—1882), in his 1882 recollections of his 1828—1831 Cambridge association with DARWIN, writes: "One day in his rooms he (DARWIN) dosed a large Beetle with a penful of Prussic Acid. The creature almost instantaneously kicked & fell over on his back apparently dead — but after a few minutes exposure to the sun it rallied and walked off as if nothing had happened to it, to our great amazement. Was this a discovery in toxicology?" (J. M. HERBERT to Francis DARWIN, 12 June 1882, DARWIN Papers, Cambridge University Library, Box 112). This was, presumably, the "Longicorn beetle" that DARWIN wrote about to PREYER. His observation of the effects of prussic acid on this beetle does not seem to have been followed up.

44 This was in reply to PREYER's request for information on Dr. Robert DARWIN's essay on vision (see note 34). This was the only scientific paper that Dr. Robert DARWIN published. DARWIN felt that, although his father was an acute abserver, and although he formed many theories, father's "mind was not scientific, and he did not try to generalize his knowledge under general laws . . ." (BARLOW, ed., "Autobiography" (n. 15), p. 42).

45 DARWIN's discoveries were that the ova of *Flustra* possess organs of locomotion; and that the small black globular body, mistaken for the young of *Fucus loreus*, were really the ovum of *Pontobdella muricate*, he was wrong on both these claims: the 'ova of Flustra' were pilidium larvae, and 'ovum of Pontobdella' was an egg case full of eggs. (Richard B. FREEMAN, "Plinian Society of Edinburgh", in "Charles DARWIN: A Companion". Archon Books, Hamden, Connecticut, USA, p. 232. Also see note 31).

46 DARWIN fails to mention that the reason he worked so hard at collecting, and then naming beetles, was that if he found a new beetle a notice of it, along with his name, was published in: James Francis STEPHENS, "Illustrations of British entomology; or a synopsis of indigenous insects", etc. This was a work published between 1827—1845 in 11 volumes, 4 volumes of Haustellata, and 7 Mandibulata. DARWIN's beetles were published in the first 5 volumes of the latter.

collecting, minerals, shells, plants, Bird-skins[47] have all had their turn. ("At" crossed out) Near the end of my Cambridge life HENSLOW persuaded me to begin Geology. I was ("always" inserted) very fond of observing the habits of Birds, & WHITE's Nat. Hist. of Selborn[48] thus had ("a great" crossed out) much influence on my mind.[49] But of all books, HUMBOLDT's Travels[50] had ("by far" inserted) the greatest influence. — I read large parts over & over again. — I had nearly managed to get a party to go to the Canary Isl[d],[51] when the offer of joining the "Beagle" was made ("to" inserted) me & joyfully accepted. I suppose, however, no one started worse prepared than I was except as a mere collector. I knew nothing of anatomy, & had never read any systematic work on Zoology — I had never touched a compound microscope & had begun Geology for only about 6 months. But I took out plenty of Books & worked as hard as I could & dissected roughly all sorts of the lower marine animals. Here I felt fearfully the want of practice & knowledge. My education in fact began on board the Beagle. I remember nothing previously which deserved to be called education except some experimental work at chemistry when a school-boy with my Brother. — No doubt collecting largely in so many branches had improved my powers of observation.[52]

I never wrote so much about myself in my life, & I hope it may be worth your reading, but I doubt.

Believe me, my dear Sir
Yours Sincerely
Ch. DARWIN

I do not know whether you will care to see extracts from my letters printed by Prof. Henslow, but I send a coppy by this Post."[53]

In the above letter DARWIN was mainly elaborating on what he had previously (in answering PREYER's question about "Additional biographical notes") written about his scientific work and experience at Edinburgh and Cambridge. In his recollections he experienced several different emotions. Regret that he had not learned anatomy and that he had not had systemic training in zoology, and embarrassment in writing about himself. He also experienced pleasure in recollecting his first scientific discovery at Edinburgh, and in his friendships with GRANT and HENSLOW; and he accurately depicted how his passion for

47 In his recollection of his early years DARWIN alwys mentioned that he had a passion for collecting diverse objects. This seems to be the only place where he mentions that he collected "Bird-skins". He was taught to skin birds, when he was a student at Edinburgh, by John EDMONSTON. Edmonston was a black, and a freed slave, who had traveled with the English naturalist and explorer, Charles WATERTON. See FREEMAN, "Charles DARWIN: A Companion" (n. 45), "John Edmonston", p.133.

48 Reverend Gilbert WHITE, "The Natural History of Selborne" (1788).

49 DARWIN recollects that: "From reading WHITE's "Selborne" I took much pleasure in watching the habits of birds . . . In my simplicity I remember wondering why every gentleman did not become an ornithologist." (BARLOW, ed., "Autobiography" (n. 15), p. 45).

50 Alexander von HUMBOLDT, "Personal Narrative of Travels to the Equinoctial Regions of the New Continent during the Years 1799 to 1804", 7 vols. (London: 1818—1829).

51 DARWIN planned to make a trip to the Canary island of Teneriffe.

52 Joseph HOOKER, DARWIN's intimate friend and scientific confidente, wrote that DARWIN recognised three stages in his career as a biologist: "the mere collector at Cambridge; the collector and observer in the "Beagle", and for some years afterwards; and the trained naturalist after, and only after the Cirripede (baracle) work." DARWIN, ed., "Life and Letters" (n. 13), 1, p. 315.

53 "Extracts from Letters Addressed to Professor HENSLOW", Cambridge, 1 December 1835, 31 pp.

collecting diverse things had developed his powers of observation, and how his "Beagle" experience had developed his capacity for hard work. His main feeling, as he "scribbled" his recollections, was pleasure — a feeling which seems to have developed suddenly and unexpectedly — and he expressed this feeling by telling PREYER, "I never wrote so much about myself in my life."

He also enjoyed writing about himself to PREYER, who he respected as a scientist and then as a German who valued him (in contrast to the Englishmen who had abused him), and who — in writing his biography in German — would cause a German audience to value him.

His assertion that he had "never" written as much about himself was inaccurate. In August 1838 he had written, for his private perusal, what he entitled "Life"; a 1,700 word recollection of himself from the ages of 4 to 11.[54] His letter to PREYER, however, may have been the longest autobiographical letter that he had ever written to another person.

After May 1870 DARWIN and PREYER remained in contact — exchanging publications, and discussing different scientific subjects — until DARWIN's death in April 1882.[55] PREYER published a second biographical sketch of DARWIN in 1879[56]; after 1870, however, he did not query DARWIN for any further biographical information. In 1896, when PREYER was gravely ill and had only a year to live, he published a third, book length, biography of DARWIN.[57] In this he printed DARWIN's letters to him , in German translation. The frontpiece of his book contained a facsimile reproduction of the first page of DARWIN's February 17th, 1870, letter.

It may be that, writing his autobiographical letter to PREYER, then motivated DARWIN to write his "Autobiography".

54 COLP, "'I was born a naturalist'" (n. 2), pp. 21—36.

55 In the spring of 1871. after the publication of DARWIN's book "The Descent of Man, and Selection in Relation to Sex", DARWIN and PREYER corresponded about some points in the anatomy of the external human ear. In July 1877 DARWIN sent PREYER his article, "A Biographical Sketch of an infant", and PREYER wrote him of his observations on human infants and on very young guinea pigs. In January 1878 DARWIN wrote PREYER a note of thanks for having been elected an Honorary Member of the Medicinisch-naturwissenschaftliche Gesellschaft zu Jena. In August 1880 DARWIN and PREYER met briefly in Cambridge, in England, and discussed the advances of Darwinism in Germany. DARWIN told PREYER that he regretted his difficulty in reading German because of the great amount of original research done in his field in Germany. In November-December 1880 DARWIN and PREYER exchanged letters about DARWIN's just published book, "The Power of Movement in Plants." In October 1881 PREYER sent DARWIN his just published book "Die Seele des Kindes", and DARWIN wrote him a warm note saying: "How I wish that I could read German easily, for the subject interests me much and I have no doubt that your work is excellent." With this book, and his observations of the psychological life of children, PREYER established himself as one of the founders of modern developmental psychology. The correspondence of DARWIN and PREYER, after 1870 is in DARWIN Papers, Cambridge University Library, and consists of the following letters. DARWIN to Preyer: April, 30 and May, 13, 1871; July, 8, 1877; January, 25, 1878; November 27, 1880. (These letters are also published in German translation in PREYER, "DARWIN" (n. 24), pp. 158—168), PREYER to DARWIN: July, 6, 1877; November, 12 and December, 2, 1880.

56 Prof. W. PREYER, „Charles DARWIN. Eine biographische Skizze", „Kosmos". Vol. 3, February 1879, pp. 339—350. This "Sketch" contained, for the time, the most complete biography of DARWIN's writings.

57 PREYER, „Darwin" (n. 24). PREYER became ill with a kidney and liver ailment in 1888, and because of this he retired as Professor of Physiology in Jena, and spent the last 9 years of his life as a private scholar in Wiesbaden. He died in Wiesbaden on July 15, 1897.

DARWIN began writing his "Autobiography" in May, 1876, and he did so because of several influences. He was 67 years old and feeling the proximity of death. He wanted to leave a record of his life and work: first for the private perusal of his children and grandchildren (he had recently learned that he was, for the first time, going to become a grandfather); and eventually for the public. Then, in the first sentence of his "Autobiography" he mentioned another influence: "A German Editor having written to me to ask for an account of the development of my mind and character with some sketch of my autobiography, I have thought the attempt would amuse me . . ."[58]

Who was this "German editor"? DARWIN does not further identify him, either in his "Autobiography" or his other writings; and his extant correspondence contains no letter from such an individual. The only German editor that he had an extended correspondence with was Dr. Ernst KRAUSE, one of the editors of „Kosmos"; and he first wrote KRAUSE in March 1877 (10 months after he had written his "Autobiography").[59] In 1870 he received a book from Oskar PESCHEL, editor of „Das Ausland". But there is no evidence that PESCHEL corresponded with him, or tried to direct him in any of his writings[60] Among all of DARWIN's extant letters (including letters to such German scientists as Ernst HAECKEL, Hermann MULLER, Fritz MULLER, Hugo THIEL, Ferdinand COHN, Victor CARUS, and Otto ZACHARIAS) there is no mention of any work which was about DARWIN's life and the development of his mind and character. There is only mention of such a work in his 1869—1870 letters to PREYER.

It seems plausible to suppose that in May 1876 — along with thoughts of death and grandchildren — DARWIN recollected, first of all, the pleasure that he had experienced in writing to PREYER 6 years previously. He was a man who long remembered certain of his thoughts and feelings, and then developed these rememberances in his writings. The recollection of this 6 year old pleasure explained what DARWIN meant when he said that writing his "Autobiography" 'would amuse' him. Since he desired to write about himself, and then was embarrased by this desire, he used his past relationship with PREYER as an explanation — it could also be called an excuse — for his "Autobiography".

PREYER was certainly not an "Editor" in 1869—1870, and he never seems to have been an editor. DARWIN, however, may have used the term "Editor" in a loose way. PREYER had edited his 1869 notes for the „Das Ausland" biography; and PREYER had then asked for more recollections, and failed to edit these when they were received. Perhaps the knowledge that his February 17th letter had not been used was another stimulus for DARWIN to write his "Autobiography". In the "Autobiography" he developed many of the events and feelings he touched on in his letter.

58 BARLOW, ed., "Autobiography" (n. 15), p. 21.

59 DARWIN to Dr. Ernst KRAUSE, March 25, 1877, Down, Beckenham, Kent. In addition to this there are 45 other letters of DARWIN to KRAUSE. Most of these letters concern the biography of Dr. Erasmus DARWIN, and KRAUSE never suggested that DARWIN write about himself. When „Kosmos" needed a biography of DARWIN it was PREYER who wrote it. PREYER seems to have been the only serious German biographer of DARWIN, during DARWIN's life. The correspondence between DARWIN and KRAUSE is held in the Henry E Huntington Library, San Marino, California.

60 When PREYER sent DARWIN his „Das Ausland" biography, in May 1870, he also sent him a copy of Oskar PESCHEL's book, "New Problems of Comparative Geography as an Attempt at a Morphology of the Earth's Surface". The book was sent at PESCHEL's request. DARWIN, in his letter of May 15, 1870, thanked PESCHEL for the book, commenting that it looked "extremely interesting", and that he was grieved by his difficulty in reading German. (Preyer, „DARWIN" (n. 24), pp. 150—151).

Darwin was a man who, quietly, methodically, and steadily — from behind his desk — wrote many letters, about diverse topics, to many correspondents. Scholars working on the DARWIN Letter Project have estimated that during his life DARWIN wrote about 10,000 letters. Many of his important letters have been lost. It is possible that letters from him to a "German Editor", about an autobiographical work, will be discovered. At present all that can be said with certainly is that he received an autobiographical impetus from his 1869—1870 correspondence with PREYER.

Acknowledgments

In the preperation of this essay I have needed help from many quarters, and it has always been generously given. It is a pleasure to thank the following: Frederick BURKHARDT and William MONTGOMERY, editors of the Collected Letters of Charles DARWIN, for answering questions about DARWIN's correspondence and directing me to sources of information; Peter GAUTREY, of the Cambridge University Library, for sending me copies of the DARWIN-PREYER correspondence and Robert HUNT's letters to DARWIN, and the Syndics of this Library for permission to quote from these letters; Stephen CATLETT, Library of the American Society, and Glen JENKINS, Howard DITTRICK Museum of Historical Medicine, for sending me copies of DARWIN documents; Lt. Col. Charles CASOLANI (retired), for searching out sources of information in The British Museum and The British Museum (Natural History); Paul COHN for researching at Columbia University Library.

I owe a special debt to my mother. Her work in biology and her copy of "The Life and Letters of Charles DARWIN" stimulated my early interest in things about DARWIN. Before her sudden and untimely death she translated much of what PREYER had written about DARWIN from German into English. She would be happy to know that she has aided me in work which will appear on the occasion of the Centenary of DARWIN's death, in Germany, at a Colloquium devoted to "DARWIN Today." I dedicate this essay to her memory.

Sektion Marxistisch-leninistische Philosophie der Martin-Luther-Universität Halle-Wittenberg

Darwinism and Marxism — Including a Consideration of the Personal Contacts Between MARX, ENGELS and DARWIN

S. KIRSCHKE

When DARWIN died nearly a hundred years ago on April 19th, 1882, Karl MARX' son-in-law, Charles LONGUET, wrote an impressive obituary for this great naturalist. This was published first in "La Justice"[1], and almost immediately re-published in „Der Sozialdemokrat". The article begins with the statement that since COPERNICUS there had not been as revolutionary an idea as that of DARWIN, and goes on to say that DARWIN's ideas expand all human knowledge and make way for a bold new philosophy of the world and its society. These ideas were confirmed and extended by discoveries in the closest and also the most remote areas of science — biology and the study of language[2]. Friedrich ENGELS also stressed the great value of DARWIN's achievement when he spoke at the grave of Karl MARX one year later. He mentioned in the same breath MARX' view of the development of human history and DARWIN's discovery of the laws of evolution of natural organisms[3]. With this comparison ENGELS repeated the view he had expressed to MARX three weeks after the publication of DARWIN's "On the Origin of Species ...", that the evolution of species is a historical process.[4] This reveals that MARX and ENGELS appreciated the scientific essence of the book and also its philosophical relevance. In ENGEL's article "Ludwig FEUERBACH and the End of Classical German Philosophy" one can see that DARWIN's evolutionary theory is one of the three great discoveries of natural sciences which have given rise to a new form of philosophical materialism, i.e. dialectic materialism[5]. In this article ENGELS stated that the "Copernican turn" which DARWIN had introduced into biological thinking made possible the subsequent revolution in philosophical thought. The revolution started not with DARWIN but with the philosophers MARX and

1 The Paris newspaper "La Justice" was founded in 1880 by G. B. CLEMENCEAU. He was editor of the paper until 1897. During this time "La Justice" was the paper of the extreme left wing of the Radical Party which represented the interests of the lower middle class and the middle bourgeoisie by defending the program of democratic and social reforms. After the amnesty of July 11th, 1880, Ch. LONGUET became a member of the editorial staff and a sub-editor of great consequence.

2 DARWIN. „Der Sozialdemokrat", Zürich, No 18, 27th April 1882, p.1.

3 ENGELS, F.: Das Begräbnis von Karl MARX (1883). In: MARX, K. und ENGELS, F.: Werke, (MEW), Vol. 19, p.335.

4 ENGELS, F.: Letter to K. MARX of 11th or 12th December 1859. In: MEW, Vol. 29, p. 524.

5 ENGELS, F.: Ludwig FEUERBACH und der Ausgang der klassischen deutschen Philosophie (1886). In: MEW, Vol. 21, p. 295. — Compare also: ENGELS, F.: Ausgelassenes aus „FEUERBACH" (1886). In: MEW, Vol. 20, p. 468.

ENGELS. Nevertheless, DARWIN's definition and criticism of teleological thinking in biology was a significant contribution to materialist philosophy[6].

The differences between biological and philosophical theories of evolution in regard of both content and scope must not be ignored[7]. Thus, in the 1860's, bourgeois sociologists confined DARWIN's theory to expressions such as "the struggle for existence", "the survival of the fittest", etc. They considered these to be the driving forces of world events and explained capitalist society in these terms[8]. The variety of biologism disparagingly known as social Darwinism was misused by the most reactionary German imperialists to justify their plans for the extermination of what they called unworthy life and inferior races. Even now there are social forces which would like to obscure the economic basis of imperialism and hold DARWIN's theory and its adherents responsible for this misuse of Darwinism[9]. In recent years there has been a revival of social Darwinism in several capitalist countries supported by developments in genetics, behavioural studies and other branches of biology.

Some time ago the theory of evolution was used to account for the development of theories, too. It was held that scientists generate ideas in great numbers, like the progeny of plants and animals, and that these compete for survival. The proponents of this Darwinistic view of the evolution of theories[10], as well as those of neosocial Darwinism[11], attempt to deceive the causal observer with the superficial similarities between qualitatively different developmental processes. Long ago MARX and ENGELS had scoffed at the use of these inappropriate and unscientific methods by some of their colleagues.

But how is one to explain that DARWIN, having grasped the ideas of natural selection[12], was pleased at the fact that Robert MALTHUS applied his principles to the development of capitalism[13]? And how is one to understand that MARX mentioned in a letter to Ferdinand LASSALLE that DARWIN's paper served as a "natural scientific background for the historical class struggle"[14]? Or how are we to interpret the remarks by MARX in

6 MARX, K.: Letter to F. LASSALLE of 16th January 1861. In: MEW, Vol. 30, p. 578. — Compare also: ENGELS, F.: Letter to K. Marx of 11th or 12th December 1859. In: MEW, Vol. 29, p.524.

7 Compare: HÖRZ, H.: Marxistische Philosophie und Naturwissenschaften, Berlin 1974, p. 387—445. HÖRZ, H. und Röseberg, U. (Eds.): Materialistische Dialektik in der physikalischen und biologischen Erkenntnis, Berlin 1981, p. 57—69, 138—143, 239—312.

8 Compare: SCHALLMEYER, W.: Vererbung und Auslese. Grundriß der Gesellschaftsbiologie und der Lehre vom Rassedienst, Jena 1918.

9 Compare: DOMIN, G. und MOCEK, R. (Eds.): Ideologie und Naturwissenschaft. Politik und Vernunft im Zeitalter des Sozialismus und der wissenschaftlich-technischen Revolution, Berlin 1969, p. 224—246.

10 DOMIN, G. et. al.: Bürgerliche Wissenschaftsauffassungen in der Krise (I), Berlin 1972, p. 199—202.

11 Compare: WERNECKE, A.: Biologismus und ideologischer Klassenkampf, Berlin 1976.

12 That follows from DARWIN's "Notebook" 1837/38 which had been finished in February 1838, seven or eight months earlier than DARWIN read the paper by MALTHUS.

13 DARWIN wrote in his "Autobiography", Leipzig und Jena 1959, p. 100—101: „Im Oktober 1838, also fünfzehn Monate nachdem ich meine Untersuchungen angefangen hatte, las ich zufällig zur Unterhaltung MALTHUS ‚Über die Bevölkerung', und da ich hinreichend darauf vorbereitet war, den überall stattfindenden Kampf um die Existenz zu würdigen, namentlich durch lange fortgesetzte Beobachtung über die Lebensweise von Tieren und Pflanzen, kam mir sofort der Gedanke, daß unter solchen Umständen günstige Abänderungen dazu neigen, erhalten zu werden und ungünsige zerstört zu werden. Das Resultat hiervon würde die Bildung neuer Arten sein."

14 MARX, K.: Letter to F. LASSALLE of 16th January 1861. In: MEW, Vol. 30, p. 578: „. . . naturwissenschaftliche Unterlagen des geschichtlichen Klassenkampfes."

„Das Kapital“ to the effect that the organ-specific division of labour contributed towards a better understanding of social technology? Adequate answers to these and other possible questions can be obtained if one takes into account the understanding of materialistic dialectics. Thus, the qualitatively different forms of motion do not exist simultaneously, but the more highly developed forms arise from the less developed ones, and contain them in the sense of dialectic negation. Therefore the laws obtaining in society, for example, cannot be adequately explained by the known laws governing Nature but have more than superficial similarities to them. MARX and ENGELS recognised the objective inner relationship between the forms of motion and the inherent laws governing them. For the first time they defined these common laws as the basic laws of materialistic dialectics. This represented a creative development of HEGEL's ideas and differed from the mechanistic interpretations mentioned above. Certainly, the basic laws of materialistic dialectics were founded on specialized scientific knowledge treated philosophically; similarly, knowledge of philosophical evolution is no substitute for further research, for instance, into biotic evolution. Understanding of the philosophical evolutionary theory has to be analysed heuristically in accordance with Marxist-Leninist philosophy and modern biology, and has to be changed in the process. This has been shown recently by the definition of philosophical categories such as necessity and chance, causality, possibility and reality, quality etc. This indicates that the new relationships between philosophy and biology which derive from the work of DARWIN, MARX and ENGELS remain productive for the further development of both fields even a hundred years after DARWIN's death. Occasionally these correlations were disturbed[15], and from this bourgeois philosophers have concluded that there is a basic inconsistency between science and philosophy.

In view of the high opinion MARX and ENGELS held of DARWIN's work and the proximity of the three contemporaries (DARWIN lived at Down in Kent, MARX in London, and ENGELS in Manchester and later in London, too), there has always been an interest in possible personal contacts. It is certain that DARWIN never met MARX or ENGELS. The public learned of the correspondence between MARX and DARWIN for the first time from the above mentioned obituary written by LONGUET for DARWIN. He wrote: "A great socialist ... some years ago sent a paper of deep economic analysis and bold historical and social criticism of DARWIN"[16]. He mentioned a copy of the second, improved German edition of the first volume of „Das Kapital“. MARX referred favourably to the ideas of DARWIN in chapters XII and XIII of this volume, suggesting that they held heuristic meaning for the understanding of human technology. This edition had been published in 9 parts between June 1872 and May 1873, and appeared as a complete book in June 1872 with the year 1872 on the title page. MARX wrote a dedication to DARWIN in one of his copies of the book, on June 16th, 1873, "Mr Charles DARWIN, from his sincere admirer Karl MARX" (Fig. 1). But the author was not able to send the gift because he did not know DARWIN's address. He asked the chemist Carl SCHORLEMMER for the address and received it in a letter dated September 25th, 1873.[17] By October 1st, DARWIN had received the book, which has

15 GEISSLER, E.: Einige Bemerkungen zum Thema T. D. LYSSENKO und die moderne Genetik. In: Philosophie und Naturwissenschaften in Vergangenheit und Gegenwart. No 12: Philosophie und Biowissenschaften in der Geschichte, Berlin 1978, p. 22—30.

16 DARWIN. „Der Sozialdemokrat“, Zürich, No 18, 27th April 1882, p. 1.

17 Compare: REIPRICH, K.: Die philosophisch-naturwissenschaftlichen Arbeiten von Karl MARX und Friedrich ENGELS, Berlin 1969, p. 128.

Das Kapital.

Kritik der politischen Oekonomie.

Von

Karl Marx.

Erster Band.

Buch I: Der Produktionsprocess des Kapitals.

Zweite verbesserte Auflage.

Das Recht der Uebersetzung wird vorbehalten.

Hamburg

Verlag von Otto Meissner.

1872.

Fig. 1. Title-page of the specimen „Das Kapital" which MARX has sent to DARWIN. Explanation, see text
(From HUXLEY, J. and KETTLEWELL, H. B. D.: Charles DARWIN and his world, New York 1965, p. 80)

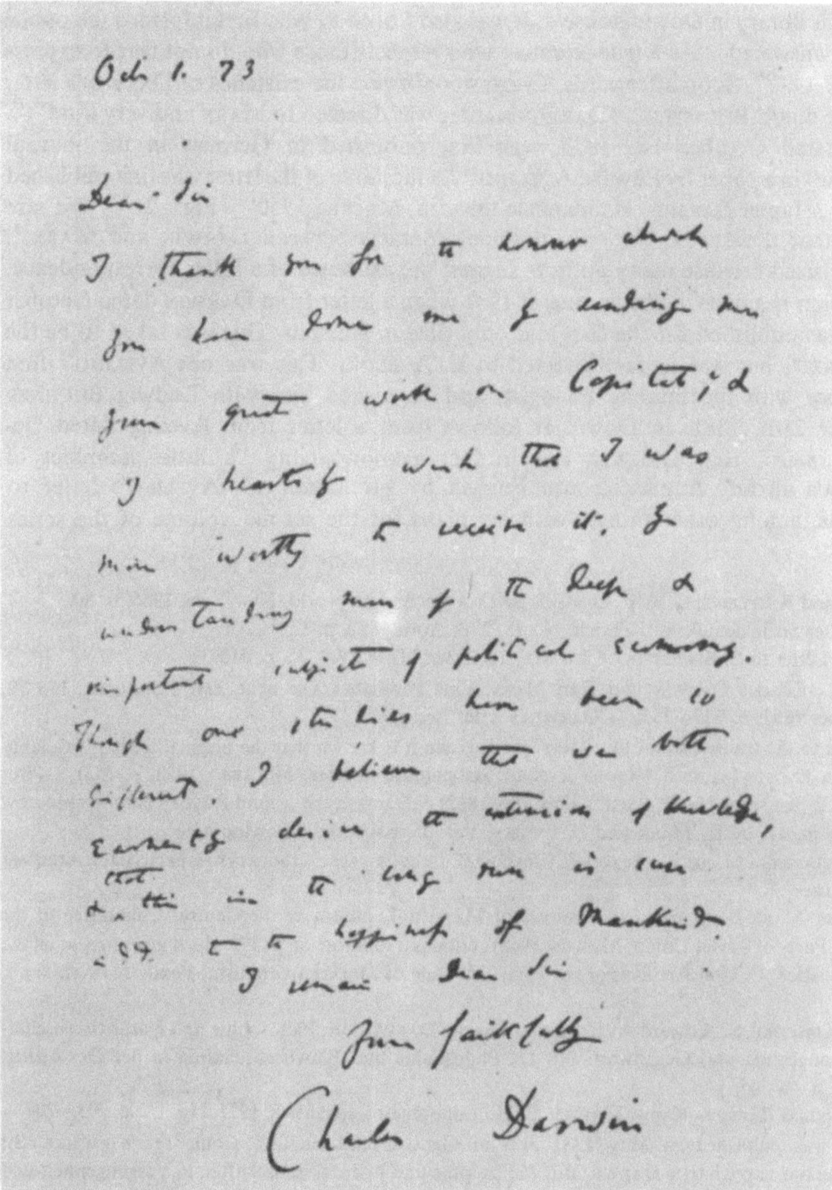
Oct. 1. 73

Dear Sir

I thank you for the honour which you have done me by sending me your great work on Capital; & I heartily wish that I was more worthy to receive it, by understanding more of the deep & important subject of political Economy. Though our studies have been so different, I believe that we both earnestly desire the extension of knowledge, & that this in the long run is sure to add to the happiness of Mankind.

I remain Dear Sir

Yours faithfully

Charles Darwin

Fig. 2. DARWIN's letter to MARX. Explanation, see text
(From ДАРВИН, Ч.: Избранные письма, Москва 1950, с. 239)

remained in his library in Down House[18]. It was also LONGUET who first informed the public that DARWIN answered: "As a true scientist, who respects those who do not fear to express original ideas . . ."[19]. Soon afterwards, ENGELS confirmed the existence of DARWIN's letter in writing to Eduard BERNSTEIN: "DARWIN's letter was directed to MARX and very kind"[20]. This letter dated October 1st, 1873, was first published in German in the journal „Die neue Zeit" in a paper by Edward AVELING[21]. A facsimile of the letter was first published in the book „Чарлз ДАРВИН, Избранные письма, Москва 1950" (Fig. 2).[22] We are now certain that these were the only personal contacts between DARWIN and MARX[23]. This is emphasized because many authors suggest the existence of a fuller correspondence.

The confusion began at the beginning of 1931 when a letter from DARWIN dated October 13th, 1880, was published for the first and only time in Russian. This was taken to be the reply to MARX[24], but was in fact directed to E. AVELING. This was not AVELING's first correspondence with the famous biologist, and he visited him with Ludwig BÜCHNER on September 28th, 1881, in Down. It follows from a letter from Aveling dated October 12th, 1880[25], that DARWIN was in fact acknowledging "a little pamphlet of Dr. BÜCHNER's already translated into English by Mrs. BESANT". AVELING's letter to DARWIN deals, among other things, with his plans for the second volume of the series

18 HUXLEY, J. and KETTLEWELL, H. B. D.: Charles DARWIN and his world, New York 1965, p. 80.

19 DARWIN. „Der Sozialdemokrat", Zürich, No 18, 27th April 1882, p. 1.

20 ENGELS, F.: Letter to E. BERNSTEIN of 3rd May 1882. In: MEW, Vol. 35, p. 315.

21 AVELING, E.: Charles DARWIN und Karl MARX. Eine Parallele. „Die neue Zeit", Stuttgart, No 50, 1st September 1897, p. 745—757. — DARWIN's letter, see p. 753.

22 From a note to the translation of this letter into Russian it is known that the original letter is available in the MARX-ENGELS-Institute (Чарлз ДАРВИН, Избранные письма, Москва 1950, p. 353). — But the letter to E. BERNSTEIN by F. ENGELS (3rd May 1882) contains in the second Russian edition (Moscow 1964) of the papers by K. MARX and F. ENGELS, Vol. 35, p.439 the following note:
„Письмом ДАРВИНА МАРКСУ, о котором упоминает здесь Энгельс, Институт марксизма-ленинизма не располагает . . ."
— The KARL-MARX-Biography (ed. Institute of Marxism-Leninism of the Central Committee of the Communist Party of Soviet Union, Moscow 1968) contains a footnote on p. 372 which gives first of all the place of the letter: Central Archives of the Party, Institute of Marxism-Leninism, Fonds 1, Inventary 5, No 3314.

23 Compare: KIRSCHKE, S.: Edward AVELING und Charles DARWIN. In: Philosophie und Naturwissenschaften in Vergangenheit und Gegenwart. No 12: Philosophie und Biowissenschaften in der Geschichte, Berlin 1978, p. 40—45.

24 Письмо Чарльза ДАРВИНА Карлу МАРКСУ. В: Под знаменем марксизма, 1931, Но. 1/2, с. 203—204. — The letter was published in May 1931 also in German and English. Compare: KIRSCHKE, S.: Edward AVELING und Charles DARWIN. In: Philosophie und Naturwissenschaften in Vergangenheit und Gegenwart. No 12: Philosophie und Biowissenschaften in der Geschichte, Berlin 1978, p. 42. — The explanations of the letter deal with that MARX had asked DARWIN to read a part of the English edition of „Das Kapital". Moreover, it was asserted that MARX had sent DARWIN a specimen of the French edition of „Das Kapital" in 1873. — Compare also: HUXLEY, J. and KETTLEWELL, H. B. D.: Charles Darwin and his world, New York 1965, p. 80: "Karl Marx venerated DARWIN and wanted to dedicate the English translation of Das Kapital to him — a request which was courteously refused". — Karl MARX. Biographie, Berlin 1979, p. 419. — KUCZYNSKI, J.: Zum Briefwechsel zwischen ENGELS und MARX, Berlin 1976, p. 162. — KUCZYNSKI, J.: Autobiographien — vornehmlich philosophische Probleme, Berlin 1977, p. 28.

25 The original letter is available in the Cambridge University Library.

"International Library of Science and Freethought" which he edited together with A. BESANT and C. BRADLAUGH. This second volume was planned to contain papers about DARWIN as a review of his discoveries and thoughts. AVELING asked DARWIN to give his written permission for this plan, to accept the dedication of the paper, and to proof-read it. DARWIN's answer dated October 13th, 1880, dealt with this request of AVELING[26]. Despite certain assertions[27], there was no correspondence between DARWIN and ENGELS, as was shown very clearly in ENGELS' reply to an inquiry by BERNSTEIN[28].

Although the correspondence between the originators of the Marxist philosophy and the founder of evolutionary theory was rather insignificant, the influence of the evolutionary theory on the development of Marxist philosophy was profound.

Discussion

BRODA: Could you please read us the full text of DARWINS letter to Marx?

KIRSCHKE: The letter from Charles DARWIN, to Karl MARX of October 1st, 1873, was taken from a facsimile. The original English text will be published together with a new German translation in the book:

KIRSCHKE, S. (Hrsg.): Darwinismus — Quellen und Wirkungen. Wissenschaftliche Beiträge der Martin-Luther-Universität Halle-Wittenberg. Halle (Saale) 1982 (in the press).

26 Letter from DARWIN to MARX. "The Times", London, May 18, 1931: "The publication of your observations on my writings, in whatever form they may appear, really does not need any consent on my part ... I should prefer the part or volume not to be dedicated to me (although I thank you for the intendend honour) ... I am sorry to refuse you any request but I am old and have little strength, and the reading of proofs (as I know from present experience) imposes a heavy strain on me."

27 Compare: KOEHLER, O.: DARWIN und wir. In: HEBERER, G. und SCHWANITZ, F. (Eds.): Hundert Jahre Evolutionsforschung. Das wissenschaftliche Vermächtnis Charles DARWINS, Stuttgart 1960, p. 16.

28 E. BERNSTEIN asked F. ENGELS (letter of 27th April 1882) to write a paper on DARWIN and the significance of the Darwinism to Socialism which was intended to be published in the news paper „Der Sozialdemokrat". ENGELS' answer, see note 20.

Institute of Physical Chemistry, Vienna University

DARWIN and BOLTZMANN

E. BRODA

Abstract

At least since 1886 BOLTZMANN came forward as a strong supporter of DARWIN. He argued for DARWIN's theory in general, but, in addition, he also made specific contributions to evolutionist theory in four fields, namely 1) bioenergetics, more particularly photosynthesis, 2) origin of life, 3) origin and evolution of the mind and 4) morality and esthetics. The general validity, but also the limits, of BOLTZMANN's ideas on ethics and esthetics are pointed out. Finally, questions about the roots of BOLTZMANN's involvement in DARWINian theory are raised. In his confidence in the evolutionary basis of all mental functions BOLTZMANN may be considered as an "absolute" DARWINist.

> If Schiller told the scientists and philosophers of his time: "Enmity be between you, still it is too early for alliance." I am not contradicting him. I just think that now the time for alliance has come.
>
> L. BOLTZMANN "On statistical mechanics", St. Louis, 1904.

Ludwig BOLTZMANN lived from 1844 to 1906 and was mostly Professor at Vienna and Graz. Along with NEWTON and MAXWELL he was the greatest theoretical physicist of classical times. Moreover he was a most interesting philosopher who forcefully argued the case of realism, occasionally also called by him materialism, against the idealism of HEGEL and SCHOPENHAUER and against the positivism of MACH. Here we are interested in his contribution to biology. But it will be seen that BOLTZMANN's views in philosophy and biology are interconnected.

General Support of Darwin

In his lecture on the Second Law to the Official Meeting (Feierliche Sitzung) of the Vienna Academy in 1886 BOLTZMANN called the 19th century, whose achievements he enthusiastically praised [1], the century not of electricity and not of steel (and not of one of the great physicists, we may add), but the century of DARWIN [5]. In an address to Graz University in 1887 in honour on the deceased physicist of Berlin, Gustav Robert KIRCHHOFF, he strongly, indeed violently, argued against the obscurantists who stood in the way of DARWIN's teachings [6]. He compared their anxious noises to those of oxen that fear the discovery of a new theorem of PYTHAGORAS. In gratitude, PYTHAGORAS had sacrificed 600 oxen to the gods. It was implied that DARWIN's adversaries were motivated not scientifically, but that they acted on the basis of prejudice and bias, even of panic.

In his later years, BOLTZMANN took any opportunity to express his admiration and support for DARWIN. The last reference to DARWIN is found in BOLTZMANN's [7] anti-SCHOPENHAUER lecture early in 1905, originally called "Proof that SCHOPENHAUER is a mindless, ignorant, philosophaster who by hollow verbiage degenerates the heads for ever and fundamentally"[1]. No doubt DARWIN was given prominence also in BOLTZMANN's [8] lecture with the curious title "Explanation of the entropy law and of love from the laws of the calculus of probability" to the Philosophical Society at Vienna University on October 28th, 1905. The text of the lecture has been lost. Thus BOLTZMANN belonged to the large number of thinkers, inside and outside biology, who in the 19th century were deeply impressed by DARWIN and his evolutionist ideas.

We may interpolate that BOLTZMANN's fellow professor at Vienna since 1895, Ernst MACH, while he was a bitter adversary of BOLTZMANN in physics and philosophy, likewise supported DARWIN [9]. In his sense physiological investigations he used evolutionist arguments. One may well ask how a positivist, who believed only in 'complexes of sensation" or "elements of experience" and who rejected the idea of an independent outer world as an object of science, could accept evolution of organisms as an objective fact. But these are the contradictions into which one inevitably runs in trying to be a positivist.

In spite of his great interest in living Nature that he showed as a boy already [10], BOLTZMANN remained a layman in biology. Nevertheless, he contributed important and fruitful ideas. All of them were inspired by DARWIN's views. We mention (1) BOLTZMANN's ideas on the struggle for the acquisition of free energy, especially in connection with plant photosynthesis, (2) his ideas on the origin of life, (3) his ideas on the evolution of the mind and of consciousness, and (4) his ideas on ethics and esthetics.

Acquisition of Free Energy

The dependence of plant life on light had been discovered in London by Jan INGEN-HOUSZ, a Dutchman mostly living in Vienna, in 1779. However, INGEN-HOUSZ did not know why exactly light is needed. Indeed he could not know it as conservation of energy was unknown in his time, and so the need for a particular source of energy did not seem to exist. The second step had been taken by the discoverer of energy conservation (First Law of Thermodynamics), Julius Robert MAYER, in 1845. He wrote [11]: "The plants absorb a force (in modern terms: energy, E. B.), light, and produce a force (energy): chemical difference".

However, in his ignorance of the Second Law MAYER could not make a distinction between useful and useless forms of energy. This distinction, in respect to photosynthesis, was left to the physicist who understood the Second Law better than anybody else, and who explained it in atomistic terms, to Ludwig BOLTZMANN [13]. In 1884 he had introduced the notion of the entropy of radiation, though the word entropy was not used explicitly. This paper was, in his famous obituary of BOLTZMANN, called a true pearl of theoretical physics by Hendrik Antoon LORENTZ [15]. Already two years later, BOLTZMANN [5] wrote about photosynthesis in his lecture on the Second Law: "The general struggle of the organisms is . . . not a struggle for the elements — the elements of all organisms are abundant in air, water and soil — nor for energy which, in the form of heat, un-

1 These words were originally used by SCHOPENHAUER against HEGEL.

fortunately unchangeably, is contained in all substances, but a struggle for the entropy (better: negative entropy. E. B.), which becomes available in the transition of the energy from the hot Sun to the cold Earth. To exploit the transition as far as possible the plants spread out the immeasureable area of their leaves and force solar energy in a way as yet unexplained, before it sinks down to the temperature level of the surface of the Earth, to carry out chemical syntheses of which one has no inkling as yet in our laboratories. The products of this chemical kitchen are the object of the struggle of the animals."

In these majestic words it was implied that plant photosynthesis arose and developed during a DARWINian struggle for improved supply of free energy. Free energy is, of course, negatively correlated to entropy. Quite generally it may indeed be assumed that the impressive improvement of the bioenergetic processes — fermentation, photosynthesis, respiration — during the ages was an essential part of evolution [12, 16]. These processes insure the capacity of the organisms to do work in the physical sense, essential in the struggle for survival. The absorption of negative entropy (negentropy, according to BRILLOUIN) as the essence of life was also, much later, argued by Erwin SCHRÖDINGER in his influential book "What is Life" [17]. A puzzle is the absence of DARWIN himself from SCHRÖDINGER's book, although SCHRÖDINGER like BOLTZMANN admired DARWIN.

Origin of Life

In 1924 Aleksander OPARIN [18] and in 1929, independently, J. B. S. HALDANE [19] began the modern investigations of the origin of life on Earth, and maybe elsewhere, on a DARWINian basis. We are speaking of "chemical" or "prebiotic" evolution, which led to the first cells. But only seldom it is mentioned in biophysical literature that similar ideas had been put forward long before by nobody else but BOLTZMANN [12]. Of course, experimental data at BOLTZMANN's disposal were meagre, and so he could not produce detailed pictures. Yet the general principles laid out by BOLTZMANN are still acceptable today. In 1886 BOLTZMANN [5] said: "We make the hypothesis that complexes of atoms evolved that could multiply by formation of similar complexes around themselves. Among the larger masses so formed those were most vital that could proliferate by division, further those that tended to move to places with favourable conditions for life". In a lecture [20] in 1904, BOLTZMANN specifically referred to DARWIN and continued: "It does not matter to us whether during the millions of years in the enormous mass of water on Earth the first protoplasm evolved 'by accident' in moist mud, whether egg cells, spores or other germs, as dust or embedded in meteorites, some time arrived on Earth from space. More highly developed individuals will hardly have fallen from the skies. Thus at first there were only quite simple individuals, simple cells or lumps of protoplasm. Constant movement, so-called BROWNian movement[2], is shown, as is well known, by all small lumps; growth through absorption of similar components and subsequent proliferation by division is fully imaginable

2 It is remarkable that BOLTZMANN was aware of the importance of BROWNian movement. Yet it was left to Albert EINSTEIN and Marian von SMOLUCHOWSKI to interpret it as a fluctuation phenomenon in a BOLTZMANNian sense. This was done in BOLTZMANN's life time, but the great master never referred to the work of these younger colleagues — notwithstanding SMOLUCHOWSKI's presence in Vienna.

mechanically[3]. Similarly it can be understood that the rapid movements were influenced and modified by the environment. Lumps where modifications took place in the sense that on an average (preferentially) they moved towards substances better suitable for absorption (better food) succeeded better in growth, and more often in propagation. Hence they soon overran the others."

Origin of the Mind

In our time the concept has much support that in evolution nervous systems and the mind arose, as a function of matter, in response to challenges in the DARWINian struggle for existence. Organisms equipped with complex and efficient information systems could maintain themselves better than their poorer brethren. At certain stages of complexity memory appeared, and still later that astonishing fruit on the tree of evolution, consciousness, with its corollary, free will. Clearly consciousness did not fall from the blue sky. We can trace it back from man to apes, lower mammals and even beyond — with decreasing efficiency [22].

Few of the contemporary authors are aware of BOLTZMANN's DARWINian thoughts in this field either. In his paper with the characteristic title "On the objective existence of the processes in inanimate nature" (1897) we read [23]: "The analogy between the sensations of man with those of the highest animals is so perfect we absolutely must ascribe objective existence to the latter sensations. But where is the limit? True, occasionally one hears doubts whether insects or divisible animals, like certain worms, have sensations. But a definite limit where sensation stops cannot be given. At last we come (going downward. E. B.) to organisms so simple that their world views and thoughts are zero. Unless we suddenly deny the predicate of existence to the sensations of animals beneath a certain level, which would be most inappropriate, we must ascribe existence even to the sensations of this unthinking organized matter where sensations are hard to demonstrate."

Morality and Sense of Beauty

In the spirit of DARWIN BOLTZMANN also derived the human sense of ethics and estetics. In 1900 he said [24]: "We still have to remember the most wonderful mechanical theory in the field of the biological sciences, namely, the theory of DARWIN. It undertakes to explain the whole diversity of the world of plants and animals through the purely mechanical principle of heredity — which admittedly in itself is dark like all primordial principles of mechanics.

The explanation of the wonderful beauty of flowers, of the richness of forms in the insect world, of the fitness of the structure of the organs in the animal and human body, all this herewith becomes the domain of mechanics. We understand that it has been useful and im-

3 The expression "mechanical" was used loosely by BOLTZMANN, and was often employed as a synonym for "scientific" or "in accordance with objective natural law". For instance, BOLTZMANN [21] spoke of the "law by whose grace the kings rule, the fundamental law of mechanics". Here I protest against superficial attempts to label BOLTZMANN as a "mechanical" thinker or "mechanical materialist". These attempts do not do justice either to the actual contents of his work or to his spirit.

portant to our species that certain sensations flattered us and were sought by us, others repelled us. We see what an advantage it has been to construct, as precisely as possible, pictures of the environment in our mind and to distinguish strongly what of these pictures agreed with experience, as true, from what did not agree, as wrong. Thus we can explain mechanically the origin of the concept of beauty as well as that of truth.

We understand, too, why only individuals could continue to exist who abhorred with all intensity of their nerve power and tried to prevent certain highly pernicious influences, but strove with equal nerve power towards others that were needed for their own preservation or that of the species. Thus we understand how the whole intensity and force of our emotional life evolved, joy and pain, hatred and love, desire and fear, delight and despair. ... Thus, in Nature and in the arts the all powerful mechanics rules, as it does in politics and in social life".

Now BOLTZMANN was surely right when he thought of morality within populations, maybe families, of animals. The devotion of the parents to each other and to the offspring within, say, the fox family, the goose family, the wolf pack or the baboon herd is an essential factor for the survival of the species. The amazing discipline of the worker bee is needed for the hive. The insect's enjoyment of bright colours and of pretty shapes is a factor in their evolution and in the coevolution of the flowers they visit. The harmony of the birds' song must also be a product of the struggle for existence, of variation and selection. Similar considerations also apply to man's ancestors. On a DARWINian basis BOLTZMANN [20] further explained, in his magnificent attack (1904) against the formalistic views of Wilhelm OSTWALD, who had proposed an algebraic equation to measure happiness, the true meaning of this fact of life.[4]

In this way, BOLTZMANN indeed arrived at the last roots of the moral and esthetic sense of all animals, including man. Thus he provided the basis for an understanding of the contents and the values of moral and esthetic concepts. Seen from this point of view, these concepts, also in humans, are not arbitrary any more. It must be admitted that Albert EINSTEIN, in practice much more concerned with social ethics among humans than BOLTZMANN, did not find the right starting point, and derived morality in an arbitrary and unconvincing way [25, 26]. This regrettable fact was a consequence of EINSTEIN's neglect of the idea of evolution among organisms.

4 The lecture against OSTWALD ends with the pertinent words in a long-drawn sentence indeed: Why does a seemingly so harmless essay ... appear to me so dangerous to science? Because it means a relapse into the pleasure with the purely formal, into the methode of the socalled philosophers that is so pernicious for progress to construct theorems (Lehrgebäude) from mere words and phrases and to emphasize only their pretty formal interconnection (called purely logical or even aprioristic foundation) without looking whether this interconnection corresponds exactly to reality and is sufficiently founded in the facts, a relapse into the method to let oneself govern by prejudiced opinions, to bend everything unter the same principles of classification, and to force everything artificially into one and the same system, not to want to see true mathematics for algebraic formulae, true logic for syllogisms that seemingly are built according to the rules (schulgerecht), true philosophy for nonsense (Krimskrams) that decorates itself philosophically, the forest for trees — a method that unfortunately will always appeal more strongly to the multitude than the method of the natural sciences that gives less scope to imagination.

Biological and Social Evolution

It must be conceded that BOLTZMANN was not quite clear about social evolution in mankind. To formulate our problem, let us first stress the mistakes which BOLTZMANN did not make. He did not try to ascribe inborn differences in mental functions to different parts of mankind, to human races or to social classes. It is implied that he kept away from the mistakes and platitudes of "Social DARWINism", i.e. he did not try to explain the success of the members of the upper social classes to superior physiological, including mental, equipment. It is also implied that the success of the Whites in conquering the world was not due to genetic superiority.

Moreover, BOLTZMANN rightly sought the reason for the intellectual differences between peoples of different historic ages in the differences between the milieus, in the differences between the influences to which the peoples were exposed. It will be seen below that BOLTZMANN, in pointing to changes of the laws of thought (Denkgesetze) during historical times, chose as examples "a naive people like the Greeks", the "mediaeval scholastics" and "the present direction". Modifications through "education, teaching and one's own experience" are mentioned. Similarly BOLTZMANN spoke of the difference between children, uneducated and educated persons. We summarize that according to BOLTZMANN the recent socio-cultural evolution of mankind had no genetic basis, and nothing in common with biological evolution. So far so good.

BOLTZMANN also noted contradictions between ethical attitudes within given human societies. For instance, he spoke of such contradictions in his anti-SCHOPENHAUER lecture [7]. In more detail the problem was treated in the lecture on the "principles of mechanics" (1900) in Leipzig [24]: "The activity of the conservatives, of pedantic, reactionary (zopfige), rigid, judges of manners (Anstandsrichter), who insist on exact observation of any traditional habit and of any rule for social relationships, on precise employment of all their titles in adresses and concession of all social privileges often appears laughable to us. Yet it is beneficial and has to be to prevent brutalization in the social relationships. On the other hand the uninhibited, the emancipated, the hommes sans gêne, see to it that it (conservatism. E. B.) does not lead to petrification of the mind. Both kinds of humans struggle against each other, and together they keep society in the right equilibrium."

After continuing by praising "one of the most magnificent, most admirable mechanisms produced by mankind, that of capital, of money" BOLTZMANN said: "Whoever failed (ist zu kurz gekommen), grumbles about mammon; the cheat who bends the rules for lust of profit is expelled like useless stuff from a living organism; but for our modern civilization the flow of money and the stock exchange are as important as the art of printing, as steam, as electricity".

Clearly BOLTZMANN was not well informed about social matters. He was unable to correlate moral tendencies with social situations. He did not see that basically conservatism is the ideology of the "haves", a progressive attitude (not necessarily "sans gêne") the natural trend among the "have-nots". Needless to say that in BOLTZMANN's time the main classes of the rich were the aristocracy and the big bourgeoisie, and of the poor the proletariat and the poor peasantry.

It follows that the moral (and esthetic) ideas serve not only to promote the success of mankind in general, but also — or instead? — the success of different classes of society, or of nations. Life and death struggles may be the wellknown consequence. Evolutionist ideas (not on a biological level) can be applied to them.

BOLTZMANN was not thinking in terms of societal change. His ignorance of social matters is evident from the lack of references to them in his writings. BOLTZMANN loved to speak not only of the sciences, but also of the arts and of other achievements of mankind. But while he was a thoroughly humane person, the problems of the poor, the struggles of the working class for the Ten Hours Day and for the right to vote, the emergence of Trade Unions and of Socialist movements remain unmentioned, though they affected dozens of millions of contemporary men and women in Austria and elsewhere. Nor were war and peace treated adequately.

The Basis of BOLTZMANN's Involvement With DARWIN

We are arriving at the question of the basic of the admiration of BOLTZMANN for DARWIN. Was it just BOLTZMANN's interest in the great science of biology and his fascination with DARWIN's meteoric successes? Did they move him, from his middle age, to give more and more thought to a second branch of science? Was it the attraction of the statistical element in DARWIN's teaching? Or was it the seeming contradiction between the general increase of entropy in inanimate Nature and the decrease in entropy in growing organisms that struck BOLTZMANN, as it strikes Ilya PRIGOGINE [27] now?

Careful reading of BOLTZMANN's writings leads to a different conclusion. In his fundamental research in physics BOLTZMANN again and again became concerned with the origins and the limits of our laws of thought. He insisted that these laws that arose in evolution must be broadly true, but not necessarily in every case. They may lose validity in extrapolations beyond the realm of experience. In a discussion with the great Heinrich HERTZ BOLTZMANN [28] wrote in 1899: "I want to raise certain objections against this proposition (that the laws of thought are faultless. E. B.), or at least explain it more clearly. Certainly we must bring with us a rich treasure of laws of thought. Without them, experience would be quite useless; we could not even fix it through internal pictures. Almost without exception, these laws of thought are inborn, but nevertheless they suffer modifications through education, teaching and one's own experience. They are not quite the same for a child, for a plain uneducated person and for a scientist. We shall appreciate this when we compare the direction of thought of a naive people like the Greeks with that of the mediaeval scholastics and this again with the present direction. Surely there are laws of thinking that proved their worth in such a way without exception that we have absolute confidence in them, that we consider them aprioristic unchangeable principles of thought. But nevertheless I think that they developed slowly. Their first source consisted of primitive experience of mankind in the original condition, gradually they became stronger and clearer through complicated experience, until finally they assumed their present sharp formulation; but I should not recognize the laws of thinking as absolutely supreme judges. We cannot know whether they will not suffer one or the other modification. On may recall how definitely children or uneducated persons are convinced that it must be possible to distinguish up and down in all places of space through mere feeling, and how they believe to be able to deduce therefrom the impossibility of antipodes. If such people wrote logic, they would certainly consider it a law of thought that is evident a priori. Likewise in the beginning aprioristic objections were raised against the theory of COPERNICUS. In the history of science there are many cases where theorems were either proved or refuted through evidence which was thought to correspond to evident laws of thinking, while now we are convinced of its futility. The sole and ultimate verdict about the fitness of the pictures is based on their representation, as simple as possible and completely correctly, of experience. . . . Precisely here is the test for the correctness of the laws of thinking." Thus according to BOLTZMANN the laws of thought evolved, in a DARWINian way, in response to need.

BOLTZMANN as an Absolute Darwinist

Thus Darwinism saved BOLTZMANN from the antinomies of Immanuel KANT. Again and again BOLTZMANN [29] referred to them: Finite or infinite space, finite or infinite time, infinite divisibility or indivisibility of matter. BOLTZMANN esteemed KANT, but he considered his views as "not tenable", as he especially made clear in his magnificant lecture against SCHOPENHAUER [7]. We are unable to deal with these antinomies because the statements and assertions involved go beyond actual experiences. For lack of such experience, the concepts in question could not be formed in evolution.

Thus we come to the conclusion that BOLTZMANN was so to speak an **absolute** Darwinist [30]. Darwinism was not a side-line in his thinking, but had a central role. The DARWINian approach enabled him to deal with the logical problems in setting up physical theories. BOLTZMANN was confident that the antinomies would disappear, as science learns to deal with the real, objectively existing, external world.

It may be noted here that K. R. POPPER, while he is an admirer of BOLTZMANN and also of DARWIN, nevertheless expressed skeptic views about Darwinism. In his important and stimulating intellectual autobiography POPPER said [31]: "I regard DARWINism as metaphysical[5] . . . It is metaphysical because it is not testable. . . . And yet, the theory is invaluable. I do not see how, without it, our knowledge could have grown as it has since DARWIN." One wonders how a theory that contributes to knowledge can be untestable. POPPER in the same book also denied "that Darwinism can explain the origin of life". POPPER wanted to restrict Darwinism to a "research programme". POPPER later recanted [33]. In view of the attacks of anti-scientific Creationism against Darwinism in the USA, which in our time have reached new heights, this recantation is doubly fortunate and welcome.

Evolutionary epistemology and gnoseology is a matter of the greatest importance in our time, and so it is surely justified to consider BOLTZMANN's convictions, his absolute Darwinism, in some detail within a symposium called "DARWIN today".

Ich danke Herrn Professor Hans KUHN, Göttingen, für seinen Hinweis, daß BOLTZMANN nicht mit „PRIGOGINES Problem" befaßt war.

References

[1] See 2), pp. 56, 198; 3, 4).

[2] BOLTZMANN, L.: Populäre Schriften, Leipzig (1905). A new edition, somewhat shortened, was published by Vieweg, Braunschweig—Wiesbaden (1979), with a foreword by the present author. All page numbers refer to the latter edition.

[3] BRODA: Ludwig BOLTZMANN, Mensch—Physiker—Philosoph, Deuticke, Vienna (1955) and Deutscher Verlag der Wissenschaften, Berlin (1956).

5 BOLTZMANN [32] called metaphysics a "mental migraine of mankind."

[4] BRODA, E.: Philosophical Biography of Ludwig BOLTZMANN, in: E. G. D. COHEN and W. THIRRING, eds., The BOLTZMANN Equation, Springer, Vienna (1973).
[5] See 2), p. 26.
[6] See 2), p. 47.
[7] See 2), p. 240.
[8] BOLTZMANN, L.: Erklärung des Entropiesatzes und der Liebe aus den Prinzipien der Wahrscheinlichkeitsrechnung, Phys. Bl. 32, (1976) 337.
[9] BLACKMORE, J. D.: Ernst MACH, His Life, Work and Influence, Univ. Calif. Press, Berkeley (1972).
[10] JÄGER, G.: Ludwig BOLTZMANN, Neue Österr. Biogr., Vol. 2, Amalthea-Verlag, Vienna (1925).
[11] MAYER, J. R.: Mechanik der Wärme, Stuttgart 1845, quoted in 12).
[12] BRODA, E.: The Evolution of the Bioenergetic Process, Pergamon Press, Oxford (1978).
[13] BOLTZMANN, L.: Ableitung des Stefanschen Gesetzes, etc., reprinted in 14).
[14] HASENÖHRL, F., ed.: Wissenschaftliche Abhandlungen von Ludwig BOLTZMANN, Barth, Leipzig, (1909).
[15] LORENTZ, H. A., Ludwig BOLTZMANN: Vhdl. D. phys. Ges. 9, (1907) 206.
[16] BRODA, E. and G. A. PESCHEK: Did respiration or photosynthesis come first? J. Theor. Biol. 81 (1979) 201.
[17] SCHRÖDINGER, E.: What is Life? Cambridge Univ. Press, Cambridge (1945).
[18] OPARIN, A. I.: The Origin of Life (Russ.), Moscow Worker, Moscow (1924). Reprinted in J. D. BERNAL, The Origin of Life, Weidenfeld and Nicolson, London (1967).
[19] HALDANE, J. B. S.: The Rationalist Annual, London (1929). Reprinted in J. D. BERNAL, see ref. 18.
[20] See 2), p. 225.
[21] See 2), p. 178.
[22] BRODA, E.: Free will and evolution, J. Theor. Biol. 69, 755 (1977).
[23] See 2), p. 94.
[24] See 2), p. 170.
[25] BRODA, E.: BOLTZMANN, EINSTEIN, natural law and evolution, Comp. Physiol. Biochem. 67B. (1980) 373.
[26] BRODA, E.: Ethics and evolution in BOLTZMANN's and EINSTEIN's thought, In: M. KAGEYAMA, K. NAKAMURA, T. OSHIMA and T. UCHIDA, eds., Science and Scientists, Japan Scientific Societies Press, Tokyo, and Reidel, Dordrecht (1981).
[27] PRIGOGINE I. and STENGERS, I.: Dialog mit der Natur, Piper, Munich (1980).
[28] See 2), p. 160.
[29] See 2), pp. 179, 213, 242, 253.
[30] BRODA, E.: BOLTZMANN and DARWIN, International BOLTZMANN Symposium, Vienna (1981).
[31] POPPER, K. R.: Unended Quest. An Intellectual Autobiography. Fontana-Collins, London (1976). German version, amended in some details: Ausgangspunkte. Meine intellektuelle Entwicklung. Hoffmann und Campe, Hamburg (1979).
[32] See 2), p. 257.
[33] POPPER, K. R.: Natural selection and the emergence of mind, Dialectica 32, 44 (1978), quoted by H. ZEISEL, POPPER on DARWINism, Science 212, 873 (1981).

Discussion

GRANTHAM: I think it is wrong to suggest that human social and cultural evolution has nothing to do with biological evolution (I am not sure whether you intended this statement as representing BOLTZMANN's position or your own). Later you said that BOLTZMANN held that the laws of thought and ideas about time and space evolved in a DARWINian way, thus seeming to contradict the first statement.

The tripling of the volume of the hominid brain in a few million years, the acquisition of speech and tool making, the upright stance and our special human behaviour seem to show interaction between cultural and biological evolution.

BRODA: Socio-cultural evolution is rapid and requires no genetic changes. Of course, genetic changes occurring during long periods in prehistory were bound to have socio-cultural consequences.

KLIX: You have mentioned that Dr. BOLTZMANN related the laws of human thinking to evolutionary theory. He assumed obviously that some aspects of such laws are linked with a pre-human period of evolution. On the other hand he mentioned (as you cited) that there is educational dependence in human thinking.
My question: How did he assume that both are related or interlinked with each other?

BRODA: The only point where I want to criticize BOLTZMANN is his failure do discriminate sharply between genetic and socio-cultural evolution. It is not always clear what was in his mind at a given instant.

NOVAK: I would agree with what Prof. BRODA said about the difference between biological and cultural evolution. Geneticists are inclined to conclude too easily that various characters are hereditary without having sufficient evidence. This applies to many characters arising during evolution and various morphological characters. The extent to which human language is conditioned to man's adapting his speech to the actual time of his individual development has been shown, for instance by VYGOTSKY in the Soviet Union. This is later no longer possible or rather restricted.

We have defined the concept of "non-hereditary phylogenetic changes" (NPC) as changes produced by various morphogenetic factors during ontogenesis without a corresponding change in the genome which may exist for many generations because the given factor continues to work for many generations. The experience with explants in tissue cultures shows that this is a very common phenomenon also at the biological level. The phenotypic variability is very broad, and by far most of the per se characters of organisms are thus "non hereditory". This seems to apply to all features of cultural evolution. It is known from many cases that Australian aborigines educated from birth in European families showed all capabilities of white children. The note of an English anthropologist that "Neanderthaler had already all the capacities to study with success the Oxford University" seems to be quite well founded.

GRANTHAM: The only phylogenetic change that is not hereditary that I can think of is extinction.

KLIX: To compare old languages with respect to their age brings some consequences regarding vowels and consonants. The symbolic fixation of consonants is relatively young. Older languages have only consonant fixations. The problem is how to fill the gaps without shifting the possible outcomes. This point is essential because you especially referred your comparison to vowels and consonants.

SCHOLZ: Is the evolution of languages an appropriate model for biological evolution of species in term of mutation, isolation, and development of new (linguistic) species?

BRODA: The Japanese have difficulties in pronouncing the letters "l" and "r". Nevertheless a Japanese baby brought to the West will learn these letters easily. No genetic, but only socio-cultural evolution is involved.

ANTONOV: There exist profound analogies in the evolution of human languages and biological evolution, and it was DARWIN who discussed this in ch. XIV of the "Origin of Species".

GRANTHAM: Perhaps we are adapted to learn a human language and if a human were placed on planet X he would not be adapted to speak the language used there.
(By this I imply that the evolution of speech is not independent of our biological evolution)

USSR Acad. Sci., Moscow, USSR

Ch. DARWIN's "The Formation of Vegetable Mould Through the Action of Worms", Its Methodological Premises

M. S. GHILAROV

The name of the great naturalist Charles DARWIN is connected in the mind of humanity in the first line with the theory of natural selection, with his strictly materialistic theory of organic evolution. Using consistent materialistic premises DARWIN has shown the possibility of the origin of multitude of plants and animals, of the rise of new species as result of the accumulation of small hereditary changes of ancestral forms. DARWIN's methodology was spontaneously dialectic one; his theoretical constructions may serve as brilliant illustrations of the philosophical law of quantitative changes transition into the new quality.

Correct methodological premise, ability to subtle observation, astonishing diligence of the great scholar enable him to reserve undisappearing traces not only in biology but also in other branches of natural history. In geology especially famous are his views on reef formation (1847, 1874) due to the slow growth and skeleton deposition of coral polyps based on the same methodological premise, on the admission that small "insignificant" quantitative changes when accumulating provoke significant qualitative change, a qualitative "jump". Certainly a "jump" in philosophical meaning, DARWIN himself used to stress that "natura non facit saltus". This methodological approach is clearly evident in the last DARWIN's work on "The Formation of Vegetable Mould through the Action of Worms" (1881), the centennary of which has been celebrated at a special international symposium on earthworm ecology, organized by Dr. J. E. SATCHELL in Grange-over-Sands (England) in August—September 1981.

This last DARWIN's book was of a paramount significance for the development of soil science. Not in vain Sir E. J. RUSSEL (1927) has written on this book evaluating it as "the most interesting book ever written on the soil". A. YARILOV, a famous Russian historian of soil science, wrote in 1936 that the author of "The Formation of Vegetable Mould" is to be recognized as one of founders of modern soil science.

Earthworms attracted the attention of the great English naturalist during all his life.

First communication on their activity was made by DARWIN in 1837 in the Geological Society of London; he showed that pieces of chalk etc. laid on the meadow surface became some years after under the turf being convered with earthworms castings. Already than he concluded that all the vegetable mould (as he called humified soil layer) had passed through earthworms intestines. Being heavily ill after his "Beagle" voyage and staying all the time in his Down, DARWIN continued observations and gathered literature upon earthworms, counted their castings on pathways and laws. I was happy to be invited in 1958 to Down by DARWIN's family and to look at the places of his systematic observations. DARWIN also reared earthworms in pots when studying more detailly their behaviour and activity. It was just as the result of all his life work the "swan song" of the great naturalist, his last accomplished work — "The formation of vegetable mould". This work was published not long before the famous Russian naturalist Vasiliy DOKUCHAEV issued his classical book "Russian chernozem" (1883) which is regarded as the very foundation of modern concept of the soil as a "fourth natural body" in addition to three discriminated by C. LINNEUS — mineral, plant, animal.

DARWIN was the first who characterized and quantitavely evaluated the role of biological agents in soil formation, this enables us to call the great naturalist as the founder of biological direction in soil science. In the preface to his work DARWIN remembered his first communication on the subject in Geological Society and mentioned that in "Gardner's Chronicle" (1869) Mr. FISH rejected his conclusions on the participation of earthworms in vegetable mould formation "merely on account of their assumed incapacity to do so much work". Mr. FISH remarked that "considering their weakness and their size the work they represented to have accomplished is stupenduous". DARWIN adds with bitterness: "Here we have an instance of that inability to sum up the effects of a continually recurrent cause, which has often retarded the progress of science, as formarly in the case of geology, and more recently in that of the principle of evolution" (pp. 5—6)[1]. This DARWIN's note reflects his profound certitude of the correctness of his methodological approach, which he has applied when examinating the reef formation and when explaining evolutionary changes under the action of natural selection resulting in formation of new species. Methodology of scientific investigation, originating from actualistic Ch. LYELL's concepts in geology, reached its extreme sharpness in studies of the "Origin of the Species".

When finishing his book on earthworms DARWIN has written: "It may be doubted whether there are many other animals which have played so important a part in the history of the world, as have these lowly organized creatures. Some other animals, however, still more lowly organized, namely corals, have done far more conspicuous work in having constructed unnumerable reefs and islands in the great oceans; but these are almost confined to tropical zones" (p. 288).

Now we can mention with real sadness that the founder of the genetical soil science DOKUCHAEV (1883), who based his soil classification on the ways and modes of soil formation underestimated the role of earthworms, though he had thoroughly studied and cited DARWIN's book on the subject. DOKUCHAEV considered countings made by DARWIN as either exaggerated or having local significance. Such DOKUCHAEV's position founds its explanation in the fact that DOKUCHAEV when studying chernozems (black soils) worked in steppe regions, where the activity of earthworms is not so evident, especially in summer, when DOKUCHAEV studied soil profiles.

Later the immediate disciple of DOKUCHAEV — G. VYSSOTSKY (1930), who worked in steppes whole years long, explained on base of his observations and experiments the high stability of chernozem's granular structural units characterizing earthworm activity as one of main factors of this washing resistant granular structure.

In DARWIN's approaches to the characteristics of earthworms soil forming activity especially essential is the fact, that he, when studying soil, draws attention not to its static state, does not describe its horizons etc., but investigates the "vegetable mould" from the view-point of its dynamic, the process of the development of the soil layer.

1 Cit. after "The formation of vegetable mould . . ." by Charles DARWIN, Thirteenth thousand, John Murray, L. (1904).

In DARWIN's "Origin of the species" time ecology as a branch of biology was not still formed. The term "ecology" and its definition were introduced later by the adept and passionate fighter for Darwinism — the famous German zoologist Ernst HAECKEL (1869); the concept of the biocenose was formulated still later by MÖBIUS (1877). But DARWIN himself in the 3rd Chapter of "The Origin of the Species" when illustrating the struggle for existence used just changes in biocenoses as examples, whereas in his "Earthworms" his attention was drawn to that aspect of ecology which in that time and many years later was still neglected by ecologists.

Ecology according to the very definition of this science studies interaction, interrelationships of organisms and their environment. Up to short time ago ecologists studied dependence of organisms on their environment; DARWIN in his "Earthworms" has brilliantly shown the other side of the medal — the influence of organisms on their environment, dependence of the milieu on their activity.

He also realized the quantitative approach to the evaluation of the environment changing activity of organisms; in particular he estimated the velocity of changes of the earthsurface under the influence of earthworms.

The broad generalization of various observations on earthworms led him to the conclusion on their geological significance, on their participation in denudation processes, in transformation of hilly relief into a plain.

The last book by Ch. DARWIN may be also regarded as one of foundations of invertebrate ethology. Interesting is that DARWIN stressed the non-stereotypic, adaptive character of earthworm activity, e.g. when closing their holes with fallen leaves and other objects.

Though some DARWIN's descriptions of the earthworm behaviour are rather anthropomorphic, they are, as all his descriptions are, very clear and detailed so that they allow us to discriminate strictly objective facts observed by him from their interpretation.

The comparison of methodological premises of such different works by DARWIN as "Origin of the Species", "Coral reefs" and "Earthworms" reveals the philosophical concepts of the great scholar.

Being an adversary of catastrophism, standing on Aristoteles position that "Natura non facit saltus" DARWIN developed in his works the principle of dialectic philosophic concepts on qualitative changes resulting the accumulation of small quantitative changes and convincingly showed it in the theory of evolution, in geology and the soil science.

Department of Ornithology and Herpetology, Zoological Institute,
USSR Academy of Sciences, Leningrad, USSR

Origin of the Unisexuality and Hybridogenous Speciation in Vertebrate Animals

I. S. DAREVSKY

In this short presentation I shall dwell on the problem of reticulate evolution. Needless to say, the main path of speciation is through divergent evolution as was brilliantly grounded by DARWIN. At present, however, some data have been obtained on new species formed not as a result if divergence of characters of ancestral form, but, conversely, as result of convergence of two ancestral species due to natural hybridization. This can be shown in the scheme that has a shape of reticle, hence the term "reticulate evolution". This type of speciation is common in the plant kingdom, and quite recently has been shown in animals, vertebrates including.

As is known DARWIN attached much importance to hybridization which he specially discussed in the 9th chapter of his book. However, he had no facts that would directly indicate the role of hybridization in the origin of species.

At present such facts are known and I shall discuss the problem further. In fish this phenomenon was observed in members of the American genera *Poecilia* and *Poeciliosis.* This described in *Poeciliopsis* are five diploid bisexual species and as minimum seven distinct unisexual ones consisting of females only.

Hybridization of two bisexual species *P. monacha* and *P. lucida* resulted in a unisexual diploid form conventionally termed *P. monodea-lucida.* In the recurrent hybridization that followed with each of the parental forms two more forms appeared that were triploid, i.e. *P. monacha* — 2 *lucida* and *P.* 2 *monacha lucida* as is shown in the Figure 1.

The centre of this complex is *P. monacha* genome that takes part in formation of five unisexual forms.

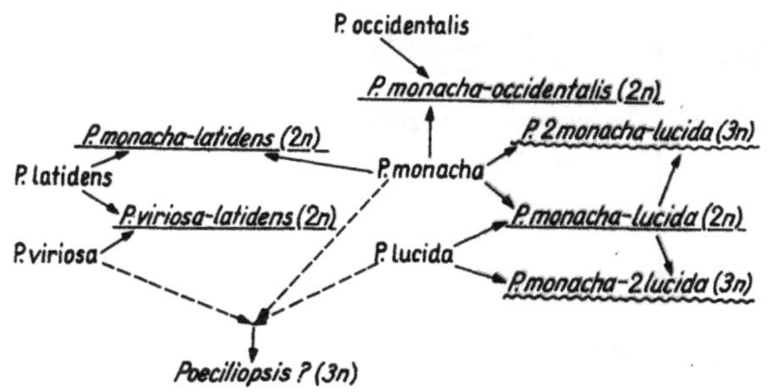

Fig. 1. The general scheme of reticulate speciation in the group of North American fishes of the genus *Poeciliopsis* (from BORKIN and DAREVSKY, 1980). Unisexual species of hybrid origin are typed in italics, diploid species reproducing hybridogenesis (creditogenesis) are underlined with a straight line, ones reproducing by gynogenesis by a wavy line.

Triploid unisexual species reproduce by gynogenesis. In diploid unisexual species a remarkable way of reproduction called hybridogenesis was discovered by SCHULZ [13,14]. Not going into details I shall only mention that in hybridogenesis spermatozoid fuses with the egg cell to form a hybrid zygote. Further in the process of oogenesis adult females lose father's chromosomes and only maternal chromosome set is left in the egg cell [14].

Since the term "hybridogenesis" is widely used in the literature, to avoid confusion we denoted the above phenomenon as "creditogenesis" formed from the latin word "creditum" meaning "money given on loan". We meant that paternal genome appears to be "given on loan" as R. SCHULTZ [14] describing this phenomenon has patly remarked.

The vast literature that deals with the problem of the origin and evolution of the unisexual species *Pecilia* and *Peciliopsis* was discussed in a recent paper by BORKIN and DAREVSKY [2].

Now let us consider hybrid speciation in amphibians. North American salamanders of the genus *Ambystoma* include a group of four related species, two of them (*A. jefersonianum* and *A. laterale*) are diploid and consist of males and females, whereas two others (*A. platineum* and *A. trimblai*) are triploid and consist of females only that reproduce by gynogenesis.

Using electrophoresis it was established that both triploid species in this complex were formed as a result of hybridization between the above bisexual species [16, 17].

Later two more unisexual, diploid and triploid forms of *Ambystoma* that were also formed as a result of hybridization [11] were found on the islands of the Erie Lake. Relations of bisexual and universal species of the genus *Ambystoma* are shown in Figure 2.

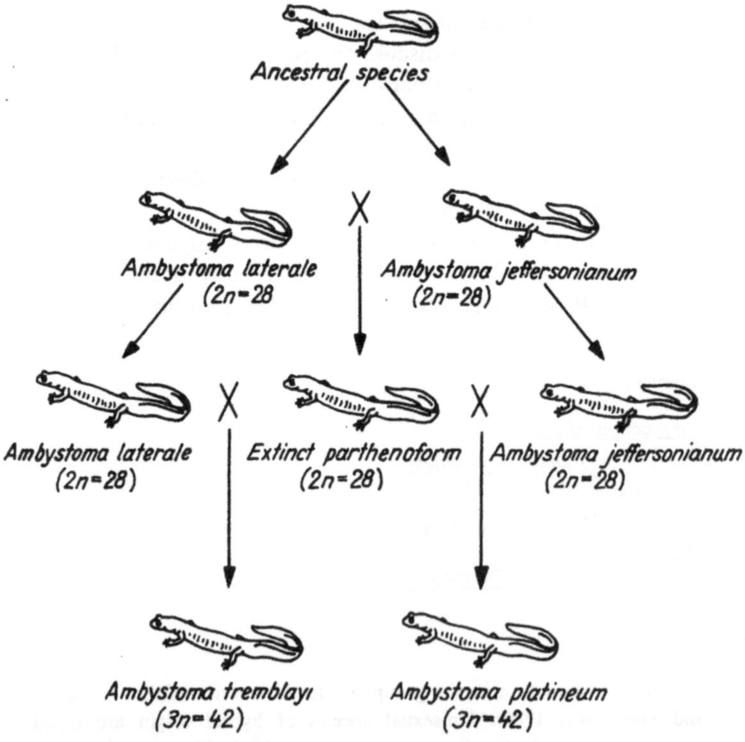

Fig. 2. Scheme of reticulate speciation in the group of North American salamanders of the genus *Ambystoma* (island forms from the Lake Erie not indicated).

I shall not touch upon the highly interesting problem bearing on hybridization in the group of European green frog as this problem now being studied has not been completely clarified.

Speciation resulting from hybridization has been established in such higher vertebrates as reptiles. Over 20 years ago for the first time 4 parthenogenetic species of Caucasian rock lizards of the genus *Lacerta* [7, 8] were described by me. Later Professor T. UZZEL and I showed that their unisexual species initially were formed as a result of natural hybridization between 4 related sympatric bisexual species [18]. It turned out that similar parental species participated in formation of different parthenogenetic species as is shown schematically in Figure 3.

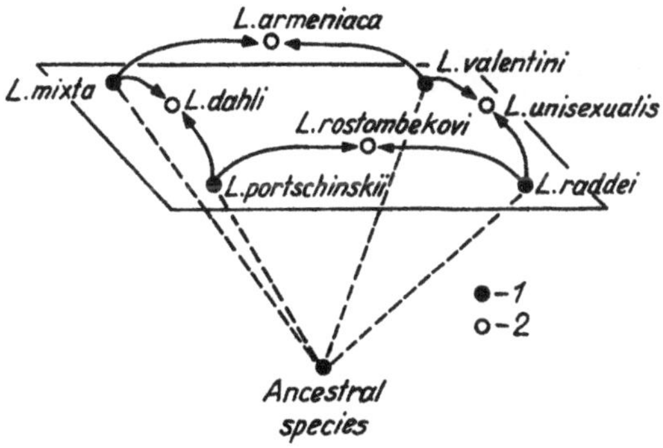

Fig. 3. General scheme of hybridogenic speciation in the group of Caucasian rock lizards of the genus *Lacerta* (from BORKIN and DAREVSKY, 1980). 1 — diploid bisexual species; 2 — unisexual species of hybrid origin.

One other well studied example of unisexual reproduction in reptiles are the so-called American whip-tail lizards of the genus *Cnemidophorus*. The genus includes about 40 species of which 15 diploid and triploid species reproduce by parthenogenesis.

The American scientists have shown that all the parthenogenetic species of this genus as is the case with *Lacerta*, appeared as a result of natural hybridization. Initially diploid parthenogenetic species were formed, later due to the recurrent hybridization with a parental form triploid species [3] were formed. This is shown in the scheme in Figure 4.

It should be emphasized that in all the above cases the hybrid origin was confirmed by such up-to-date methods as electrophoresis, skin grafting, and karyological analysis. Mitochondrial DNA analysis method was also used in one case. One of *Lacerta* parthenogenetic species has recently been "synthesized" by way of artificial hybridization in the laboratory by DANIELJAN [6]. Some gynogenetic fish species earlier were "synthesized" in a similar way [15].

Theoretically speaking even a single parthenogenetic female is capable of originating a unisexual population. Special research using the method of skin grafting have proved it in some cases. The rejection reaction or the lack of it as well as its type allow to judge of the genetic relationship between the donor and the recipient.

CUELLAR [4] has done 175 skin graftings in 20 individuals of two isolated populations of the parthenogenetic species *Cnemidophorus uniparens* in New Mexico and in the southwestern USA. Acception of grafts was observed in 98,8% cases. This led CUELLAR to a conclusion about genetic identity of all the investigated individuals which as he suggestets had

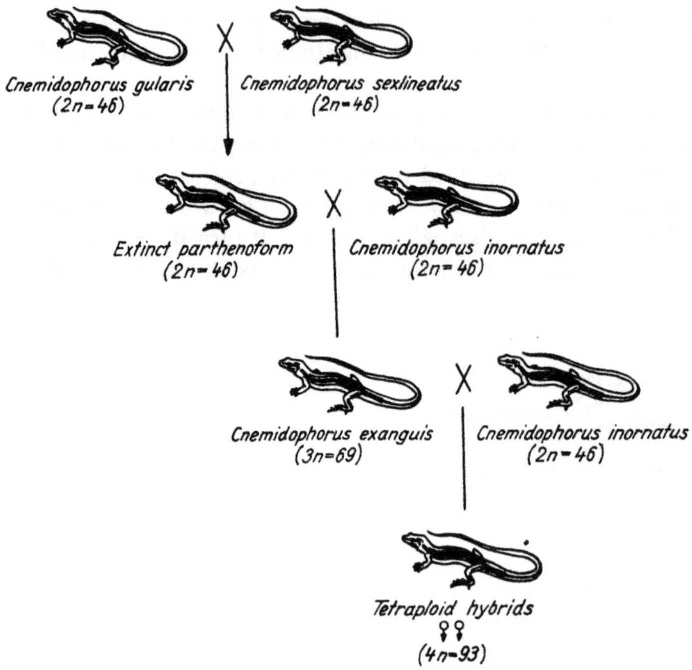

Fig. 4. Scheme of successive hybrid origination of one diploid and one triploid species of American lizards of the genus *Cnemidophorus* and of single male and female tetraploid individuals.

originated in a single parthenogenetic female. Subsequently, using the same method CUELLAR estimated the degree of genetic similarity between several populations in unisexual *Cnemidophorus neomexicanus* and *C. velox*. Genetic homogeneity of the former was established in New Mexico throughout 160 miles. In experiments on the much more widely distributed species *C. velox* rejection was observed when skin grafts were performed in individuals of different isolated populations. It turned out that this species includes at least three isolated clones in the states Colorado, New Mexico, and Utah. Every such clone has apparently arisen irrespective of various female founders.

Similar experiments using skin grafts have been recently performed by me and DANIELJAN [10] in an unisexual species of *Lacerta unisexualis*. The animals (35 females) were taken from 5 isolated populations located on an average 25 km from each other. The experiment was so planned that every lizard was a donor and a recipient at the same time. Thus 150 patches of skin were grafted. Of them 145 (96%) were completely accepted. As in the experiments of CUELLAR this shows a genetic homogeneity of the studied species and we assume that initially it originated from a single female fonder.

This is actually a peculiar case of manifestation established by E. MAYR, the so-called Founder Principle and its consequences. This question requires a special consideration and I shall not discuss it now.

Though the hybrid nature of parthenogenesis in some case was proved a few problems bearing on this theory remained obscure. The matter is that natural hybrids between different species of reptiles occur rather frequently in nature whereas formation of hybrid parthenogenetic species is observed comparatively seldom. To explain this disparity WRIGHT and LOWE [20] have suggested an original theory relating the process of transition of single hybrid to unisexual species to the changes in the environmental conditions. Working out this theory they noted that American parthenogenetic *Cnemidophorus* inhabit extremely changeable habitats that were formed as a result of destruction of natural communities, e.g. forests as a result of being cut down, deserts replacing steppes, steppes replacing forests, sand beaches being overgrown, etc. According to the terminology used by the authors unisexual hybrid species became a sort of "weeds", filling the weed habitats. Disturbance of primary habitats results in destruction of ecological barriers between the parental species that having come into contact begin to intensively hybridize. It is the initial presence or lack of weed habitat that is responsible for the success or failure in origination of hybrid individuals, as regards their ability to transform into independent unisexual species. It is possible to understand the cause if we recall the rule formulated by the Soviet biologist G. F. GAUZE. According to that rule two species with similar biological requirements are incapable of a long co-existence in one and the same ecological niche. The hybrid individuals that sometimes appear in the zone of contact co-exist with males and females of the both parental species and inevitably begin competing with them. Therefore to become confirmed ecologically they should inhabit their own ecological niche for which they choose the weed-habitat. It is under those conditions as a result of natural selection that the potential tendency of female hybrids to parthenogenesis is revealed.

This tendency is a manifestation of hybrid heterosis. The advantage of a unisexual species at the stage of formation is in particular the two-fold increase of its reproduction rate, since 100% of the individuals in populations are females. As a result of quick increase in number and prompt distribution the new species occupies an independent area and becomes gradually separated from the bisexual forms from which it originated.

It should be emphasized that origination of unisexual species in such a way is influenced by natural selection. This completely conforms to one of the principal propositions of DARWIN's theory of the origin of species.

The theory gives a satisfactory explanation of a transition to unisexual reproduction in Caucasian parthenogenetic lizards. At present they all inhabit the territory that about 10,000 years ago was covered by mountain glacier. Even in the epoch of maximum glaciation some ice-free areas remained in some regions of Transcaucasia with their thermophilic flora and fauna that evidently included also different bisexual species of rock lizards. Retreat of the glacier in the post glacial period resulted in a gradual release of extensive "weed-habitats" and settling of young unisexual species whose areas are still overlapping the zone of parents distribution. In the newly formed zones of sympatry parthenogenetic females obtain the possibility to couple with males of sympatric species. As a result peculiar sterile triploid hybrids are formed whose sterility is due to almost a complete reduction of ovaries [8].

Though the hybrid origin of parthenogenetic *Cnemidophorus* and *Lacerta* has now been explicitly proved some authors criticized the hybrid theory of the origin of parthogenesis. In particular it was criticized by the American zoologist Orlando CUELLAR. Even a discussion "Parthenogenetic Lizards" was started in the "Science" [12]. Without going into details I shall mention that CUELLAR suggested a hypothesis of spontaneous origin of parthenogenesis but not the "hybrid" hypothesis. According to that theory some unfertilized females can start reproducing by diploid parthenogenesis. Later coupling with males of their own or some other species they may produce triploid individuals that reproduce by parthogenesis. Some facts are known when single occasional triploid specimens were observed among normal bisexual species of lizards. It is probable that some parthenogenetic species of lizards have been formed in such a way, however, it is hybridization that plays a decisive role in the formation of unisexuality in reptiles.

A study of majority of known cases of unisexuality among vertebrates has shown that there exists a specific way of evolution when natural hybridization, unisexual reproduction and polyploidy are three successive interrelated steps. On the first stage as a result of hybridization of two diploid species a diploid form appears which later begins to reproduce by parthenogenesis. Diploid gynogenetic fishes of the genera *Poecilia* and *Poeciliopsis*, diploid salamanders of the genus *Ambystoma* and parthenogenetic *Lacerta* and *Cnemidophorus* stopped at that stage. On the second stage hybridization of such unisexual diploid species with males of parental or some other species leads to formation of triploid form also reproducing by gynogenesis of parthenogenesis.

Triploid fishes, triploid *Ambystoma* and triploid whiptail lizards of the genus *Cnemidophorus* are on the same level. As to Caucasian *Lacerta*, they form as mentioned above. triploid sterile hybrids that are an evolutionary deadlock (Fig. 5).

Subsequently as a result of coupling of triploid females with diploid males of their own or foreign population the third stage begins, i.e. formation of tetraploid species (males and females). The tetraploid level suggests returning to bisexual reproduction, though on

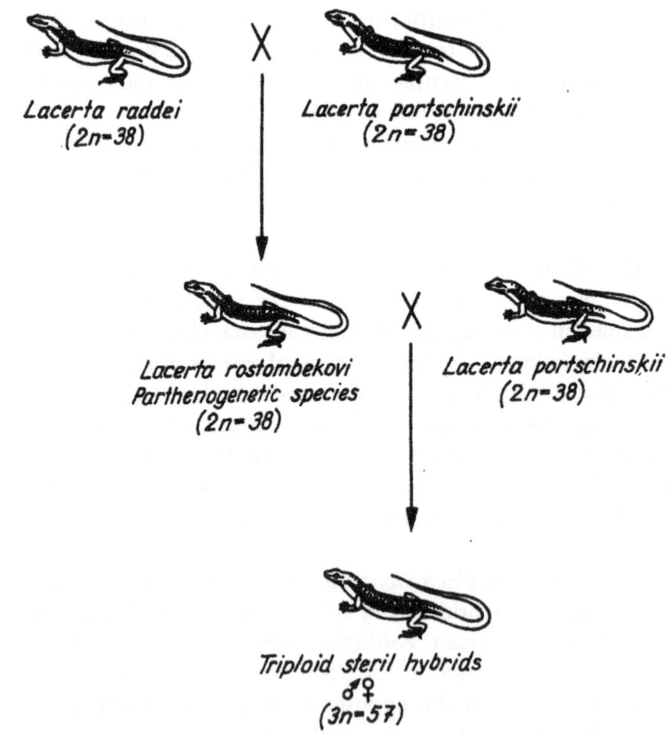

Fig. 5. Scheme of successive hybrid origin of parthenogenetic species *Lacerta rostombekovi* and of triploid steris hibrids.

the level of polyploidy. Such tetraploid species in reptiles are not yet known, however single tetraploid individuals (males, females) have been discovered in some *Cnemidophorus* that are in the state of being new species "in statu nescendi" (Fig. 4). This question calls in for further study. The principal scheme of hybrid speciation in vertebrates is shown in Figure 6.

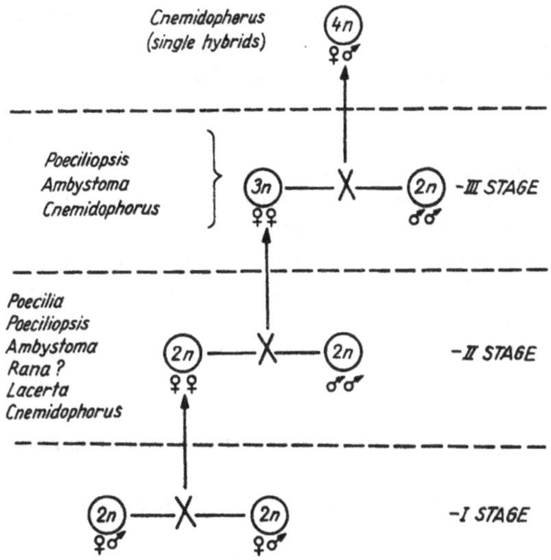

Fig. 6. Universal scheme of hybridogenic speciation in vertebrates (See the text).

The above said conforms to the hypothesis of the so-called indirect formation of natural polyploidy in bisexual animals, that was developed by the eminent Russian scientist Academician B. L. Astaurov. These stages including bisexual tetraploids were obtained by B. L. Astaurov [1] in the experiments on silk worm *Bombyx mori.* A similar view point on the interrelation of hybridization, parthenogenesis and polyploidy was expressed also by Schultz [13]. As is known the present so-called biological definition of species can be applied only to bisexual animals in whose populations panmixy is developed. Therefore it appears to be hardly applicable to the unisexual species just mentioned.

Hence, Simpson's definition of species would be more suitable if applied to unisexual species. The proposed by Simpson term "evolutionary species" implies a complex of populationts or clones that evolve irrespective of other such complexes. With regard to this definition we can assume that hybrid forms are species only in the case if they are 1) genetically independent on parental species, 2) capable of independent existence and do not require a constant inflow of hybrid individuals, 3) possess a peculiar ecology and area. A hybrid unisexual species appears to possess all these characteristics.

References

[1] Astaurov, B. L.: Experimental polyploidy in animals with special reference to the hypothesis of an indirect origin of natural polyploidy in bisexual animals. Genetica, Moscow, 5 (1969), 129—149.

[2] Borkin, L. J. and I. S. Darevsky: Reticulate (Hybridogenous) speciation in vertebrates. J. General Biology, Moscow, 41 (1980), 485—506.

[3] COLE, C. J.: Evolution of parthenogenetic species of reptiles, In: Intersexuality in the animal kingdom. R. REINBOTHE (ed.) Berlin—Heidelberg, New York, Springer-Verlag, (1975), 340—355.

[4] CUELLAR, O.: Genetic homogeneity and speciation in the parthenogenetic lizard *Cnemidophorus velox* and *C. neomexicanus*, evidence from intraspecific histocompatibility. Evolution, **31** (1977), 24—31.

[5] CUELLAR, O.: On the origin of parthenogenesis in vertebrates: The cytogenetic factors. Amer. naturalist, **108** (1974), 625—648.

[6] DANIELJAN, F. D.: A study into relationship of bisexual and parthenogenetic species of rock lizards using the methods of skin grafting and natural hybridization. The problems of herpetology. Abstracts of communications presented to the Fifth All-Union Herpetological Conference, Leningrad (1981), 46—47.

[7] DAREVSKY, I. S.: Natural parthenogenesis in a polymorphic group of Caucasian rock lizards related to *Lacerta saxicola Eversmann*. J. Ohio Herpet. Soc. 5 (1966), 115—152.

[8] DAREVSKY, I. S. und V. N. KULIKOVA: Natürliche Parthogenese in der polymorphen Gruppe der Kaukasischen Felseidechse (*Lacerta saxicola Eversmann*). Zool. Jb. Syst. **89** (1961), 119—176.

[9] DAREVSKY, I. S. and V. N. KULIKOVA: Natural triploidy in a polymorphic group of rock lizards (*Lacerta saxicola Eversmann*) resulting from hybridisation between bisexual and parthenogenetic forms of this species. Doklady Acad. Nauk SSR, Moscow, **158** (1964), 202—205.

[10] DAREVSKY, I. S. and F. D. DANIELJAN: A study of the degree of genetic homogeneity in the unisexual lizards *Lacerta* unisexualis DAREVSKY using skin graft technique, Proc. Zool. Inst. USSR Acad. Sci., Leningrad (1979), 65—70.

[11] DOWNS, F. L.: Unisexual *Ambystoma* from the Bass Island of the Lake Erie, Occas. Pap. Mus. Zool. Univ. Michigan, **685** (1978), 1—36.

[12] Parthenogenetic lizards. Comments by P. E. VANZOLONI, J. W. WRIGHT, Ch. J. COLE and O. CUELLAR. Science, **201**, N 4361 (1978), 1152—1155.

[13] SCHULTZ, R. J.: Hybridization, unisexuality, and polyploidy in the teleost *Poeciliopsis* (Poeciliidae) and other vertebrates. Amer. Natur., **103** (1969), 605—619.

[14] SCHULTZ, R. J.: Reproductive mechanism of unisexual and bisexual strains of the viviparous fish *Poeciliopsis*. Evolution, XV, (1961), 302—325.

[15] SCHULTZ, R. J.: Unisexual fishes: laboratory synthesis of a "species". Sciences, **179** (1973), 180—181.

[16] UZZEL, T. M.: Relations of the diploid and triploid species of the *Ambystoma jeffersonianum* complex (Amphibia, Caudata). Copeia (1964), 257—300.

[17] UZZEL, T. and S. M. GOLDBLATT: Serum proteins of salamanders of the *Ambystoma jeffersonianum* complex, and the origin of the triploid species of this group. Evolution, **21** (1967), 345—354.

[18] UZZEL, T. and I. S. DAREVSKY: Biochemical evidence for the hybrid origin of the parthenogenetic species of *Lacerta saxicola* complex (Sauria, Lacertidae) with a discussion of some ecological and evolutionary implication. Copeia, N 2 (1975), 204—222.

[19] VRIJENHOEK, R. C.: Genetic relationship of unisexual-hybrid fishes to their progenitors using lactate dehydrogenase izozimes as gene markers (*Poeciliopsis*, Poesiliidae), Amer. Natur., **106** (1972), 754—766.

[20] WRIGHT, J. W. and C. H. LOWE: Weeds, polyploids, parthenogenesis and geographical and ecological distribution of all female species of *Cnemidophorus*, Copeia (1968), 128—138.

Discussion

VOGEL: The concept discussed by DAREVSKY on speciation by hybridization might, according to DUTRILLAUX, have a counterpart in the speciation of man, Chimpanzee, and Gorilla. First separation of Gorilla from the common ancestor; then, separation of Chimpanzee; then, extensive hybridization between Proto-Gorilla and Proto-Chimpanzee, and finally, separate development of Gorilla and Chimpanzee.

BRODA: It should be added that the possibility of reticulous evolution is now being much discussed in connection with primitive organisms, namely prokaryotes.

Invited Discussion Remark

M. S. GHILAROV

The present stage of Darwinism's development is closely connected with concepts of so called new systematics, with concepts of biological species (instead of typological ones) developed by E. MAYR, J. HUXLEY, G. G. SIMPSON and others connected with view-points of S. CHETVERIKOV, R. A. FISHER, Th. DOBZHANSKY and others.

According to these concepts the basic unit of evolution is a population.

Species as a unit may be regarded as subdivided into inferior groups on various principles. One of them is well known subdividing into taxonomic categories of inferior ranks — subspecies, forms, races and so on. But from the ecological view-point a species as an entity consists always of more or less isolated populations, nevertheless whether representatives of various populations are morphologically different or similar.

Geometrical progression of propagation, struggle for existence, natural selection are well known factors of speciation and evolution. The same factors determine the course of another process on the Globe surface — formation of biocenoses.

The formation of biocenose is in the first line the result of interspecies competition on a concrete arena of life.

In a biocenose, in a biogeocenose, in an ecosystem the populations of those species persist, which are adapted to this given ecosystem to its abiotic factors, to an effective coexistence with other species populations composing this biocenose.

Thus both species and biocenose consist of the same units — of populations. It can be illustrated by the following scheme.

	biocenose				
species	1	2	3	4	... n
A	A_1	A_2	A_3	A_4	A_n
B	B_1	B_2	B_3	B_4	B_n
C	C_1	C_2	C_3	C_4	C_n
D	D_1	D_2	D_3	D_4	D_n
N	N_1	N_2	N_3	N_4	N_n

If we place various concrete biocenoses of the given type in horizontal row (1, 2, 3, 4, ... n) and species in a vertical row (A, B, C, D, ... N) every species, e.g. A, consists of populations entering different biocenoses ($A_1 + A_2 + A_3 + A_4 + \dots A_n$), whereas every biocenose (e.g. 3) consists of populations of various coexisting species ($A_3 + B_3 + C_3 + D_3 + \dots N_3$).

But every population, every biocenose of a given ecosystem adapted to its environment changes this environment in such a manner that the further existence in it of representatives of the given species, thenafter of the whole biocenose becomes impossible. In such a way proceeds so called ecological succession, the process of ecosystem's change. When in an ecosystem due to the activity of the biocenose environment becomes unfavourable for the given species, its population either extincts or representatives of this population have to migrate into new habitats, where the ecosystem is favourable. Therefore dispersal is one of the most significant functions of species.

Dispersal is a factor of evolutionary changes of a species — its dispersing individuals get new condition something differing from previous ones; this explains the progressive significance of dispersal, as it was stressed still by Herbert SPENCER. But in the same time dispersal insuring the finding habitats corresponding to existing requests of the given species has also the dialectically opposit conservative significance, as already LAMARCK has pointed.

Some indiviuals in an ecosystem undergoing ecological succession adapt themselves to these changes.

As the direction of succession of similar ecosystems in a given locality is similar, succession becomes the significant factor influencing the direction of natural selection; in this way speciation may proceed being e.g. connected with changes of food plants as it is observed in bark beetles and Chermes plantlice in coniferous forests.

But the main adaptation to changing conditions is the ability to migration, to dispersal.

In the case of animal active dispersal its effectiveness depends on the ability of the individual to oriented locomotion, connected with the progressive development of the musculature,

Thus active locomotion is performed by a highly differentiated organism, realized at the stage of accomplished development, by adult animal (as it is in fishes, in terrestrial of sense organs, of central nervous system, of the effective metabolism etc.
vertebrates, in insects) and the course of the ontogenese in actively dispersing animals is progressive (aromorphous, sensu A. N. SEVERTZOFF).

When dispersal is passive (anemochorous, hydrochorous, zoochorous) it is necessary for the species that the number of dispersing individuals would be great but the biomass of each dispersing individual would be minimal. Therefore in passively dispersing animals the sexual productivity is high, dispersal is performed at the stage of egg or larva, evolution of the ontogenese becomes regressive.

So the mode of dispersal determines the evolution of species in a progressive or regressive way.

The more strict are the demands of an organism at all stage of its ontogenese, the more significant becomes the function of dispersal.

E.g. in the case of field crop insect pests in contemporary fields undergoing crop-rotation, their survival depends not only on adaptations to feed on cultivated plants (direct factor of natural selection in these changing conditions) but also on the ability to active dispersal, to flight (an indirect factor influencing direction of natural selection).

Department of Evolutionary Biology, Czechoslovak Academy of Sciences, Praha

The Relation Between Darwin's and Lamarck's Concepts in the Light of the Today Biology

V. J. A. Novák

The publication in 1859 of Darwin's main work, "The Origin of Species", is rightly regarded as the beginning of scientific biology. This does not in any way mean, on the other hand, that it put an end to all attempts to deny or to cast aspersions on the scientific truth of evolution. On the contrary, the remaining decades of the last century saw a bitter struggle against Darwin's views. Even now, when no real scientists or any thinking person doubt the reality of the gradual development of organisms from the origin of life on Earth to the appearance of modern Man, not all attempts to discredit the evolutionary idea have ceased. Today, of course, they can no longer be made directly and openly, at least not in a scientific forum, so we are wittnesses to all manner of disparagement and attempts to mislead readers and to cast the shadow of doubt on these established scientific facts.

One of the most successful of the ways to discredit the idea today has been to play off against each other the views of individual scientists, and then so assert that there is no agreement on the question, and that therefore evolution is merely one of the hypotheses on the origin of life and man, no more likely, at least, than creationism. The manner in which this can be exploited in the struggle against scientific truth is demonstrated by the present so-called fundamentalists in the USA, to whom even the present American president, Ronald Reagan, lent his support during his election campaign (cf. La Recherche, 1981). A typical example is the exaggeration of the contradiction between so-called Lamarckism and Darwinism, or, as they are more correctly called today, between neolamarckism and its different varieties like psycholamarckism or mechanolamarckism and neodarwinism.

Jean Baptist Lamarck (1744—1829), one of the most outstanding scientists of the early part of the last century, whose significance is only just beginning to be fully appreciated, formulated the idea of evolution in his "Philosophie Zoologique" (1809), fifty years before Darwin. At that time, however, during the ebbtide of the French revolution, in the Napoleonic era preceding the Bourbon restoration and the temporary renewal of the power of the Church, the situation in France was not nearly as favourable for new and revolutionary scientific ideas as that fifty years later in liberal England, when Darwin put forward his conclusions. Lamarck's work, which concerns all evolution of life from its origin on Earth to the appearance of man, had, seen in the light of present knowledge, inevitable shortcomings owing to the state of knowledge and the way of thinking of his time. But compared with the approach and the views of his contemporaries, including the standpoint of his main opponent, once his junior colleague at the Musee National in Paris, the well-known paleontologist Georges Cuvier, with his "catastrophism", the Lamarck's work was a whole epoch ahead of its time.

It might be said without exaggeration that we can find in it all the most significant ideas of biology and related disciplines of the century [8]. LAMARCK, e.g. is also to be regarded as one of the discoverers of the cell theory. The chapter 5 of his "Philosophie zoologique", part II: "On cellular tissue as the medium in which the organization of all living bodies was formed", and the following citation from the same chapter: "As far as I am aware, nobody has previously mentioned that cellular tissue is the general foundation of all organization and that without it here could not be a single living body and no such body could be formed". Even SCHLEIDEN and SCHWANN's much later formulation is not so undisputable. At the same time it enhances the significance of PURKYNĚ's discovery of the cell nucleus in 1824. This has been written in 1809, i.e. long time before SCHLEIDEN (1839) and SCHWANN (1839) and before PURKYNĚ (1824). He also developed the idea of actualism in geology long before Charles LYELL etc. It would therefore naive to imagine, as has more than once been stated in the literature (of e.g. ARTEMIEV, 1970), that it was LAMARCK's scientific shortcomings which engendered the aversion and attacks against his work and his person on the part of his contemporaries and of many figures from later generations. The main cause of the vehement opposition to LAMARCK's views and to LAMARCK himself as described, inter alia, by his contemporary and successor E. Geofrey St. HILLAIRE, is clear from a remark by the English entomologist William KIRBY [5], written shortly after LAMARCK's death: "LAMARCK's great error, and that of many other of his compatriots, is materialism; he seems to have no faith in any thing but body, attributing every thing to a physical, and scarcely any thing to a metaphysical cause. Even when, in words, he admits the being of a God, he employs the whole strength of his intellect to prove that he had nothing to do with the works of creation. Thus he excludes the Deity from the government of the world that he has created, putting nature in his place." (Quoted from BURKHARDT [2], p. 189).

While in France itself LAMARCK's politically experienced opponents masked their intentions by would-be scientific arguments and personal invectives, here in a foreign, not engaged country a spade was called a spade. It would be difficult to find more satisfactory post-mortem justification and appreciation of the views and aims of LAMARCK, a proponent of revolutionary rationalism, than these words from his religious opponent.

In spite of all the opposition of his contemporaries and of subsequent generations, Frederik ENGELS in his Dialectics of Nature (cf. 1966) was able to state with full right that: "All the mentioned scientists (paleontologists L. von BUCH, D'ORBIGNY, MÜNSTER, KLIPSTEIN, HAUER and QUENSTED) were attracted by the power of facts and nearly against their own will to the LAMARCK's hypothesis on the descent of all living things". "The descendence theory has thus established itself nearly unobservably in the views of the scientists engaged with the study of fossil organisms as soon as 1832 and 1848" (it means 11 years before the appearance of Origin of species by DARWIN).

In this connection DARWIN's attitude to LAMARCK and his work is very interesting. Unlike LAMARCK, who was a child of the revolution who always said and wrote what he thought, DARWIN, drawing a lesson from LAMARCK's gate and lack of success, invested considerable energy in avoiding a refusal of his work. He was able not only to hide his atheist views, as we known them from his personal correspondence, to the extent of being an active member of the (Anglican) Church in Down, the town where he lived "to do not cause his wife pain", as he himself puts it, but, apart from a single very guarded sentence in his "Origin of Species", also his views on the origin of Man, which he brought out into the open after the full success of his main work only, in his Descent of Man, twelwe years later. This also influenced his attitude to LAMARCK.

When, after the publication of the Origin of Species in 1859, his friend, the excellent geologist of DARWIN's level — Charles LYELL (whose geological concept has been also anticipated by LAMARCK as mentioned above), wrote to him that his book was a continuation of Lamarckism, DARWIN replied touchy that to put the thing so would compromise the acceptance of his work. But it is typical of the honesty and depth of DARWIN's approach that, several years later, in the epilogue to the third edition of the Origin of Species when the success and appreciation of the Origin of species were assured; he describes LAMARCK in the epilogue to the IIIrd edition of his book as a "righty famous naturalist". In the conclusion to the fifth chapter of the Origin of Species he also expresses doubt as to whether he has paid the due attention to the question of the direct action of the environment.

It must be borne in mind that, as is becoming ever more clear today, not only all the main ideas of LAMARCK's work, including his three laws, are not incorrect, and if properly interpreted are not in contradiction with the basis of DARWIN's idea. Not only that, but they are not even contradictory to any of the findings of modern genetics. So, e.g., it is recognized generally now that all adaptations are determined at the same time by the genetic mechanism (genome) of the species in question and by the given environmental conditions (the gene expression). They are thus only hereditary within a specific range of both internal and external conditions. If these are changed, the „gene expression" changes and so far the changed conditions continue, also the new change expressions repeats for unlimited number of generations. Nothing more is said by the "second law" of LAMARCK, so far we do not give it a new meaning never forseen by LAMARCK.

On the other hand, none of the "evidence" against the "heredity of acquired characters", i.e. the inheritance of ontogenetic adaptations stands up a thorough analysis. Out of many of them, let us just mention the DARWIN's example of the workers of social insects. The fact that their specific adaptations looking very complicated are the first sight, also occur in solitary bees of relative species (*Andrena*, *Helictus* etc.) shows clearly that they are actually no new acquisitions reached on the basis of natural selection, but that they were inherited from solitary ancestors. New are the properties of queens, which are ontogenetic changes on the basis of neoteny being not at all hereditary [8]. Equally serious objections can be raised against WEISMANN's famous mouse's tail cutting experiments, or, against the many experiments in which *Drosophila* were kept for tens of generations under changed temperature, humidity and light conditions.

The whole antithesis between "Darwinism" and "Lamarckism" must therefore be considered as a misunderstanding and in some cases even a figment of the adversaries of the idea of evolution aimed at its discrediting.

It is my opinion that a deeper knowledge of the whole course of the struggle for the truth of the evolutionary idea is not only in itself very interesting and important from the point of view of the history of science but also of significance for a better understanding of the difficulties with which new ideas and knowledge meet even today, especially those which are associated with certain social consequences.

References

[1] ARTEMIEV: Priroda, Moscow (1970).

[2] BURKHARDT, R. W., jr.: The spirit of System. LAMARCK and Evolutionary Biology. Harvard Univ. Press, Cambridge, Mass., (1977) 285 pp.

[3] DARWIN, Ch.: Origin of Species, Oxford Univ. Press, London (1859).
[4] ENGELS, F.: Dialektik der Natur. Moscow—Leningrad (1935).
[5] KIRBY, W.: On the Power, Wisdom and Goodnes of God. Philadelphia (1837).
[6] LAMARCK, J. B.: Philosophie zoologique. Paris (1809).
[7] LYELL, Ch.: Principles of Geology. 3 vols., London (1830—33).
[8] NOVÁK, V. J. A.: J. B. LAMARCK and his work from the view of the present biology. Ças. nár. mus., Praha, (1980) 149, 98—106.

Museum für Naturkunde der Humboldt-Universität zu Berlin;
Zentralinstitut für Philosophie der Akademie der Wissenschaften der DDR, Berlin

On the Formation of the "Synthetic Theory of Evolution" at the Beginning of the 20th Century

K. SENGLAUB and K. WENIG

If one opens any book on Evolution published during the last twenty or thirty years, be it on biogeographic, ecological or ethological aspects, on elementary processes or on population genetics, one will notice that DARWIN is appreciated and praised not only from the standpoint of the science historian. The contemporary Theory of Evolution centres (much more) on DARWIN's concept: his discovery that populations evolve due to an interaction between variability and selection constitutes the backbone of the theory. Freed from the slogans of classical Darwinism, the essence of DARWINian knowledge states that differential adaptation of variants to the environment in nummerically restricted population leads to selection; the measure of adaptation to the environment is given by reproductive success, i.e., the proportion of offspring left for the following generations.

In 1959 many meetings took place to celebrate the 100 years that had elapsed since the publication of "On the Origin of Species". There was an almost unanimous appraisal of DARWIN, and acceptance of the Synthetic Theory of Evolution which is based on his fundamental ideas was almost complete.

All this gives the impression that development of theoretical evolutionary thought since DARWIN has been rather continuous, that its growth has led to firmer convictions, thanks to an increase in our knowledge, the accumulation of experimental evidence and the incorporation of new research levels. But the contrary is true. Probably no other natural scientific theory has existed that has been submitted to such confusing situations and manifold conflicts as the DARWINian theory.

It is often said that a critical period set in for Darwinism at the turn of the century. This formulation may lead to the belief that the situation was more convenient before, but this would also be incorrect: the ageing DARWIN yielded increasingly to the pressure of certain factors and on the role of use and disuse. And the overwhelming majority of „Darwinists" who celebrated DARWIN as the one who had led the Theory of Descent to triumph considered his theory on variability and selection, his "populational thinking", to be of secondary importance. The "old-Darwinists", acting with the best of intentions, did not tire of insisting before and after the the turn of the century that not only variability and selection but also Lamarckist mechanisms are undoubtedly necessary for evolution to occur. Since DARWIN, even as a young man did not completely exclude a few Lamarckist factors although he totally rejected psycholamarckism he was constantly cited in this respect.

A. R. WALLACE (1823—1913) and the clear sighted German zoologist A. WEISMANN (1834—1914) can be regarded as consistent Darwinists. The former, independently but somewhat later than DARWIN, also advanced the selection principle. Since all subsequent development has confirmed the consistent DARWINian standpoint it is not difficult today for the evolutionary biologist to avow that he is "more DARWINian" than DARWIN himself.

DARWIN recognised that variability is the source of evolution. That this variability existed, and what happend to it as result of selection, was decisive for this theory. He knew nothing about the causes of genetic variability, but this was not very important. On the other hand, what was important for his theory was the assumption that genetic differences are maintained from one generation to the next and do not increasingly disappear by dilution due to blending processes. Widespread contemporary notions on the fluid nature of heretary traits required such a dilution. DARWIN was never able to cope with this question, which was raised by one of his greatest critics who DARWIN took very seriously: Henry Fleeming JENKIN — not a biologist but an engineering researcher [10].

To think this question had already been solved during DARWIN's lifetime! But neither DARWIN nor his followers knew that in the „Verhandlungen des Naturforschenden Vereins" from Brno, Volume 4, published in 1866 [13], the laws of heredity had been expressed with extreme clarity, and that experimental proof had been given to show that heredity was not mediated by fluids but by discrete constitutional units.

When in 1900, the discreteness of the hereditary units and the statistical laws of their transmission were rediscovered and spread worldwide by de VRIES, CORRENS and TSCHERMAK, it did not lead to the triumph of the DARWINian variability-selection theory but, on the contrary the previously mentioned critical phase began. In 1966 G. L. STEBBINS remarked that the fate of the Theory of Evolution during the first part of the twentieth century constitutes one of the most remarkable paradoxes in the history of science [16].

Most of the representatives of the young science of genetics turned away from DARWIN. "Populational thinking" practically disappeared. All attention centred on single mutant individuals. Selection was only seen as the means by which totally detrimental mutants were eliminated. Many geneticists looked to macromutations as promising evolutionary agents, whereas DARWIN had always stressed the importance of small or minute changes within nummerically large populations.

In 1901, H. de VRIES (1848—1935) stated, "that species do not arise but disappear in the struggle for existence and by natural selection" („Daß die Arten durch dem Kampf ums Dasein und durch die natürliche Auslese nicht entstehen, sondern vergehen") ([17], p. 150). For years he stressed that mutations are totally different to individual Darwinian variations. Most of de VRIES' interpretations of DARWIN are so wrong, that it is difficult to imagine that he ever read DARWIN. It seems that at that time DARWIN was considered to be such a fossil entity that hardly anyone read his original work.

1903 and later, W. JOHANNSEN (1848—1935) remarked, "Selection produces nothing", it "only eliminates" („Selektion produziert nichts", sie „rottet nur aus") [11]. The meritorious English geneticist, W. BATESON (1861—1926) developed the idea — stated clearly in a lecture delivered in 1914 — that evolution consists of a loss of inhibitory factors [2]. Even in 1916, J. P. LOTSY (1867—1931) saw the basis of evolutionary change only in new combinations, thus accepting the notion of evolution through constant genes [12].

These are only a few examples. In those days there existed a great variety of misunderstandings, one-sided standpoints and manners of thinking that can nowadays hardly be imagined. In 1972, B. L. ASTAUROW (1904—1974) in his now famous paper on theoretical biology attributed the dispute between the young and rather selfreliant Mendelism and the ageing and rather complacent Darwinism mainly to difficulties in mutual understanding. In this connection he speaks of Darwinists as the fathers and of the Mendelists as the sons [1].

But there were not only two parties in dispute. In 1913, L. PLATE (1862—1937), who was a contemporary of the conflict, made a distinction between old DARWINians, who considered both Lamarckist as well as Darwinist factors to be valid, neodarwinists of different tendencies and strict Lamarckists [15]. Many paleontologists belonged to the latter group, thus making it clear that it was frequently the discipline which was determinant in establishing the theoretical evolutionary conviction of the scientists. The student of symbiosis, P. BUCHNER (1886—1978), who in the nineteen thirties swore by Lamarckism together with the zoologist J. W. HARMS (1885—1956) (7; 8), the anatomist H. BÖKER (1886—1939) [3] and others, remarked in 1938 that biologists were divided into two camps in relation to questions of Lamarckistic mechanisms: "Those — mainly geneticists — who absolutely reject them, sometimes with disdain, while the others, one may be trempted to say, the real zoologists — declare that they can always recognize the validity of the inheritance of acquired traits" („Die einen — zumeist Genetiker — lehnen sie entschieden ab, ja manchmal mit Spott, ab, die anderen — man ist versucht zu sagen, die eigentlichen Zoologen, — erklären, daß sie auf Schritt und Tritt in der Natur das Walten der Vererbung erworbener Eigenschaften erkennen" ([4], p. 352).

The new evolutionary models proposed by the geneticists encountered decisive resistance on the part of "real" biologists. To them the genetic theories seemed enstranged from nature, and they regarded them as ideas born in laboratories, greenhouses and on the desks of biometricians. These "real" biologists did not return to DARWIN but turned to Lamarckism. In both enemy camps it became fashionable to declare DARWIN to be dead.

Among the numerous testimonies to this critical situation, one is worthy of special mention as its author did not agree with the general opinion but felt he must justify himself under its pressure. In 1921, Carl W. NEUMANN published a new translation of DARWIN's "The Descent of Man" („Die Abstammung des Menschen") as a Reclam paperback; in the preface he proclaimed that the triumph of the Theory of Descent was accompanied by the defeat of the Theory of Selection. He spoke of crumbling pillars and cracked walls, and concludes with small confort: "Even though the strong and proud building given to us by DARWIN is to old to be reconstructed, it still remains a towering monument that fills us with awe of the time when modern biology was born" („Und wenn selbst das große und stolze Gebäude, das DARWIN uns schenkte, zum Umbau zu alt ist, so bleibt es doch ein ragendes Denkmal, das ehrfurchtgebietend die Zeiten zurückruft, da die moderne Biologie überhaupt geboren wurde") [5, p. 8].

For dozens of years we see in DARWIN's work not a frail and only historically revered monument. In 1949, the geneticist and Nobelprize winner H. J. MULLER (1890—1967) described DARWIN's masterful arrangement of observation material to show the regulatory action of natural selection and his acute recognition of selection's multiple aspects as an "intellectual monument that has no parallel in the history of human thinking up to the present day" [14, p. 459].

If one considers books on Evolution written during the nineteen thirties, forties or even later and remembers that complex scientific insight often requires an incredibly long time to be incorporated into text-books, it becomes very clear that many botanists, almost all zoologists and the paleontologists without exception were devotees of Lamarckism.

The great breakthrough is well known. It was specifically prepared by a paper published by S. S. ČETVERIKOV (1880—1959) in Russian in 1926 under the title "On certain aspects of the evolutionary process from the standpoint of modern genetics". The breakthrough is represented mainly by a book by Th. DOBZHANSKY (1900—1975) that appeared in 1937 with the title "Genetics and the Origin of Species" and had a conscious connection to DARWIN's "On the Origin of Species".

It showed that the new knowledge of genetics can be applied to natural populations and is compatible with DARWIN's evolutionary concepts. This book was immediately accepted by taxonomists, ecologists and zoogeographists, who had been in conflict, and marked an immediate end to Lamarckist arguments in ornithology, theriology (mammalogy) and in other fields. Although "nongenetical" endeavors to develop the theory did not contribute further towards solving the questions of causality, a large amount of data was accumulated, specially in the field of intraspecific geographical variation, that could be immediately interpreted in the light of the newly acquired knowledge.

Thus the synthesis was accomplished by a geneticist. But soon this new synthetic theory was expanded in taxonomic biogeographical, ecological and other respects, leading to a more comprehensive meaning of the expression "synthetic".

In regard of the causes for the critical period, there are still many open problems. STEBBINS pointed out that Darwin was a "naturalist" while the rediscoverers of MENDEL's laws and their followers were not. In German, this word has no equivalent, and the translation „Naturforscher" found in dictionaries is not approriate to describe scientists like de VRIES. JOHANNSEN and BATESON certainly were „Naturforscher" (students of natural sciences), but they turned more typically to individuals and individual mutants rather than to natural populations, or even to populations at all.

In his preface to DOBZHANSKY's "Genetics and the Origin of Species", L. C. DUNN evaluates this book as an expression of a "back to nature movement". M. HARTMANN (1876—1962) reacted in a similar fashion in his preface to the German edition that was published in 1939 under the misleading title "The Genetic Basis of Speciation" („Die genetischen Grundlagen der Artbildung"). He wrote that it may be seen "that geneticists are not such narrow laboratory and garden specialists as they have been frequently considered by other disciplines" („daß die Genetiker durchaus nicht so enge Laboratoriums- und Gartenspezialisten waren, wie sie von anderen biologischen Forschungsrichtungen vielfach angesehen wurden") [6, p. III].

We — the authors of the present communication — are not, as has already been mentioned, of the opinion that the multiple aspects of the critical period and their causes have yet been sufficiently and thoroughly clarifield. But we believe that the truth may lie in the above mentioned direction. The attitude towards what specially E. MAYR, in a broad sense, has called "population thinking" has surely played an important role.

Research into this problem of understanding how the "Synthetic Theory of Evolution" came about is very important because in many respects it leads to philosophical and methodological problems, for instance the idea of thinking in terms of populations instead of individuals, that led to the knowledge that the unit of evolution is the population. Evolutionary thought is inextricably linked with statistical considerations to include the element of chance that the mechanical materialists denied.

The biologists, similar to the physicists, have been helped to adopt a dialectical view of evolution. It is possible to give other examples, such as the relationships between elements (individuals) and system, chance and law [9]. This knowledge can help us to understand and interpret the mutual influence of philosophy and biology during the early 20th century.

References

[1] Astaurow, B. L.: Die theoretische Biologie und ihre nächsten Aufgaben, In: Sowjetwissenschaften, Gesellschaftswissenschaftliche Beiträge **6** (1972).

[2] Bateson, W.: Mendel's principles of heredity, inauguraladress, In: Nature **93** (1914), 635—642.

[3] Böker, H.: Einführung in die vergleichende Anatomie der Wirbeltiere, Bd. 1 (1935), Bd. **2** (1937) Jena.

[4] Buchner, P.: Allgemeine Zoologie, Leipzig (1938).

[5] Neumann, C. W. (Ed.): Darwin, Ch.: Die Abstammung des Menschen, Leipzig (1921), preface.

[6] Dobzhansky, Th.: Die genetischen Grundlagen der Artbildung, Jena (1939), prefatory note.

[7] Harms, J. W.: Wandlungen des Artgefüges, Tübingen (1934).

[8] the same: Lamarckismus und Darwinismus als historische Theorien — ein Kampf um Überlebtes, In: Jenaische Zeitschrift für Naturwissenschaften, **73**, 1—27.

[9] Hörz, H.; U. Röseberg (Ed.): Materialistische Dialektik in der physikalischen und biologischen Erkenntnis, Berlin 1981.

[10] Jenkin, F.: Review on Ch. Darwin's "On the Origin of Species", In: North British Review, Edinburgh (1867).

[11] Johannsen, W.: Über die Erblichkeit in Populationen und reinen Linien, Jena (1903).

[12] Lotsy, J. P.: Evolution by means of hybridization, Hague (1916).

[13] Mendel, J. G.: Versuche über Pflanzenhybriden, In: Verhandlungen des naturforschenden Vereins zu Brünn, **4** (1866), 1—47.

[14] Muller, H. J.: The Darwinian and modern conceptions of natural selection, In: Proc. Amer. Phil. Soc., Vol. 93 (1949).

[15] Plate, L.: Selektionsprinzip und Probleme der Artbildung, (4th Ed.), Leipzig and Berlin (1913).

[16] Stebbins, G. L.: Processes of Organic Evolution, New Jersey (1966).

[17] Vries, H. de: Die Mutationstheorie, Vol. I, Leipzig (1901).

Université Lyon I, Equipe Evolution Moléculaire, 69622 Villeurbanne, FRANCE

The Genome as Unit of Selection: Evidence from Molecular Biology

R. GRANTHAM, C. GAUTIER and M. GOUY

Introduction

We seem to still be the only group in the world that studies all sequenced genes at the same time. This is like aerial mapping, you get only a gross overall view and the details must be worked on later. However, the first time through the results may be interesting; an overall pattern is often useful. One advantage of our approach is that we don't have to worry about defining a sample, that has already been done by the laboratory workers who determined all these sequences. Another advantage is that we needn't try to estimate the probability that what we find in our analysis applies to the other sequences genes not included in our sample. A final advantage is that we have the largest sample possible and no one can criticize us for taking too small a sample — or one tailored to prove our preconceptions on evolution. Of course the problem of whether what we find in our analyses holds for unsequenced genes remains, but this can only be resolved by determining more sequences! By the way, we don't always work with **all** the sequences at once, we sometimes analyze only those from one species or group of species — or one kind of gene, for example the immunoglobulins. But then we do take all sequences in a given category.

Aside from this sampling philosophy I would like to mention our general orientation in studying evolution. In saying this I am glad to see several philosophers of science present and I think UNESCO is to be congratulated for sponsoring such a colloquium in which experimenters and theoreticians confront people who reflect on why we are working and what we are really contributing to scientific knowledge and progress. Man has been studying nature for over 2500 years but the way he explains his observations varies. There appear to be 3 kinds of explanation of nature:

Adaptation described by ANAXIMANDER in the 6th century B. C., who said man descended from fish. This is not adaptation by natural selection, but the idea of the accommodation of organisms to their environment seems to be there, although all direct writings of ANAXIMANDER on the subject have been lost.

Randomness of DEMOCRITUS and his teacher LEUCIPPUS, who saw the universe as a product of chance collisions of particles moving in the void. MAUPERTIUS and MENDEL accepted this approach in imagining, respectively, how the first organisms came into being and how genetic combinations are formed. On the modern scene neutralists are descendants of DEMOCRITUS.

Finality or fixism, also called essentialism and now creationism. This notion of a plan in nature, that everything has a **purpose**, was elaborated by PLATO and ARISTOTLE; they claimed there is no evolution in the sense of transformation of species, only imperfect realization of the original plan.

These 3 kinds of explanation are still used. Their peaceful interaction may lead to increased understanding of ourselves and our relation to nature — if the interaction stays peaceful. One way of keeping the controversy peaceful is to make it as scientific as possible. Hence I think scientific journal editors are right to give space to this topic. As for my own research group, we are clearly sons and daughters of ANAXIMANDER and DARWIN — although we sometimes do calculate probabilities and occasionally think like the ancient atomists. So you see that the here present Dr. KEYNES (a direct descendant of Charles DARWIN) has many relatives!

Hominid and Mitochondrial Evolution

Now as background for our analyses, a fast look at hominid evolution. Not everyone agrees but there is some consensus that *Homo sapiens sapiens* (HSS) derived from *Homo sapiens neanderthalensis* (HSN) 40 to 50 thousand years ago, that HSN evolved out of *Homo erectus* (He) *possibly* one-half to one million years ago and that around 2 million years ago HE came from *Homo habilis* (HH), who in turn descended from *Australopithecus* (AUS in Figure 1). Several australopithecines existed, the oldest may have had *Ramapithecus* as *ancestor*. This is an active and controversial field; the LEAKEY's and JOHANSEN are not ready to agree [1].

We are trying to learn something about human evolution by studying not only sequences of human nuclear genes, but also those of the mitochondrial (mt) genome in humans and other species. Mt are symbionts in our cells, several thousand per cell. The mt DNA is only 5 parts per million as long as our nuclear DNA but we cannot live without mt, which perform a number of essential functions for us. Incidentally, mt are not inherited by a MENDELian mechanism; they come exclusively from the mother (mt inclusions in spermatozoa are unknown in all mammals investigated).

The mt genome shows a considerable amount of genetic variability Curiously, this appears to be 2 to 10 times higher in ape mt and 3 to 30 times higher in those of other

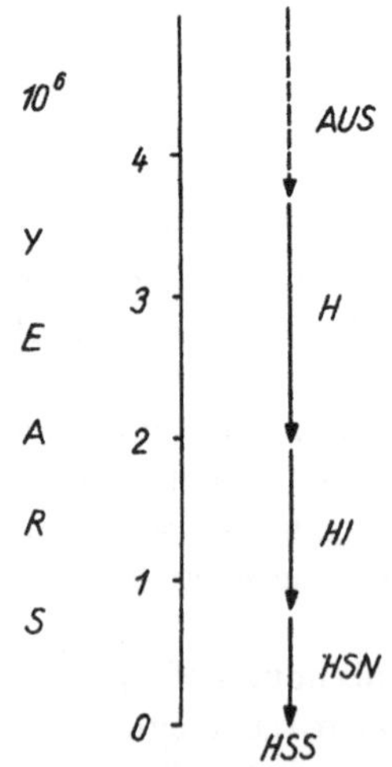

Fig. 1. Hominid evolution. This is the succession of hominid species accepted by some paleoanthropologists, although there is no general agreement on the phylogeny (1—5). AUS *Australopithecus*, HH *Homo habilis*, HE *Homo erectus*, HSN *Homo sapiens neanderthalensis*, HSS *Homo sapiens sapiens*.

mammals than in human mt [2]. Recent workers found 6 morphs — that is 6 different sequences of the human mt genome — among 235 individuals from 5 ethnic groups [3]. Their results are consistent with an Asian origin for modern man, probably within the last 50,000 to 100,000 years [3]. The evolutionary scheme deduced from such analyses seems compatible with the tree for hominoids themselves [4] that is, the closest relatives to HSS are the chimpanzee and gorilla but there is a common ancestor between the group *Homo* and the group chimp + gorilla. Differences remain, however, on hominid phylogenies [5]. All this work on primate mt must be taken as preliminary because of the small samples and gross methodology.

These studies suggest a harmonization of the evolution of human nuclear and human mt genes, but a difference exists as shown later. To demonstrate this our methodology must be introduced.

Use of the Codon Catalog by different Species

First, we shall use simple arithmetic to show that the genetic code (Figure 2) is exploited differently by different species to code proteins. We find that all synonymous codons or, what amounts to the same thing, all degenerate bases (those in the third

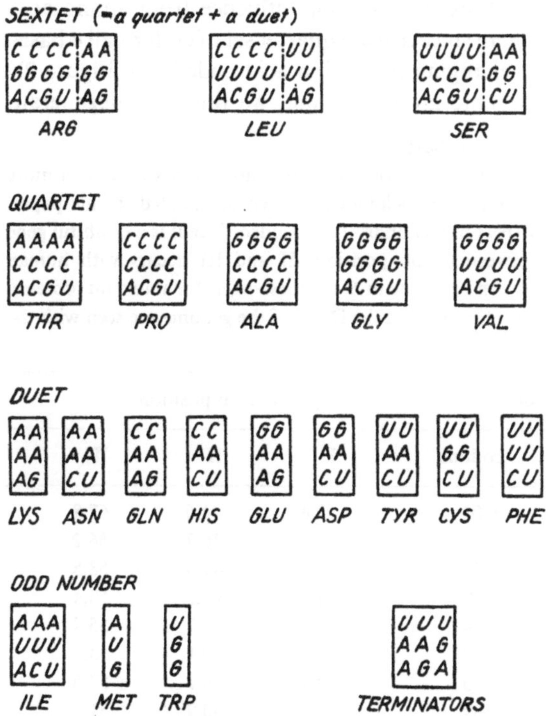

Fig. 2. The genetic code. Codons read vertically. For each of the 20 amino acids there corresponds a set of from one to six codons. Quartet sets contains 4 synonymous codons, duets contain 2 synonymous codons. Note that a sextet is in fact composed of a quartet and a duet, thus half of the total of 64 triplets is made up by the 8 sets of quartet codons. A, C, G and U represent the nucleotide bases of mRNA, adenine, cytosine, guanine and uracil. By definition, the single codons of Met and Trp contain no degenerate bases.

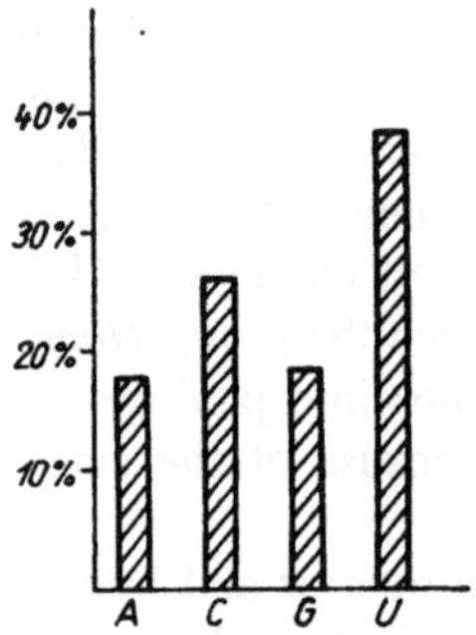

Fig. 3. Degenerate base usage. Frequencies of the 4 bases in position III of codons in the 8 quartet sets (a total of 32 different codons, see Figure 2) are given for the combined 161 mRNAs in our nucleic acid sequence bank in January 1981 (7). Clearly, for most genes U is the preferred degenerate base.

codon position) are not employed to the same extent. Figure 3 reveals that combining all the protein genes in our computerized bank [6] gives unequal frequencies of purines and pyrimidines as third bases. In the total sample [7], pyrimidines were favored, making up 64.0% of all degenerate bases in quartet codons. No possible influence from amino acid frequencies in the proteins coded can occur here since in each quartet set of codons all 4 bases are possible in codon position III without changing the amino acid coded (see Figure 2). If we look at what happens in each taxonomic group, though, we find that degenerate

Table 1 Taxa ranked by G + C content of codon third bases
Here the 61 different codons are analyzed for % G + C in position III. Only genome types with 12 or more genes appear except for yeast and SMT (*Saccharomyces* mitochondria), which are treated in this paper. URF 6 of HMT (human mitochondria) and BMT (beef mitochondria) is excluded since it is a doubtful gene [15, 16]. The greatest GC content difference between codon positions II and III occurs with human genes, between positions I and III with SMT and between I and II with *E. coli* messengers for abundant proteins. Greatest differences between % G + C in codon position III and the DNA of the genome are seen with the chicken, human and SMT.

	Number of		% G + C in position		
Genome type	genes	codons	I	II	III
Chicken	13	3401	58.4	44.6	67.8
Human	22	3662	53.3	36.7	66.2
Mammal excluding human	36	6729	54.8	41.9	58.5
Bacteria without *E. coli*	24	3509	59.9	41.1	58.4
E. coli, high expression	20	4434	63.1	39.7	55.4
E. coli, weak expression	37	8947	58.6	40.1	53.1
Adenovirus	18	3985	57.4	46.7	52.6
dsDNA phage	15	2589	54.2	40.9	51.0
HMT	12	3515	48.9	38.5	49.4
Yeast	9	2396	47.5	41.4	48.5
ssDNA phage	45	7903	51.8	41.4	38.6
BMT	12	3513	47.0	36.5	37.4
Papova virus	12	3795	48.2	39.6	32.6
SMT	6	1475	36.2	32.2	10.0

base use varies. In Figure 4 we show relative frequencies of the 4 bases in the third position of quartet codons for different types of species. Clearly, degenerate base use changes from one type of genome to another. Extreme examples are: fungal mRNAs employ 3 times as much U in position III as do animal messengers; bacterial codons contain 10 times as much degenerate G as do those of yeast and slime mold. Also we note that animal messengers have an average G + C content in quartet degenerate bases of over 60% while most of the other groups show less than 50%. By the way, this is not to match the G + C content of the first 2 bases in the codon (the nondegenerate bases) or of the whole genome, which is around 40 to 42% for animals. We see the same thing with human genes taken alone, the 22 human mRNAs now sequenced (immunoglobulins excluded) have 66.2% G + C in codon position III, taking all 61 codons (see Table 1).

We believe that variation of the internal cell environment is not sufficient explanation for the differences in coding strategy. Consider the bacterial cell and 3 kinds of parasites, ssRNA, ssDNA and dsDNA phages. Do mRNAs of the 4 genome types use of codon catalog similarly? Here we need a new figure since in Figure 4 prokaryote and eukaryote viruses were combined in the case of ssRNA and dsDNA genomes. In Figure 5 we compare only phages and bacteria for degenerate base frequencies, from the same sample of 161 genes as in Figure 4 [7]. No common pattern emerges; each of the 4 genome types has its own degenerate base use schema in spite of functioning in the same cell.

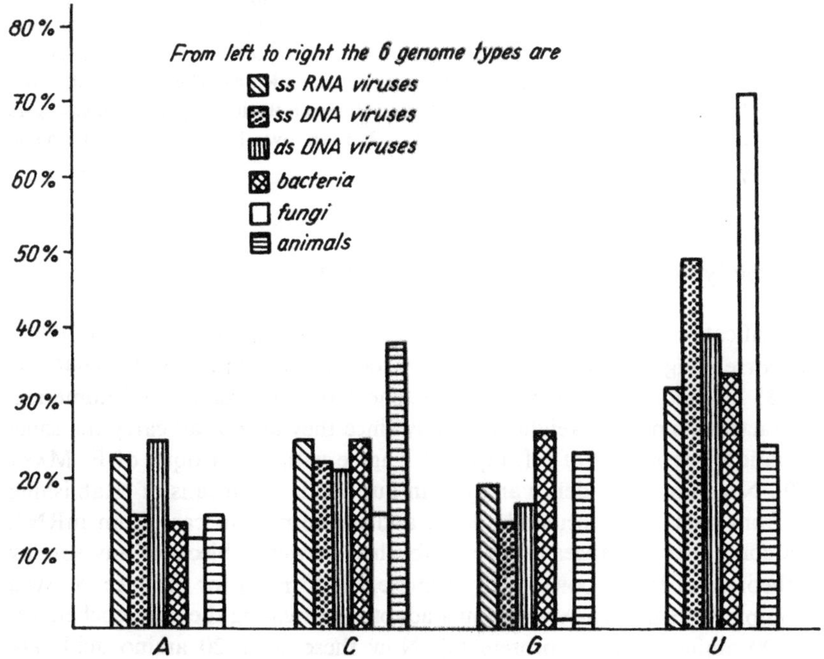

Fig. 4. Degenerate base use by genome type. The total sample of 161 genes is the same as in Figure 3 [7]. Here third base frequencies in quartet codons are shown for 6 genome types. Animal mRNAs are seen to prefer C to any of the other bases for filling position III. Incidentally, this preference is extreme in the antifreeze gene of the winter flounder (not included in present sample), in which the composition of position III of all codons is 25% A, 6% U and 69% C; no codon has G as degenerate base (Y. Lin and J. Gross, Proc. Natl. Acad. Sci. USA **78**, 2825—2829 (1981)).

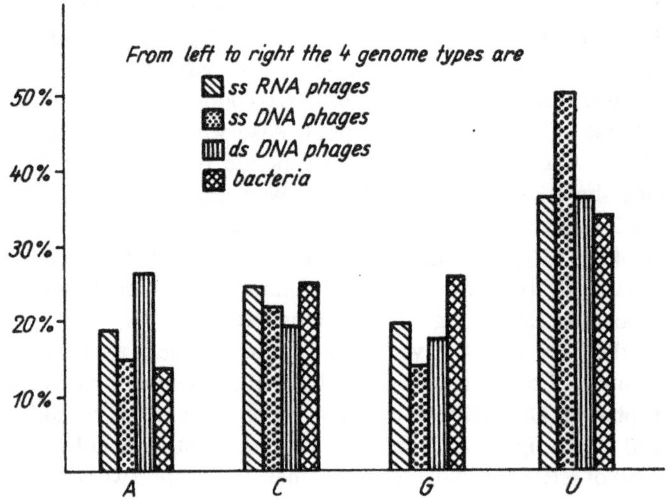

Fig. 5. Third base frequencies in phage and *E. coli* genes. These 4 genome types are taken from the same sample as in Figures 3 and 4 [7]. Considerable variation in degenerate base usage occurs among the 4 genome types but they do all prefer U to any other base for filling position III of quartet codons. However, as seen further on, *E. coli* protein genes can be separated into highly and weakly expressed mRNAs. When this is done we find that the 20 highly expressed messengers now sequenced have 39.7% U in position III of quartet codons while the 37 weakly expressed *E. coli* genes have only 24.1%. Thus all 3 phages have third base uses nearer the highly expressed bacterial mRNAs. There is less difference between highly and weakly expressed genes with the other 3 bases (for each the low expression messengers use 4 to 6% more than do the high expression mRNAs). For an overall comparison of coding strategies in phage and bacterial genes see Figure 13 of [7].

Correspondence Analysis and Automatic Classification

Two immediate clarifications are necessary. First, we are not doing numerical taxonomy. Our own work in correlating mutation rate with physicochemical properties of exchanging protein residues [8] has indicated the inappropriateness of the premises of numerical phenetics. All characters cannot be weighted equally since they do not all carry the same amount of information (see note 9 of ref. [8]) and I agree with the critique of E. MAYR on this method [9]. Next, correspondence analysis in our hands is a means of establishing genetic distances from data of molecular biology, and in the present case from mRNA sequences. In fact conceptual distances can be established between objects of any sort by extension of the theorem of PYTHAGORAS. For example, the formula for distance between amino acids is a simple Pythagorean-type formula adapted to treating the physicochemical properties of the 20 amino acids of protein [8]. Now these same 20 amino acids are encoded in the genes by 61 different trinucleotide codons (Figure 2). Hence there is a choice on the average of about X synonymous codons for each amino acid in constructing a gene. If you wanted to chemically synthesize an artificial gene to produce a protein of one species in cells of another species (examples of this are known — see ref. 10 for an attempt at producing human interferon in *E. coli*), you would find many ways, perhaps **too many** ways for convenience, of doing this. On the average there are 9 ways of coding a dipeptide, over a thousand for coding a decapeptide, etc. For coding an alpha or beta globin chain more mRNA are possible than you could make a copy of with all the matter in the universe.

This brings us to our purpose in studying all sequenced genes simultaneously: we wanted to know if there was a pattern in the biological exploitation of all the potential information available in coding proteins. Could part of this information relate to something other than specifying amino acids? Does each species synthesize the same kind of messenger for all its proteins as far as choices among synonymous codons are concerned? And if we do find patterns of codon usage, what do they mean, how are they related to the evolution of taxonomic groups?

We do correspondence analysis on frequencies of the 61 codons in each mRNA to create a multidimensional scatter diagram in which each messenger is represented by a point. Neighboring points are mRNAs with similar uses of the codon catalog (the machine places the points by a simultaneous comparison of the codon frequencies in each messenger with those in each other messenger, no judgement on our part is involved in situating an mRNA in this "space cloud", thus the result is completely reproducible anywhere in the world). We supplement this analysis with automatic classification which delimits classes of mRNAs having more or less similar coding strategies. This is done by minimizing the dispersion (the multivariate analogue of variance) in each class of a given number of classes while maximizing the overall dispersion [11—14]. This allows us to define classes of coding strategy, each class being made up of a list of messengers having similar codon choices. Automatic classification is done before projection of the space cloud into 2 dimensions, which inevitably involves some distorsion. In fact the whole analysis is done before the projection and the quantitative data from the analysis can always be studied to refine the impression given by the plane figure resulting from the projection.

The Data Base

Before exhibiting some results with these multivariate methods we describe our nucleic acid sequence bank. Our work sample is composed of nucleic acid sequences — of all published sequences of complete genes or of 150 or more continuous and fully determined nucleotides. The only thing excluded from the bank are tRNA sequences as individual molecules since these have been extensively studied by other workers. Everyone now knows that nucleic acids are the long molecules that contain all the genetic information. For example, the human nuclear genome is a linear DNA molecule that if straightened out would be nearly a meter long. This may be hard to believe since it is enclosed inside a gamete, which only measures 3×5 microns in the case of the male gamete, the spermatozoon. This sequence contains 3 billion (3×10^9) nucleotide base pairs, which are organized into 23 chromosomes. Each chromosome probably carriers several thousand genes; the whole genome may have 50,000 to 100,000 genes, we don't know just how many. The sequences in our bank are usually those of a gene or region of a genome. In about a dozen cases, however, we have a sequence of a **whole genome** — this is so for several viruses, mt, etc. For instance, the longest sequences in the bank so far are those for the HMT (human mitochondria) and BMT (beef mitochondria) genomes, which are completely sequenced DNA molecules of around 16,500 base pairs each [15, 16].

We offer our computerized bank of nucleic acid sequences as a scientific gift to all scientific laboratories upon demand. It has been published as a handbook and has over 1100 sequences of all kinds in it [6]. We furnish the bank on magnetic tape, either 800 or 1600 bpi, without charge.

Multivariate Results

Using correspondence analysis and automatic classification, we have repeated the study of use of the 61 codons several times as the total sample of sequenced genes has increased [6, 17—20]. We have consistently found a stable overall pattern of codon usage among the major taxa. This fact we have expressed as the genome hypothesis: all genes in the same genome, or same genome type, tend to have the same coding strategy. Two examples of the use of these methode appear in Figures 6 and 7. Each type of

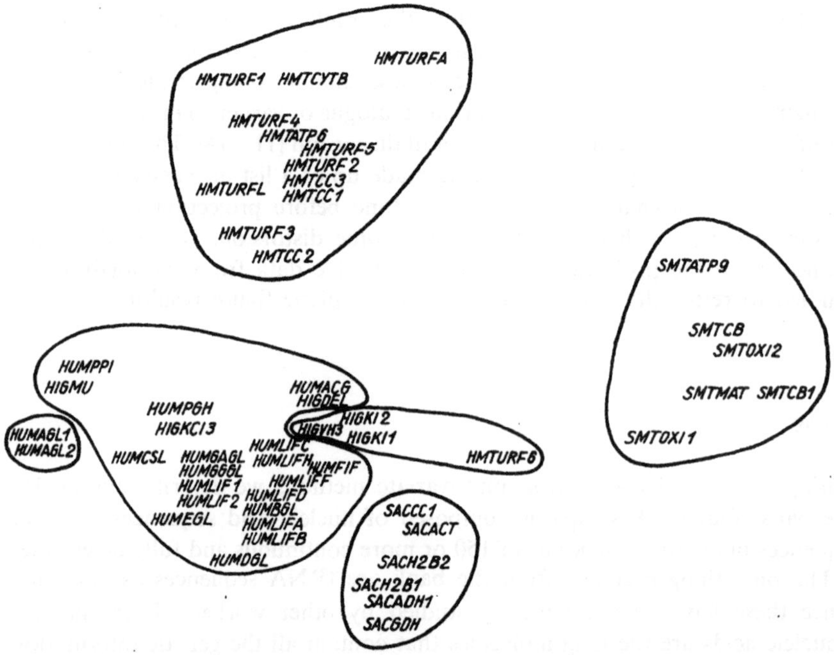

Fig. 6. Correspondence analysis on codon frequencies in yeast and human mitochondrial and nuclear genes. As known, there is some variation of the genetic code dictionary between mitochondrial and nuclear genes; for example, UGA codes Trp in yeast mt but is a terminator in nuclear genes. In this correspondence analysis, in order to avoid mixing such non-homologous cases with those without sense variation, we have eliminated all codons of doubtful or changing assignments. This removes AUG; UGG; AUA, AUC and AUU; CGA, CGC, CGG and CGU; AGA and AGG; ACA, ACC, ACG and ACU; CUA, CUC, CUG and CUU; UUA and UUG (see Figure 2 for amino acid assignments in nuclear genes). Forty homologous codons remain and their frequencies in each gene have been taken for the correspondence analysis which was followed by automatic classification into 6 classes [6, 7, 11—14]. All sequenced human and yeast nuclear genes and all sequenced HMT and SMT genes have been included in the analysis. The horizontal axis corresponds approximately to the G + C content of the degenerate bases [17 and 18].

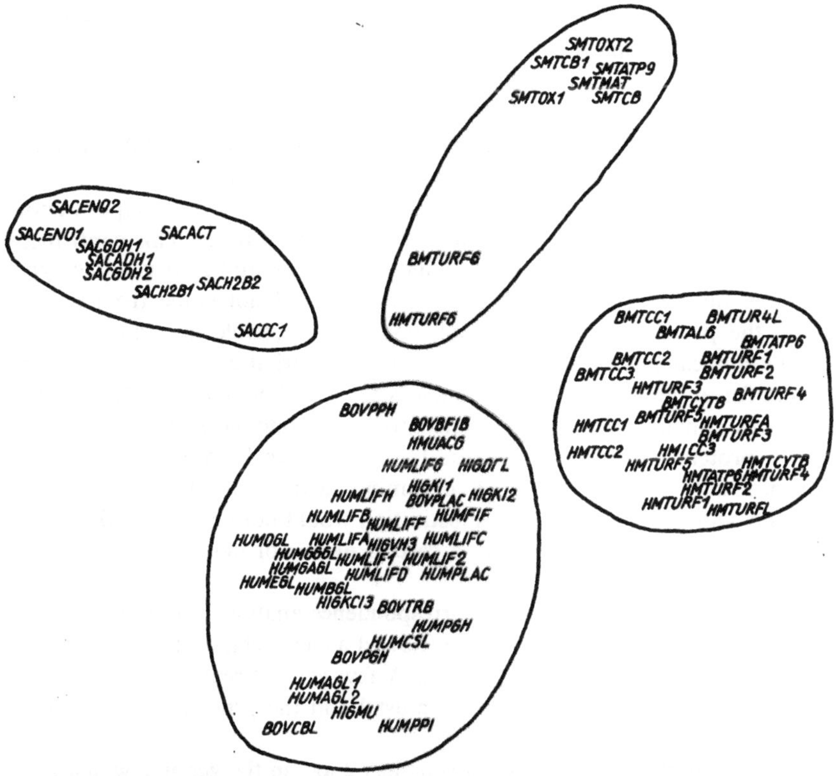

Fig. 7. Correspondence analysis on codon frequencies in beef, human and yeast mitochondrial and nuclear genes. This correspondence analysis has been done on the 50 codons now considered homologous between mt and nuclear genes, codons for Thr and Leu have been included (see legend Figure 6). Automatic classification for 4 classes gave nearly perfect separation of the genes according to genomic origin, excepting the two URF6 sequences. These 2 unidentified reading frames may not be genes [15, 16]. The scale change between Fig. 6 and 7 has no significance since these are relative distances. Fig. 7 is more "condensed" than Fig. 6. At this level of comparison (75 total genes) all human and bovine nuclear genes possess the same kind of coding strategy, as indicated by their falling in one class. Likewise, BMT and HMT genes have similar uses of the genetic code, although this is not identical as is reflected by the fact that BMT genes occupy the top half of the class to the right. Orientation of this figure has no meaning; here the vertical axis correlates with the G + C content of the degenerate bases (note the G + C content of codon position III of HMT and BMT in Table 1).

genome in Figure 7 shows practically all of its sequenced genes falling in the same class. Both figures strongly support the genome hypothesis. I shall now give some background for the formulation of this hypothesis.

In 1976 when our bank only contained 6 mRNA sequences we did correlations of the base composition in each codon position between messengers. We thought we saw pattern in the results [21]. In 1978 the bank had grown to 29 protein genes which we analyzed by some "gene indexes" we had imagined [22]. A grouping of mRNA according to species seemed to be emerging; for instance all 5 SV 40 genes were separated from the other 24 genes by our indexes. But we wanted a more elegant method and so went to computerized correspondence analysis. This methodology change was a bound ahead for us both qualitatively and quantitatively and has permitted us to keep abreast of analyzing the exponentially increasing sequences in the bank. In a few months we were ready to announce the genome hypothesis based on use of the codon catalog [20].

In the first 2 publications using correspondence analysis we worked on 80 and then 90 mRNAs, obtaining separation of genes according to species type [19, 20]. A little later we analyzed the 119 genes then in the bank and verifiied that the principal factor separating mRNAs was simply the G + C content of codon third bases. This confirmed previous studies and gave a clear segregation of messengers according to genome type [17, 18]. At the same time we showed that the segregation did not come from amino acid frequencies in the proteins coded. More information is possible in a nucleic acid sequence than in the colinear protein sequence. Note that a priori at each site in a protein there are 20 choices for an amino acid but 61 choices for a codon. We have found that nucleic acid sequences contain information related to the taxonomic status of a gene; this is not so for a protein sequence, at least as far as we have been able to discover [17—19]. It is conceivable that an exchange of amino acids could be neutral at a particular protein site while the required base substitution would not be neutral. Selection at the levels of protein and nucleic acids have a certain degree of independence, as our results imply.

More recently [7] we submitted 161 genes to correspondence analysis, this is the sample from which Figures 3—5 are drawn. This time we showed a clear relation between coding strategy and expressivity of bacterial genes [23, 24]. A few general remarks are in order on the expressivity of genes in proteins. The key must be in the gene sequences but at least 3 control levels are possible:

Number of loci. For example the alpha globin gene has 2 loci in the genome which give identical proteins. Also, ribosomal proteins in mammals appear to possess up to 20 loci each [25]. Finally, a small fraction of *Drosophila* protein genes have copies repeated several times in the genome [26].

Transcription. The number of copies of RNA run off is what counts here; this is believed to be regulated in the 5′ region before the translation initiator codon. The Cambridge group has proposed a transcription attenuation type of regulation to account for the fact there are 60 time as many rRNA molecules as of any mRNA species in human mitochondria [15, 27].

Translation. Here a triple possibility exists. The first is ease of initiation, which again is assured by the sequence. Next comes accuracy of translation, which could relate to functional expression. A mistranslated protein may not have the same half-life as the correct protein, which could have greater thermodynamic stability and resistance to proteases. Finally, speed of translation may be determined by the sequence. This is where we enter, our indications are that speed (and probably accuracy) are related to codon usage.

Messengers for abundant proteins consistently use the codon catalog differently than do other bacterial mRNAs. The most abundant cell proteins of known gene sequence are ribosomal proteins, major membrane proteins and elongation factors. A more profound analysis showed that the differences between coding strategies of the weakly and highly expressed mRNAs relate to 2 factors in codon choice. First, messengers for abundant proteins preferentially use codons decoded by the most abundant tRNA species in the cell [23]. Second, these messengers choose between the pyrimidines U and C in codon position III so as to avoid codons giving extreme (either very high or very low) energies in pairing with the anticodon of the cognate tRNA [7, 23]. It appears therefore that coding strategy can vary somewhat within a taxon to modulate messenger expressivity. Furthermore, these results suggest that coding strategy is linked to speed and fidelity of translation [24]. We shall see if this explanation holds in other taxa as data on abundance of different proteins becomes available.

Concluding Remarks

You have to do some of these analyses yourself to believe them. We are hoping to find more collaborators in the study of nucleic acid sequences. Thus, we have published our bank as a handbook [6]. Also, we offer it on magnetic tape to any scientific laboratory in the world, free. The Handbook is, of course, sold by the publisher, but any royalties from the sale are plowed back into our group's research funds; we make no individual profit from publishing the Handbook. All these sequences are available on magnetic tape without charge to all scientific laboratories; the format is the same as in the Handbook.

Why all this orthoselection that we find evidence of by the systematic separation of taxonomic groups according to codon usage? A partial answer is harmonization of codon and anticodon distributions. But this in fact begs the question — for why does one kind of species have a certain kind of distribution and another kind of species a different kind of distribution of both codons and anticodons? What is it in a species that determines what kind of distribution of codons or anticodons will characterize the species? It is not only the external environment; mammals live in all sorts of environments yet they all (so far) have rather similar uses of the codon catalog.

Our working hypothesis is that the overall nucleotide metabolism of an organism determines its use of degenerate bases. This would imply a metabolic adaptation of the whole genome and not of one gene. I am sorry to say that the experimental data on nucleotide pool compositions are not yet sufficient for testing this hypothesis, but some data are beginning to appear and we hope that a test will become feasible . . . Perhaps a synthetic theory of physiology and genetics is in order . . .

Our results so far are compatible with the idea that the human nuclear and mitochondrial genomes are being selected in concert in spite of each having its own coding strategy. That is, all the genetic information taken together is what counts.

References

[1] Johansen, D. and M. Edey: **Lucy** (Granada, London, 1981).

[2] Ferris, S., W. Brown, W. Davidson and A. Wilson: Extensive polymorphism in the mitochondrial DNA of apes; Proc. Natl. Acad. Sci. USA **78**, 6319—6323 (1981).

[3] Denaro, M., H. Blanc, M. Johnson, K. Chen, E. Wilmsen, L. Cavalli-Sforza and D. Wallace: Ethnic variation in Hpa I endonuclease cleavage patterns of human mitochondrial DNA; Proc. Natl. Acad. Sci. USA **78**, 5768—5772 (1981).

[4] Ferris, S., A. Wilson and W. Brown: Evolutionary tree for apes and humans based on cleavage maps of mitochondrial DNA; Proc. Natl. Acad. Sci. USA **78**, 2432—2436 (1981).

[5] CRONIN, J., N. BOAZ, C. STRINGER and Y. RAK: Tempo and Mode in hominid evolution; Nature **292**, 113—122 (1981).

[6] GAUTIER, C., M. GOUY, M. JACOBZONE and R. GRANTHAM: **Nucleic Acid Sequences Handbook**, vols. 1 and 2 (Praeger, New York, 1981).

[7] GRANTHAM, R., C. GAUTIER, M. GOUY, M. JACOBZONE and R. MERCIER: Codon catalog usage is a genome strategy modulated for gene expressivity; Nucleic Acids Res. **9**, r43—r74 (1981).

[8] GRANTHAM, R.: Amino acid difference formula to help explain protein evolution; Science **185**, 862—864 (1974).

[9] MAYR, E.: Biological classification: Toward a synthesis of opposing methodologies; Science **214**, 510—516 (1981).

[10] EDGE, M., A. GREENE, G. HEATHCLIFFE, P. MEACOCK, W. SCHUCH, D. SCANLON, T. ATKINSON, C. NEWTON, and A. MARKHAM: Total synthesis of a human leukocyte interferon gene; Nature **292**, 756—762 (1981).

[11] HILL, M.: Correspondence analysis a neglected multivariate method; J. Roy. Stat. Soc., Ser. C **23**, 340—354 (1974).

[12] BENZECRI, J.: **L'analyse des données** (Dunod, Paris, 1973).

[13] DIDAY, E. and L. LEBART: L'analyse des données; La Recherche **8**, 15—25 (1977).

[14] FAGES, R.: La notion de disperion en classification automatique; Journée de classification automatique Soc. franç. Nice (1978).

[15] ANDERSON, S., A. BANKIER, B. BARRELL, M. de BRUIJN, A. COULSON, J. DROUIN, I. EPERON, D. NIERLICH, B. ROE, F. SANGER, P. SCHREIER, A. SMITH, R. STADEN, and I. YOUNG: Sequence and organization of the human mitochondrial genome; Nature **290**, (1981), 457—465.

[16] ANDERSON, S., M. de BRUIJN, A. COULSON, I. EPERON, F. SANGER, and I. YOUNG: Bovine mitochondrial genome sequence (submitted for 1982).

[17] GRANTHAM, R.: Workings of the genetic code; Trends Biochem. Sci. **5**, (1980), 327—331.

[18] GRANTHAM, R., C. GAUTIER, and M. GOUY: Codon frequencies in 119 individual genes confirm consistent choices of degenerate bases according to genome type; Nucleic Acids Res. **8**, (1980), 1893—1912.

[19] GRANTHAM, R. and C. GAUTIER: Genetic distances from mRNA sequences; Naturwissenschaften **67**, (1980), 93—94.

[20] GRANTHAM, R., C. GAUTIER, M. GOUY, R. MERCIER and A. PAVÉ: Codon catalog usage and the genome hypothesis; Nucleic Acids Res. **8**, (1980), r49—r62.

[21] GRANTHAM, R.: Composition et variabilité des bases en positions dégenerées et non dégenerées dans les codons de quelques mRNA; Comptes Rendus Paris Ser. D **283**, (1976), 1667—1670.

[22] GRANTHAM, R.: Viral, prokaryote and eukaryote genes contrasted by mRNA sequence indexes; FEBS Letters **95**, (1978), 1—11.

[23] GOUY, M.: Etude des aspects dynamiques de la protéosynthèse chez Escherichia coli; thèse de 3e cycle, Université Claude Bernard Lyon I (1981).

[24] GOUY, M. and R. GRANTHAM: Polypeptide elongation and tRNA cycling in *Escherichia coli*:: A dynamic approach; FEBS Letters **115**, (1980), 151—155.

[25] MONK, R., O. MEYUHAS, and R. PERRY: Mammals have multiple genes for individual ribosomal proteins; Cell **24**, (1981), 301—306.

[26] CRAIG, E., H. BIESSMANN, and B. MCCARTHY: Cloned complementary deoxyribonucleic acid of *Drosophila* cells. Relationship of genome copy number to messenger ribonucleic acid abundance; Biochemistry **20**, (1981), 2651—2656

[27] OJALA, D., J. MONTOYA, and G. ATTARDI: tRNA punctuation model of RNA processing in human mitochondria; Nature **290**, (1981), 470—474.

Discussion

GHILAROV: You mention in the title of your lecture that the genome is a unit of selection. But the very unit of selection is the organism as a whole, the characters depending on dominant genes, whereas the genome includes also recessive ones. The recessive genes of the genome do not influence the result of selection. Selection does proceed according to features of the phenotype and not of the genotype.

GRANTHAM: I agree that the unit of selection is the organism as a whole. The whole genome through its expression, the phenotype, is selected — rather than any single gene. This is the picture we get from studying coding strategy and which we summarize as the genome hypothesis. However, many mammalian genes appear to be codominant, both alleles are expressed in cases of heterozygosity. Since such alleles ordinarily differ only by a few nucleotide bases, their coding strategies as found by our methods are nearly identical.
Regarding the genotype — phenotype dichotomy: A large part of natural selection may actually take place before the "phenotype" in the usual sense ever comes into existence. Many genetic combinations may not be viable at the stages of zygote formation and early growth. Also, gamete selection may be important and may prevent certain genetic combinations from ever forming. The gamete is a phenotype, but not the one we often think of. The common conception is that genotype is the blueprint and phenotype the house. In fact genotype meant the type species of a genus in the typology (essentialist) thinking of the beginning of this century (JOHANSSEN, 1909). The term now indicates the genetic constitution of an individual or group, the totality of genes (alleles) possessed.

In any case we say genome hypothesis and not genotype hypothesis. It would be inappropriate for us to say "phenotype hypothesis" or "phenotype as the unit of selection" for 2 reasons. First, this is what DARWIN was talking about and has already been accepted by most of us. Second, we are studying molecular evolution and nucleic acid sequences and we want to express our hypothesis in terms of the material we have worked on. Since the genome is the ensemble of all genes of a species and we find that all genes in a genome tend to use the codon catalog similarly, we speak of the genome as the unit of selection by opposition to the gene as that unit. When people talk of the selfish gene they mean the expression, the effect of the gene.

H. A. ROSENTHAL: As far as I got your point you have included into your calculations also the viruses showing a frameshift reading frame. How can you then put them at a given place? One should find them everywhere in your picture.

GRANTHAM: We included all sequenced genes in the analysis I showed. In fact, as discussed in our publications [19, 20], frameshift genes do show a somewhat different use of degenerate bases. But frameshift genes are not a general phenomenon, they are only a minor fraction of the genes in certain genomes such as ΦX174, SV40, etc.; they make up less than 10% of all sequenced genes and they are often weakly expressed at the translation level. By the way, we do not put a gene at a given place on the correspondence analysis figure, the analysis itself does this. We feed in codon frequencies and the method determines (by the distance formula on page r47 of ref. [7]) placement of each mRNA with respect to each other messenger.

HAILA: In your talk you made use of an interesting concept, "the coding strategy" — could you explain the concept?

GRANTHAM: Coding strategy means systematic use of the codon catalog by a species or group of species, that is, choices among synonymous codons such that regular patterns of frequencies of degenerate bases result. For example, in human nuclear genes 2/3 of all degenerate bases are C and G, but in human MT genes the degenerate G + C content is much lower, as shown in table 1. Hence we say their coding strategies are different. The same coding strategy tends to be exhibited by all genes of a given species.

PASTERNAK: You have shown the spot where the SV40 DNA fits. What about the retroviruses, I mean the DNA copy of the viral RNA. Can you comment on this?

GRANTHAM: Papoviruses SV40 and BKV fit well with our concept of a genome strategy for use of the codon catalog: we have complete genome sequences for both and all the genes on both genomes have similar compositions of degenerate bases. However, we so far only have a few gene sequences from retroviruses, and they indicate a coherent coding strategy that is different from mammalian genes. This is an interesting question because where did retrovirus genes come from? If they originated as fragments of the mammalian genome it must have been long ago and the virus has developed its own coding strategy since. Of course the sample of retrovirus sequences is still small and more work must be done.

KEYNES: You said that the mitochondrial sequences were derived exclusively from the mother. Does this mean that mitochondria are determined by the X chromosome?

GRANTHAM: No, mitochondria are determined by their own genome, which lives in symbiosis with the nuclear genome of humans. However, MT are transferred with female gametes and are not included in sperm. Of course there may have been evolution in concert between the female genome (hence the X chromosome) and the MT genome to favor this situation. While we are speculating let me say that the Y chromosome of the male genome can be imagined to have taken part in this concertation by evolving to disfavor the inclusion of MT in spermatozoa.

S. ROSENTHAL: Do you see any differences between high and low expressed genes?

GRANTHAM: Yes, we feel sure that coding strategy varies in the *E. coli* genome according to the protein expression of each gene. We have indications that this is true in bacteria in general and in their phages [7, 23] but we need data on the abundance of different proteins in the same cell of eukaryotes before we can proclaim anything on modulation of coding strategy as an expressivity control in animals (or plants). The same holds for yeasts, slime molds and protozoa. We need protein gene sequences and data on the abundance of each of their proteins to settle this question.

HOFEMEISTER: I am wondering that you find genes of various viruses, phages or bacterial species within a single coding group. When we compare the sequence of the **recA** gene from *E. coli* and *Proteus mirabilis* (EITNER et al. 1981 and unpublished) we find striking differences in codon usage, i.e. coding strategy. Did you analyse genes of different species?

GRANTHAM: The **recA** gene is particularly interesting since it is inducible and its expression is high when induced in *E. coli*. The recA gene of *E. coli* has a coding strategy of a highly expressed messenger [7]. What this variation in coding strategy that you report between *E. coli* and *Proteus mirabilis* means I can't say; we have not yet seen your sequence and so do not know. Just how much codon use varies between the 2 species. It would be interesting to know if the expression of **recA** is equivalent in the 2 species.

KRÜGER: Dr. GRANTHAM, when you are talking about "The genome as a unit of selection" you cannot concentrate on the coding parts of the genome only. In the case of RNA and protein coding sequences Dr. GHILAROV might be right since the evolution of the genome proceeds more or less through selection of the organisms phenotype. On the other hand, there are a lot of regulatory sequences in the genome which are non-coding but have essential significance for the interaction with proteins (e.g., recognition sites for polymerases, repressors etc.) or the genome structure. In this way the genome is a direct subject of selection and evolution. However, you have not looked at these regulatory non-coding sequences which are a direct target for selective pressures.

GRANTHAM: There is only a terminology problem between Dr. GHILAROV and myself. I agree with you that there is more to selection than codons or amino acids. Indeed, it seems that 98 or 99% of the human genome is not translated but has other functions, which probably are subject to selection. We are studying noncoding sequences for their nucleotide organization (see R. GRANTHAM, FEBS Letters **121**, 193—199 (1980)).

F. VOGEL: Even if the different coding strategies at the DNA level are due to varying selection pressures, this does not exclude entirely a strong random element at the protein level. Here, ZUCKERKANDL's concept of "evolutionary sufficiency" might offer better explanations.

GRANTHAM: We agree there *could* be a stronger random element in proteins than in nucleic acids, but the burden of proof is on those who proclaim randomness. We do not find evidence in our work of randomness in either proteins or nucleic acids — of course we are not looking for randomness. We are seeking structures (in the large sense) at the sequence level that relate to biological functions and evolutionary patterns and mechanisms. Hence we could be overlooking some neutral elements — but somebody else will tell us about them! The role of chance in evolution has been subject to controversy since 1859 but it is not the aspect that interests my colleagues and me mostly.

S. ROSENTHAL: Do you see any regularities ahead the initiation codon in high and low expressed genes in prokaryotes?

GRANTHAM: No, we haven't found anything yet in the 5'regions that would indicate whether the codons following would be highly or weakly expressed. One problem with these 5' non-translated sequences is that they run from zero to over 2 thousand nucleotides in length. We are attempting to study them for this and other reasons, however.

RAPOPORT: 1. Is there any relation between overall GC content and the location on the distance map of correspondence analysis?

2. Is there any correlation between cardinal temperature and preference for G + C, have you compared psychrophilic, mesophilic and thermophilic bacteria?

3. Isn't the spread of your areas a measure of the randomness of the results of selection?

GRANTHAM: The response to all 3 questions is:

1. The main factor separating mRNAs in our correspondence analyses is the G + C content of degenerate bases, but this does not appear to be related to the overall G + C content of the genome or of the nondegenerate bases (see table 1).

2. Although this seems logical we have seen no such correlation so far — the sample of sequences from these organisms is still too small to resolve this question.

3. We think it is a demonstration of different kinds of molecular adaptation at the level of use of the genetic code by various taxa. The eye of natural selection seems to permit many different adaptations. I see no reason to consider this permissiveness, this liberty of adaptation, or the biological patterns realized within it, as randomness. Each major taxonomic group appears to be characterized by a particular coding strategy. We don't believe this is a chance association but that coding strategy has evolutionary significance.

NOVAK: Dr. GRANTHAM said he knows no other non-hereditary phylogenetic change than extinction of a species. It is, of course, to some degree a question of what we understand by hereditary change. Accepting, however, what geneticists understand by it, i.e. changes caused by a change in DNA structure, we see that there are two principally different types of non-hereditary change: that caused by temporary, exceptional actions resulting in a change in a given individual in just one, the given, generation and that caused by actions which continues for a series of generations and which we have called non-hereditary phylogenetic changes (NPC). My claim is that these changes are very common and very important to evolution. This concerns not only a broad variety of modifications and cases of polymorphism, but also all ontogenetic morphogenesis which is of non-hereditary character (being at the same time genetically conditioned). This being so, it is not possible to decide without a detailed genetic analysis which changes of a phylogenetical character are hereditary and which are not. This seems to be probably the case with most changes in man in connection with cultural evolution. In this way their extremely high rate (from geological point of view) as suggested by Prof. BRODA might be understood.

GRANTHAM: If by nonhereditary phylogenetic changes you mean non-genetic changes associated with cultural evolution, then there is no argument. Cultural evolution is much more rapid than genetic change in part just because it does not have to be incorporated in the genes. But I think you would introduce a terminology problem by calling cultural evolution **phylogenetic change**, which is now accepted, by definition, following the modern synthesis of Darwinism and Mendelism, with population genetics and molecular biology included, as being due to heritable changes in DNA sequences. Cultural changes are indeed very common and I believe they will be important for the future of our species. They may interact with our own biological evolution because we are creating the conditions for the future selection of humans by the great innovation we are effecting in the biosphere. But, I insist, polymorphism and ontogeny are inherited genetically, their exact realization being conditioned by the evironment, as always. Incidentally, you could remove the contradiction from the title of your concept (NPC) by letting the "P" stand for phyletic instead of phylogenetic.

Zentralinstitut für Molekularbiologie, Akademie der Wissenschaften der DDR, Berlin-Buch, DDR

Movable DNA Elements and Evolution

E. GEISSLER

As we have already pointed out in the introductory lecture, the question "Is evolution gradual **or** saltatory?", which is posed primarily, by palaeontologists but was already put by GOLDSCHMIDT [57, 58] and other geneticists, is undialectical. In fact the results obtained by present day genetics show unambiguously that "**or**" has to be replaced by "**and**": evolution proceeds gradually **and** saltatorially.

Gradual changes in genetic material by base substitution and its underlying mechanisms have long been well documented. (In our context we are unable to further substantiate that base substitutions, deletions, and insertions respectively can also produce saltatory changes because of the dialectic difference between genetic material and genetic information [54]: Minor changes in the genetic material can create completely new genetic information. This evolutionary mechanism played a major role in the evolution of such viruses as ΦX174, polyoma and SV 40.)

Mechanisms leading to gross structural aberrations and/or to an increase in the amount of genetic material have been less well understood so far, although as early as 1951 Barbara MCCLINTOCK suggested the existence of "controlling elements" in maize which are responsible not only for gene regulation but also for the induction of chromosomal aberrations [115]. Twelve years later TAYLOR [167] discovered mutatorphage Mu. Later, several other transposable elements were detected in both prokaryotes and eukaryotes.

One common property of these elements is their mutagenic activity. Results relating to this property will be summarized in this paper. Being mutagenic, the movable elements contribute to genetic variability. But this is not the only reason to regard them as evolutionarily important: as we will summarize below, transposable elements may also cause amplifications and other gross structural aberrations in the genetic material. Some of them even seem to transmit genetic information horizontally, sometimes between organisms of species that are only distantly related. Movable elements may therefore be regarded as significant

Abbreviations

ALV	avian leukemia virus	mdg	mobile dispersed genetic element
ASV	avian sarcoma virus	ME	movable element
bp	base pair	M-MuLV	Moloney murine leukemia virus
ev	endogenous provirus	M-MuSV	Moloney murine sarcoma virus
GaLV	Gibbon ape leukemia virus	MTV	mammary tumor virus
HDR	host direct repeat	RAV	Rous associated virus
HSV	herpes simplex virus	R-MuSV	Rauscher murine sarcoma virus
IS	insertion sequence	SNV	(avian) spleen nekrosis virus
Kb	kilobase = 1000 base (pairs)	tk	thimidine kinase
LTR	long terminal repeat	Tn	transposon

evolutionary factors which, among other things, may cause evolutionarily important "macromutations".

The involvement of transposable elements in bacterial evolution is discussed by CAMPBELL [31a] and KOPECKO [94a]. It should be mentioned in this connection that ANDERSON [3] and ZHDANOV and TIKCHONENKO [191] wondered whether viruses might be evolutionary factors even before the discovery of movable elements in eukaryotes and before the demonstration of striking similarities between DNA proviruses of retroviruses and movable elements. Furthermore participants at a meeting on "Genome evolution and phenotypic variation" discussed the role of mobile DNA sequences in June 1981 as reported in reviews of this meeting [101, 142]. Eric DAVIDSON, for instance, commented, "from an evolutionary point of view there is growing conviction that mobile sequences in the genome may have played an important part in moving genes around" ([101] p. 635—636).

Transposable Elements in Prokaryotes

In 1963, TAYLOR described a novel bacteriophage, designated phage Mu-1, that causes severe genetic modifications in its natural host, *Escherichia coli* K-12 [167]. Bacteria lysogenized by phage Mu contained 1—3 per cent of auxotrophic mutants. All represented one-step mutations in a minimum of 34 different loci.

It did not escape TAYLOR's notice that "phage Mu-1's dual ability to occupy many chromosomal sites and to suppress the phenotypic expression of genes with which it becomes associated resembles the 'controlling elements' of maize more closely than any previously described bacterial episome" [167].

Phage Mu nowadays is classified as a "transposable element". These are "specific DNA segments that can repeatedly insert into a few or many sites in a genome" [32]. Three classes of transposable elements are defined in prokaryotes [32]:

1. "IS elements. IS (simple Insertion Sequences) elements contain no known genes unrelated to insertion function and are generally shorter than 2 kb."
2. "Tn elements (transposons). Tn are more complex transposable elements, often containing two copies of an IS element. They behave formally like IS elements but are generally larger than 2 kb and contain additional genes unrelated to insertion function."
3. "Episomes. Episomes are complex, self-replicating elements, often containing IS and Tn elements."

There are no strict borders between elements of these classes: Most Tn elements are flanked by known IS elements or IS-like sequences, and similar segments of different types can be also found in episomes. Some "classical" episomes such as phage lambda are different to "classical" transposable elements (see below).

Movable elements (MEs) of prokaryotes share the following properties (for reviews, see [25], [31], [160]).

- IS sequences terminate in perfect or nearly perfect inverted repeats involving 20 to 40 bp.
- Most Tn elements terminate in long (800 to 1500 bp) IS-like sequences.
- Integrated MEs are flanked by short duplications (4 to 12 bp) of host DNA (table 1).
- Elements generating host repeats of the same size share more or less pronounced homologies which indicate some evolutionary relationships between them.
- In the limited number of cases studied thus far transposition involves copies of the element while the parent sequence stays at the original donor site.

Table 1
All movable elements generate short duplications of target DNA

ME	size of repeat (bp)
IS3[1]	3
IS3[1], M-MuSV[2], Dm297[3]	4
IS2, Tn1, Mu, etc.[1] Ty[4], *copia*[3], SNV[2]	5
RSV, MTV, RAV-0[2]	6
IS1, Tn5, etc.[1], FB[3], GaLV[2]	9
IS4[1]	11
IS4[1]	12

1 prokaryotic
2 retrovirus
3 *Drosophila melanogaster*
4 yeast

- Transposed elements are not circularly or cyclically permutated, presumably because the ends of the elements are involved in the integration process.
- MEs can be inserted at multiple sites by nonhomologous (illegitimate) recombination. Nevertheless, some insertion specificities have been noted for different elements.
- Transposable elements contain promoters and other transcription signals.
- If a transposable element inserts into a structural gene of the bacterial chromosome or of an extrachromosomal genetic element, gene mutations result as a consequence of "gene splitting".
- Furthermore, different chromosomal aberrations such as deletions, duplications, inversions and translocations are induced by (inserted) transposable elements.

As I have already mentioned, the concept of transposable genes originally stems from the observations of McClintock on the presumed existence of "controlling elements" in maize. It is therefore not surprising that similar elements have been detected not only in prokaryotes but in different eukaryotic systems too: elements sharing structural and functional properties with prokaryotic transposable elements have been found in yeasts, *Drosophila melanogaster* and other Diptera, and avian and mammalian cells infected with retroviruses. Ironically enough, the controlling elements of *Zea mays* have not yet been identified although initial experiments by Burr and Burr [26] in this direction seem to give promising results. The possibilities now available to use the methods of genetic engineering also in studies with plant systems should reveal them soon, however.

Transposable Elements in Eukaryotes

Transposable Elements in *Drosophila*

Different classes of transposable elements have been identified in *Drosophila melanogaster* as well as in other Diptera by genetic and/or biochemical means [64].

Biochemical experiments indicated that dispersed middle repetitive DNA sequences move from one chromosomal site to another. These sequences, called "mobile dispersed genetic elements" (mdg) by Soviet scientists, belong to different "families" of middle repetitive DNA which can be distinguished by **in situ** hybridization. At present about 20 families of mdg elements have been described [122, 146]. Different mdg elements are represented quite diversely in the *Drosqphila* genome. The copy number varies from 15 to 100 [169]. FB elements can be found at about 30 different loci [177]. YOUNG and SCHWARTZ [189] calculate that **copia** will be found at about 200 chromosomal locations. About 250 copies of the mobile dispersed element Dm 225 are found at 20—30 sites in polytene chromosomes. Other, similar repetitive and dispersed genes, which are also actively transcribed, can be found in the close neighbourhood of these loci [168].

Several elements have been studied in detail, including their cloning and sequencing. This has revealed structural features which remind of those of other MEs. They are terminally redundant and carry direct repeats of several hundred bp [39, 135] which contain imperfectly matched inverted terminal repetitions consisting of several bp.

The terminal repeats have been designated LTRs by some authors due to their similarity with structures of DNA proviruses of retroviruses. Like these, at least the LTRs of mdgl and mdg3 contain transcription signals (V. G. CHEMELIAUSKAITE, cf. [122]).

POTTER et al. [136] reported characterization of a mdg family of *Drosophila melanogaster* that carries large inverted terminal repeats which fold back after denaturation and re-annealing. They were therefore designated the foldback (FB) family [177]. In contrast to **copia** and other mdg's the members of the FB family are heterogeneous both with respect to the size of the inverted repeats and the total size of the element. Nevertheless, the ends of the inverted repeats represent closely conserved sequences.

Direct repeats of 5 bp host DNA can be found (table 1) on both sites of integrated **copia** DNA. These differ from insertion site to insertion site [39]. Elements of the 297 family generate a 4 bp duplication upon insertion (E. YOUNG and G. RUBIN, unpublished, quoted by [39]), while FB elements are flanked by 9 bp repeats of host DNA [177].

Each mdg codes for an abundant type of mRNA. At least one **copia**-encoded RNA is a functional mRNA [146]. Interestingly enough, mdg elements are transcribed symmetrically, although one strand is read much more efficiently than the other [56]. This phenomenon can be used to detect mdg elements. Two new mdg elements (mdg-3 and Dm11) were first detected on the basis of just this property [80].

The elements of these families obviously are MEs because they can be located at different positions not only in different isolates of *D. melanogaster* wild type strains [165] but also in different genetic stocks [1, 79]. In tissue culture cells the amount of certain mdg elements increases drastically the additional sequences being found at new sites as intact elements [135, 169].

Their sequences are closely conserved and are not circularly permuted. In these respects they differ markedly from certain other, moderately repetitive DNA sequences of *D. melanogaster* [184]. Mdg elements seem to be highly instable in the course of evolution, however, and they seem to be species-specific: *Drosophila virilis* seems to harbor no copies (or at most single copies) of mdg-1, mdg-3, Dm34 and Dm47. On the other hand, mdg elements found in the genome of *D. virilis* are not present in the *D. melanogaster* genome [169].

Transposable Elements in Yeast

CAMERON et al. [29] found in *Saccharomyces cerevisiae* dispersed repetitive sequences which seem to be capable of moving around the genome. They were designated as Ty elements. Ty1 is a 5,6 kb sequence possessing direct repeats of 338 bp each and designated delta (δ)

sequences. Delta sequences can occur in the yeast genome independently of Ty1. Various species and strains show significant differences with respect to the presence of Ty1 elements. While *Saccharomyces norbensis* harbors very few, if any, Ty1 sequences, *S. cerevisiae* has 30 to 40 copies [46]. Each Ty1 element is flanked by a duplication of 5 bp host DNA [44, 52] (table 1).

Fink et al. [46] found a second transposable element, Ty2, which is 6 kb in length and which contains direct terminal repeats (delta sequences) that are not identical but differ by four substitutions and the deletion of one bp and differ from the delta elements of Ty1.

Proviruses of Retroviruses as Transposable Elements

The genomes of retroviruses possess common features [cf. 37, 89, 106, 114, 156, 157, 158, 166]. They contain unique sequences at their 5′ and 3′ ends (U_5, and U_3, respectively) with short terminal repeats at the very ends (fig. 1A) (and with a 5′ cap and a 3′ polyA sequence which are of no interest in our context). After infection the RNA is transcribed by the viral enzyme, reverse transcriptase, to a double-stranded linear DNA molecule which contains an additional copy of sequences from the 3′ and 5′ termini of the viral RNA molecule (fig. 1B). These termini have been designated LTR (long terminal repeats). They range from 300 to 1,350 bp and contain inverted terminal repeats of 5 to 23 bp. LTRs of different retrovirus DNAs differ in size but have identical overall structures.

The DNA molecules are circularized by fusion, homologous recombination involving the LTRs or other processes generating circles of several sizes which differ in the presence or absence of one LTR (fig. 1C) or in the loss of a certain number of bp. The integrated provirus (deriving from the linear or from circular proviral DNA) is colinear with the nonintegrated linear genome and is flanked by the LTRs (fig. 1D). Two nucleotides of the

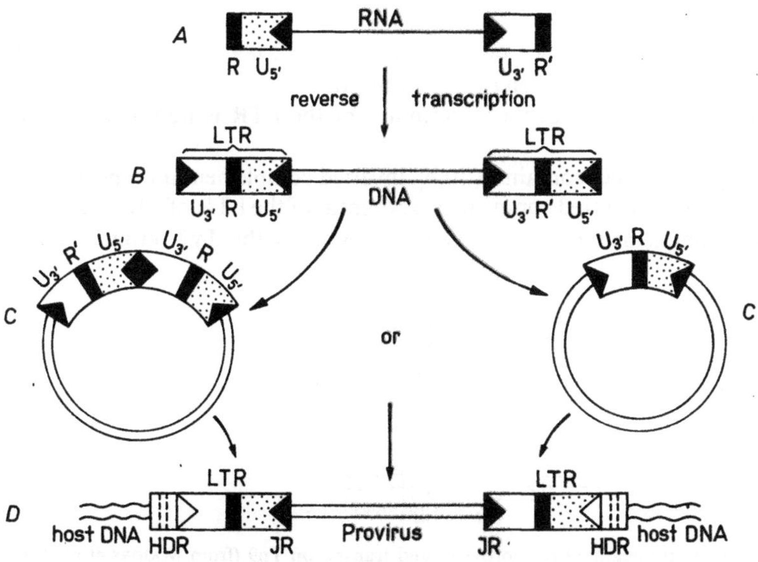

Fig. 1. Structure, reverse transcription and integration of a retrovirus genome (see text)

termini of the unintegrated DNA are absent, however. The provirus is flanked by a direct repeat of four to six bp host DNA ("HDR"). This sequence is unique at the target site before integration and becomes duplicated during insertion of the provirus.

The structure of integrated retroviral proviruses is strikingly similar to that of eukaryotic movable elements and to at least some transposable sequences of prokaryotes [37, 89, 106, 114, 155, 158]:

- Integrated sequences are not cyclically permuted, i. e. they are integrated at a fixed site of the element.
- They can be found at several sites in the genome.
- They are flanked by LTRs, direct repeats of 600 to 1300 bp.
- They are flanked by inverted repeats of three to twelve bp like Ty and **copia**.
- These are flanked by 4-bp to 6-bp direct repeats of host DNA: 4 (M-MuSV), 5 (SNV; Tn3; $\gamma\delta$; Ty1; **copia**), 6 (MTV; RAV-0) (table 1).
- Like δ sequences of yeast, MEs LTRs can be found independently of most of the retro-provirus at least in tissues of birds suffering from ALV-induced proliferative diseases [125].
- All proviruses of retrovirus end like prophage Mu, Ty1 and **copia** with the dinucleotides TG ... CA.
- Bacterial transposable elements do not exist as unintegrated DNA free of host DNA. Even unintegrated DNA of phage Mu is flanked by bacterial DNA. Retrovirus-DNA actually loses two terminal bp during integration and is thus also not integrated via free ends (as stressed by SHIMOTOHNO and TEMIN [155].

The structural homology of retro-proviruses and transposable elements is especially obvious if a provirus is compared with transposon Tn9 [106, 114] (fig. 2), and if one considers that FLAVELL and ISH-HOROWICZ [47] recently succeeded in isolating extrachromosomal circular copies of **copia** in cultured *Drosophila* cells. Most of them included a complete copy of the internal section of **copia** combined with either one or both copies of the flanking direct repeats (LTR), thus resembling circular extrachromosomal DNA copies of retroviral genomes.

Comparison of the nucleotide sequence of LTR of λ RAV2-2 [89] with the nucleotide sequence of several elements revealed among other things that

- the inverted complementary repeat at the 3′ terminus of the LTR is homologous with one terminus of IS2—6,
- the terminal direct repeat of **copia** contains a 10 bp inverted complementary repeat which is homologous to an internal inverted complementary repeat of the LTR of λ RAV2-2, and
- half of the inverted complementary repeat is also present in the Ty1 terminal direct repeat.

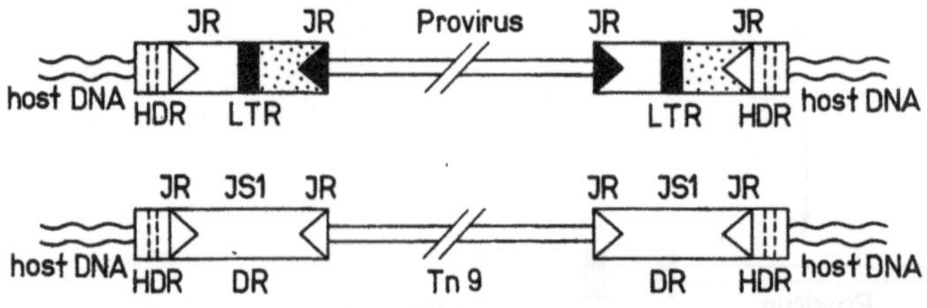

Fig. 2. Structural homology of an integrated retroprovirus and transposon Tn9 (from MAJORS et al. 1981, modified).

Thus, there actually are many similarities betyeen MEs of prokaryotes, retrovirus provirus and transposable elements of yeast and *Drosophila*. "The similarity in nucleotide sequence organization is so great that, if one is given the nucleotide sequence of a genetic element in cellular DNA, it is not possible to tell whether the element is a provirus or a transposon, i. e., had an RNA intermediate or not" [155, p. 727]. This applies, for instance, to sequences detected in the mouse genome and resembling MEs in structural and/or functional respects.

Presumed Movable Sequences in the Genome of Mammals

GEORGIEV et al. [56], by hybridization to long stretches of double-stranded RNA, discovered sequences in the mouse genome which are obviously transcribed symmetrically like mdg elements of *Drosophila* and like the provirus of M-MuSV. As these sequences detected in the mouse genome are also multiple and dispersed they might be related to movable elements. Furthermore, multiple copies of the dispersed mouse gene family VL30 [81] could be shown to be "retrovirus-like". The family consists of dispersed 5,2 kb sequences whose 30 s RNA transcripts can be packaged into pseudovirions of retroviruses and transmitted to other cells [153]. This RNA has therefore been designated virus-like **30** s RNA. The VL30 RNAs are encoded by uninterrupted 5 kb DNA sequences reiterated 20 to 50 times in the mouse genome [81]. Like movable elements, cloned LV30 genes carry direct terminal repeats of 400 bp [82] which participate in recombination processes (see below).

Furthermore, endogenous proviruses (**evs**) structurally related to the genomes of exogenous retroviruses are commonly found in the germ line of many animals.

In white Leghorn chickens at least 11 different **evs** have been physically mapped [9, 76, 77, 106].

Chinese hamster DNA contains a major interspersed repeat which is related to the human **Alu** sequence. Like mdg elements it is transcribed symmetrically. As their flanking sequences are direct repeats unique to individual family members, HAYNES et al. [69] discuss that they are transposable elements.

Finally, FRANCKE [185] has found a polymorphic short sequence in the human insulin gene which can be inserted either in a variable number of copies or in a number of discrete sizes at a single position of the chromosome. This is an indication for the possible existence of movable elements in the human genome.

Oncogenic DNA Viruses Are Different

At least some oncogenic DNA viruses, miopapovaviruses (e.g. polyomavirus, Simian virus 40) and adenoviruses resemble movable elements in some respects (table 2) [176]:

- They are mutagenic (see below).
- Their proviral DNAs are integrated at multiple sites of host cell genomes by inhomologous recombination.
- As a consequence of incorrect excision recombinant molecules are produced which comprise both viral and host cell DNA.

These viruses might therefore be of evolutionary importance, too. The differ, however, in important respects from movable elements **sensu stricto**:

Table 2
Comparison of movable elements with phage λ, an episome, and papovaviruses

	ME	λ	Miopapovavirus	Papilloma virus
Integration	+	+	+	−
Episite	+	+	−	−
Chromosite	−	+	−	−
Host duplication	+	(+)[1]	−	−
Mutagenicity	+	(+)[2]	+	?
Integration by homologous recombination	−	+	−	−
nonhomologous recombination	+		+	−

1 Because of existence of "common core" both in chromosite and episite the integrated prophage is flanked by identical sequences which in contrast to host DNA repeats caused by MEs are not induced by integration but are preexisting.
2 Gene splitting might result if λ is integrated not into the chromosite but into secondary sites.

- Integrated viral chromosomes are circularly permutated. That means, they are not integrated at a specific site of the virus chromosome ("episite").
- Integrated viral chromosomes are not flanked by host cell DNA repeats.
- At least SV40-induced mutations in the **hgprt** locus are not caused by gene splitting [172].
- Integrated SV40 genomes are not transposed. The final arrangement of SV40 DNA in the host genome is extremely stable: no alteration is observed in the arrangement of either the integrated viral DNA or its flanking DNA sequences when transformed cells are grown in culture for hundreds of cell generations [22].

Genetic Variation Caused by Eukaryotic Transposable Elements

It was already noted by McClintock [115—117] that transposable elements are not only involved in the regulation of certain genes but are capable of inducing mutations, inversions, deletions, duplications, translocations, and chromosome breaks. Furthermore they induce gene mutations by gene splitting, as first observed by Taylor [167].

Splitting of scriptons and operons may be one cause for the regulatory effects of MEs. Other means by which gene expression is regulated by movable elements are the introduction of new regulatory sequences that directly specify trancriptional switches or impacts on certain gene-folding configurations of the chromosomal region in which they integrate [188]. Some type of positional effect might happen, for example, when an actively transcribed eukaryotic transposable element or some heavily transcribed gene varried by a transposon integrates into an inactive chromosome region [146].

Gene and Scripton Splitting

Gene splitting (fig. 3) was first observed as a consequence of integration of phage Mu DNA [167]. All transposable elements cause such effects. Scripton splitting can lead to separation of promoter region and structural gene (as observed first by Bertani [16] in the

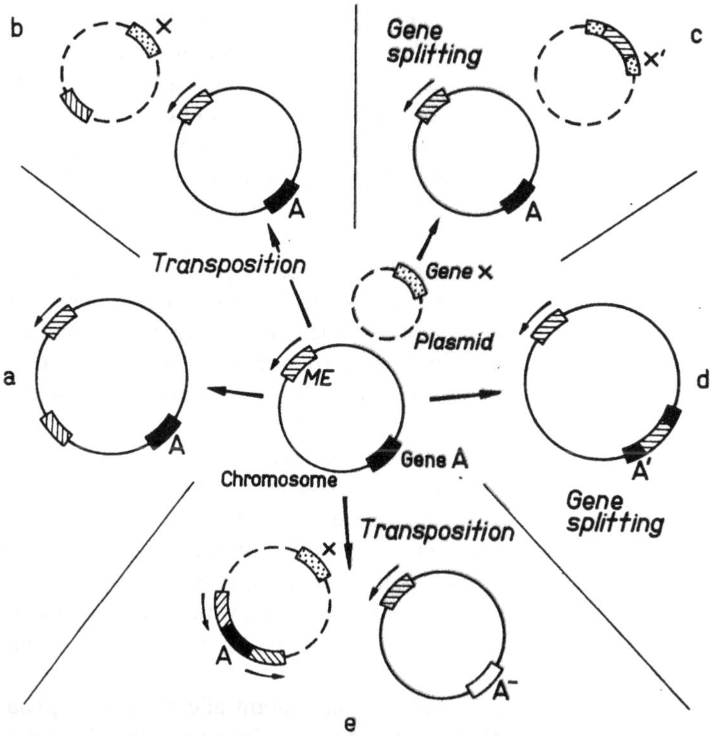

Fig. 3. Genetic effects caused by movable elements. Copies of a movable element (ME) are transposed to another site a) in the same genome or b) of a plasmid. The ME may insert into a structural gene causing its inactivation (c, d, e) Transposition gene(s) may be transposed from chromosome to plasmid and vice versa.

case of P2 prophages) or can separate genes in polycistronic scriptons. Scripton splitting is also the reason for the well known polar activity of prokaryotic transposable elements: their integration into a structural gene belonging to a polycistronic scripton acts pleiotropically and provokes not only inactivation of the affected gene but a marked reduction of expression of the genes belonging to the same scripton but distal of the insertion site.

All eukaryotic MEs should be able to induce gene and scripton splitting. It would be exciting if in addition polar effects could be observed because such effects would be a first indication for the presence of polycistronic messengers in eukaryotic nonviral systems.

Salamini [149] obtained, in appropriate crosses, indications that spontaneous mutations found at the **opaque-2** locus of maize were induced by insertion of some element.

The **white apricot** allele, **w**^{a}, of *Drosophila melanogaster* discovered some 60 years ago is due to insertion of a **copia** element into the **white** locus [18]. As the **w**^{a} mutation obviously does not completely interfere with the function of the **white** locus, Bingham and Judd [18] wonder whether the **copia** element is integrated in an inexpressed segment of this locus.

Jenkins et al. [84] determined the chromosomal location of the DBA/2J ecotropic murine leukaemia virus (MuLV) provirus, **Dbv**, and found that this provirus is closely linked to the dilute (**d**) allele on chromosome 9 which produces one of the coat colour mutations. Analysis of DNA from a spontaneous DBA/2J **d** revertant showed that these mice lack ecotropic-specific MuLV DNA sequences. These results suggest that the dilute mutation resulted from insertion of a MuLV provirus into the mouse genome. The authors wondered therefore whether other mutations of the mouse are also caused by insertion of ecotropic proviruses. They obtained preliminary results indicating that at least one further coat colour mutation is closely associated with unique ectotropic MuLV DNA sequences [84].

Physical analysis of spontaneous mutations in the **his4** gene of yeast has shown that they are due to Ty1 or Ty2 being inserted into the noncoding 5′ region of that gene and thus preventing its expression. In one of the mutants studied, the element is integrated **distally** of the "HOGNESS box" necessary for initiation of transcription [44, 46]. In that case the effect might be a consequence of an effect of Ty integration on DNA supercoiling, which is known to affect gene expression at least in *E. coli* [159].

GREER et al. [65] described two further transposable elements in yeast. One of these actually transposes a copy of the **his4**C region into an essential gene of another chromosome, thus causing recessive lethal mutations. The strains where the **his4**C gene is transposed spontaneously produce chromosomal rearrangements as well as auxotrophic mutations. The transposing element is not linked to the latter. The other element found by GREER et al. [65] had been inserted into the regulatory region of **his4**A. The element, which was designated **Ins10**, is different to Ty1 and Ty2 in size (Ins10 is about 5 kb), physical map and behavior.

Possible correlations of the insertion of mdg and instable mutant alleles of *Drosophila melanogaster* were studied by RUBIN et al. [146] by in situ hybridization with ^{3}H-labeled DNA of **copia**-like elements to polytene chromosomes of individuals carrying stable and instable alleles of the **white**, **forked**, and **yellow** loci. Several correlations have been observed, but they call for further study.

Other genetic experiments suggest that discrete DNA segments transpose from unknown sites in the genome of *Drosophila melanogaster* and integrate into specific genes, thereby generating instable gene mutations which revert to wild type with a high frequency and generate deletions in the target gene [42]. The processes involved are under control of two different mutator systems, the MR mutator system [62, 63] and the I—R mutator system [24].

Also LIM [102] has evidence that a certain instability of the X chromosome of *D. melanogaster*, indicated by the generation of X-linked recessive lethal mutations with a high frequency and their frequent reversion, is caused by the action of a still unidentified transposable element.

Insertion mutants in rat cells infected by Moloney murine leukemia virus (M-MuLV) and previously transformed by Rous sarcoma virus (RSV) were obtained by VARMUS et al. [179]. M-MuLV DNA may be integrated into the RSV provirus of the tranformed cells. In two cases studied the M-MuLv provirus is integrated on the 5′ side of the transforming gene (**src**) of RSV. This interferes with the expression of **src** so that morphological revertant cells result. One of these revertants may again revert to the transformed phenotype by excision of most of the M-MuLV DNA.

Some oncogenic DNA viruses, which are not MEs, nevertheless induce gene mutations in at least six loci of at least three different species: SV40 [108, 171], SV40 GBM [55], Hamster papovavirus (Michael THEILE et al., submitted), Bovine adenovirus type 3 [105], and human adenovirus type 2 [107]. At least mutations in the **hgprt** locus of Chinese hamster and man are not caused by splitting of the affected genes [172].

On the other hand the biological activity of adenoviral transforming genes cloned in the plasmid pBR322 is diminished if the bacterial transposon Tn5 is integrated into an exonic region of these genes (McKinnon et al. [117a].

Chromosomal Rearrangements

A transposable element studied by Ising and Block [81] designated TE1 carries the genes **white** and **roughest** and encompasses four polytene chromosome gands (3C2-6) of the distal part of the X chromosome. Interestingly, genes adjacent to an earlier site of insertion were found to be transposed together with TE ("hitchhiking").

The transposition of these elements seems to involve mdg's: Rasmuson et al. [139] obtained evidence that certain instabilities of this locus are associated with transposition of Dm225. The transposition of TE studied by Ising and Block [81] seems to involve **copia** [53]. Furthermore a correlation between instability in the **white** locus and insertion of mdg1 was found by Belyaeva [cf. 122]. Also Bingham [17] obtained genetic evidence that the mutant allele $\mathbf{w}^{DZL}$ is caused by a transposable element. This allele is mutable and causes chromosomal rearrangements: deletions, inversions, and a reciprocal translocation, all sharing a common break point.

Transposable elements also cause a large number of genomic rearrangements (fig. 4), including translocations, deletions, inversions and transposition in yeast (33, 143). One of the **his**4 mutants caused by insertion of Ty1, **his**4—912, is highly instable and gives rise to different rearrangements, such as deletions, translocations, and inversions. The most common rearrangement involves recombination between the two delta elements leaving one δ element at the original insertion site.

In the yeast mutant CYC1-T1 which overproduces iso-1-cytochrome c a chromosomal region COR containing the loci **CYC1**, **OSM1** and **RAD7** was transposed to chromosome VII while the original COR region was retained at its normal position on chromosome X [163]. The transposed region contains approximately two copies of the CYC1 gene. Thus, overproduction of a gene product is the consequence of gene amplification by transposition.

Endogenous proviruses of white Leghorn chickens are structurally related to ASV but display some aberrations which interfere with their expression [106]. In three cases (ev4, ev5, ev6) the lefthand LTR has been deleted and in two further cases only the U_3, and U_5, regions of the LTR seem to be present, possibly as a consequence of homologous recombination between the LTRs at both ends of the endogenous proviruses.

When murine sarcoma viruses are cloned in phage λ and replicated in *E. coli*, specific genome rearrangements involving LTR are observed [34, 88, 113]. For example, a specific deletion occurs when λ M-MuSV hybrids carrying the integrated provirus are replicated in **rec**$^+$ or **recA** *E. coli* [113, 114]. The deletion occurs by recombination between two LTR elements, thus removing the intervening M-MuSV sequence. Similar events have been observed with transposons.

Accordingly, one of the M-MuLV-induced revertants of RSV transformed rat·cells described by Varmus et al. [179] and mentioned above "back-mutates" by excision of most of the M-MuLV-DNA, probably by homologous recombination between both LTRs of the M-MuLV provirus.

During the propagation in bacteria of recombinant phages carrying the retrovirus-like VL30 mouse genes mentioned above, deletion mutants arise as a result of intramolecular recombination involving the LTRs of these elements [82].

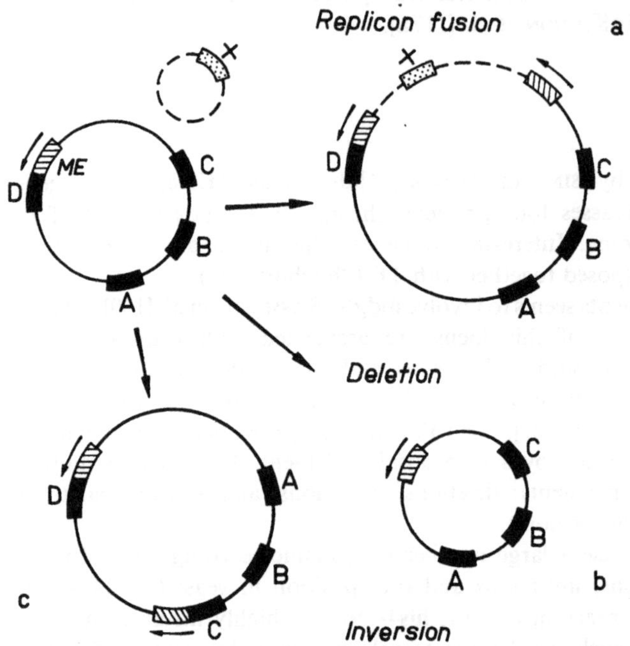

Fig. 4. Further genetic effects caused by movable elements. MEs may cause a) replicon fusions e.g. between a chromosome and a plasmid, b) deletions of adjacent gene(s) or c) inversions.

On the other hand, in M-MuLV-infected cells SHOEMAKER et al. [156, 157] found not only 8.2-kb and 8.8-kb full-length viral DNA circles containing one and both LTR copies respectively but also several circles containing delections or an inversion. The endpoints of the inverted DNA are LTR sequences which occurred in some cases also at one end of a deletion. Quite similar observations were made with IS elements and transposons. SHOEMAKER et al. [157] wonder, therefore, "whether integrated retroviral DNA can transpose", an event which might be difficult to recognize. They speculate that the loss of the terminal base pairs from the retrovirus during integration "may 'fix' the retrovirus into the host-cell DNA" although the lost base pairs are regenerated during transcription.

SV40 is no ME, but it does cause chromosomal rearrangements in nonpermissive cells where no DNA replication does in fact occur: SAGER et al. [147, 148] found rearrangements of SV40 proviral DNA and flanking host sequences which occurred with high frequencies in mouse cells carrying an integrated defective SV40 genome and in progeny produced by cell fusions with nontransformed mouse cells. Whether other parts of the mouse genome had also been rearranged could not be studied with the methods used so far. The rearrangements consist primarily of tandem duplications of part of the viral DNA, often extending into the right flanking host region, Most of them [16 of 18] include the viral origin of replication in the duplication sequence.

A characteristic feature of these arrangements is that they arise with high frequencies in some subclones and with low frequencies or not at all in others. SAGER et al. [148] wonder, therefore, whether the occurrence of these arrangements itself is under genetic control. They discuss these findings with reference to the controlling elements suspected by B. MCCLINTOCK in *Zea mays* and "consider it likely that the events . . . result from the induction of an instable transposition system, induced perhaps by the integration of the viral DNA into a particular genomic sequence" [148].

Furthermore the integration of SV40 can be associated with rearrangements involving at least 600 bp in the neighbouring host sequences [23].

Control of Gene Expression

Controlling elements in maize can transpose to any site in the genome [cf. 117, 126, 133]. Genes that come under their control produce mosaic phenotypes consisting of apparently normal sectors on a phenotypically mutant background. The mutant phenotype is caused by an association of the controlling element with a gene, whose expression is thereby inhibited. The normal sectors of the chimeric tissues are formed by clones of cells in which the controlling element has posed away from the gene which allows its expression.

In at least five different systems gene action is controlled by a specific two-element system. The controlling element designated Ds (for dissociation) or Rcp (receptor element) actually transposes only under the influence of an activating element designated Ac or Rg (regulatory element) acting in **trans**. If Ac is inactivated, the instable phenotype caused by the Ds-associated gene is stabilized.

Single-element controlling systems also exist. Transposing elements obviously do not act only in maize but at least also in *Antirrhinum majus* [151].

Attempts to characterize the molecular mechanisms involved have failed thus far because of our general lack of knowledge regarding the basic structure and function of plant genes and because the controlling elements themselves have not been isolated but followed only by indirect means.

So called ROAM mutants (**R**egulated **O**verproducing **A**lleles responding to **M**ating signals) have been found in yeast which overproduce iso-1-cytochrome c, arginase, ornithine transaminase, and urea carboxylase, respectively [43, 154]. In contrast to the mutant CYC1-T1 mentioned above, the alteration causing overproduction of iso-1-cytochrome c in mutant CXC7-H2 is a consequence of integration of a Ty1 element into a site 140 to about 260 bp in front of the 5′ end of the CYC7 gene. SHERMAN et al [154] suggest that other ROAM mutants may be caused by insertion of Ty1 elements.

WILLIAMSON et al. [186] found that insertion of MEs near the 5′ end of the yeast gene ADR2 coding for the glucose-repressible alcohol dehydrogenase (ADHII) induces **cis**-dominant constitutive mutants which express the normally glucose-repressed enzyme constitutively. For out of seven mutants carry a Ty1 insertion. In the other three the inserted elements differed from Ty1 at one or more restriction sites.

Similar to other movable elements retroviruses also have significant effects on the expression of some genes. Retroviruses include acute transforming viruses (sarcoma and acute leukaemia virus) and slow transforming viruses (ALVs and murine leukaemia viruses) (fig. 5) [183]. Slow transforming viruses differ from the acute transforming viruses in at least three important properties:

1. they induce macroscopic neoplasms in infected animals only after a prolonged latent period (4—12 months in comparison to 2—3 weeks),
2. they do not transform in vitro,
3. they do not seem to contain a transforming gene.

The chronic transforming retrovirus avian leukosis virus (ALV), for instance, induces lymphomas in most cases studied by HAYWARD et al. [70] and NEEL et al. [124] by activating cellular genes like the c-**myc** gene (fig. 6).

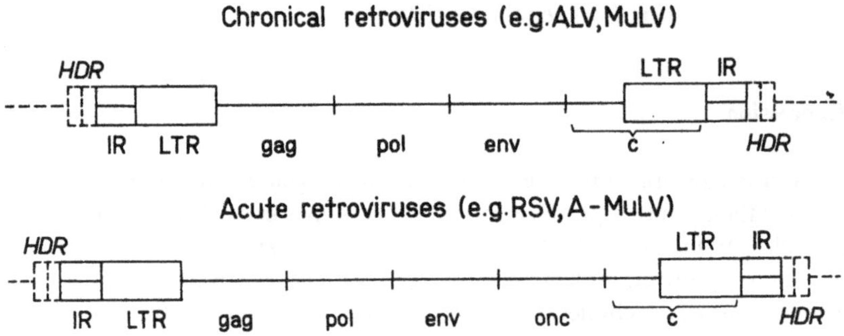

Fig. 5. Chronical and acute retroviruses and their proviruses differ in the presence of the (v-)onc region.

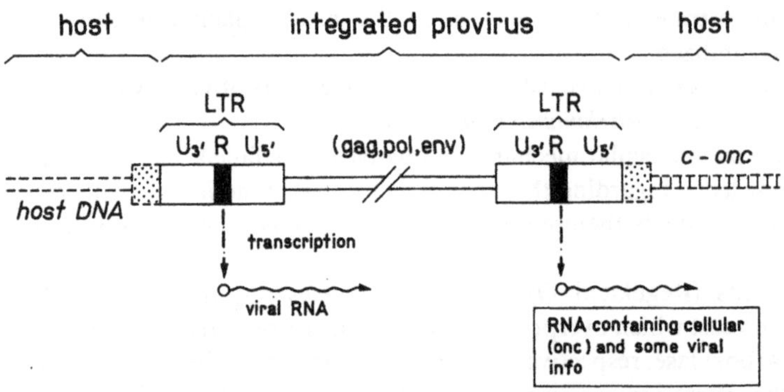

Fig. 6. Chronical retroviruses seem to activate cellular onc genes (c-onc) by transcription starting in the promoter of LTR (from HAYWARD et al., 1981, modified).

C-**myc** is one of at least ten transforming genes detected in avian and mammalian cells as normal counterparts to onc genes of acute transforming retroviruses, being the cellular equivalent of the oncogene (v-**myc**) of myelocytomatosis virus-29 (MC29). This "oncogenesis by promoter insertion" bases on the finding that the LTR of ALV contains a putative promoter sequence [cf. 89] and leads to

- stimulation of synthesis of **myc**-specific RNA (30-100-fold above normal tissues),
- synthesis of **myc**-specific RNA containing short (~ 100 nucleotides) viral 5′ sequences,
- covalent linkage of viral and c-**myc** DNA sequences (fig. 6).

The provirus of mammary tumor virus, on the other hand, does influence the expression of glucocorticoid regulated host genes [188].

In this context it is interesting that CAIRNS [27] recently proposed that most human cancers are not caused by conventional mutagenesis but are the result of transpositions.

This is in line with another observation: tumors induced by RAV-2, a typical ALV, contain a single RAV-2 provirus. These proviruses have a striking aberrant structure, most of them harboring extensive deletions. Obviously one LTR is enough to initiate activation of a cellular gene [106].

Similarly, the cellular gene c-**mos** — the cellular counterpart of the transforming gene of M-MuSV — after cloning in lambda can be activated by an LTR element of M-MuSV, so that it transforms cells with the same effiency as an analogous subgenomic clone containing LTR and V-**mos**^{M}, the viral oncogene [114, 130].

One wonders, therefore, whether endogenous retroviruses or parts of them containing at least one LTR and/or other MEs might activate, after transposition, not only cellular oncogenes but "new" genes created by duplication or by translocation of exons etc.

ROTWEIN et al. [144] found insertions of 1.5 to 3.4 kb in the 5′ flanking region of the human insulin gene especially in patients suffering from type 2 diabetes (where some ability to synthesize insulin is retained in contrast to type 1 diabetes). The authors suggest therefore that these insertions may interfere with the normal expression of the insulin gene.

Furthermore there are transposable ("cassette") elements which control the mating type in yeast [41, 71, 164]. Another element controls phase transition in *Salmonella* by inversion [192]. It contains 995 bp and is flanked by a 14-bp inverted repeat [193]. Finally, both phages Mu and P1 contain homologous 3kb-elements capable of inversion which, at least in Mu, also regulate gene expression [36]. The inversion systems seem to operate in the same way [78].

Are Movable Elements Involved in Evolutionary Processes?

Finally we should discuss whether the different results mentioned above regarding presence, structure, distribution and action of transposable elements justify speculations on their possible role as evolutionary factors.

Nonhomologous Recombination

People have discussed recombination mainly in terms of **homologous recombination** involving not only specific enzymes but identical base sequences in the DNA molecules involved. Homologous recombination gives rise to recombinants whose nucleotide sequences (at least of the region involved) are the same as those in the "parent" molecules except for the genetic markers studied (fig. 7). Homologous recombination also covers unequal crossing over that result from asymmetric pairing between homologous structures containing duplicated segments. CAMPBELL, for example, proposed in his most stimulating model of integration of phage lambda as early as 1962 that identical segments of both host and phage chromosome, designated chromosite and episite, respectively, are the targets of integrative recombination [30]. Base sequence analysis revealed that this holds true for this very type of integration: both the bacterial chromosite and the phage's episite include a common sequence of 15 bp [95]. The crossover event of integrative recombination takes place within the limites or at the boundaries of this "common-core" sequence.

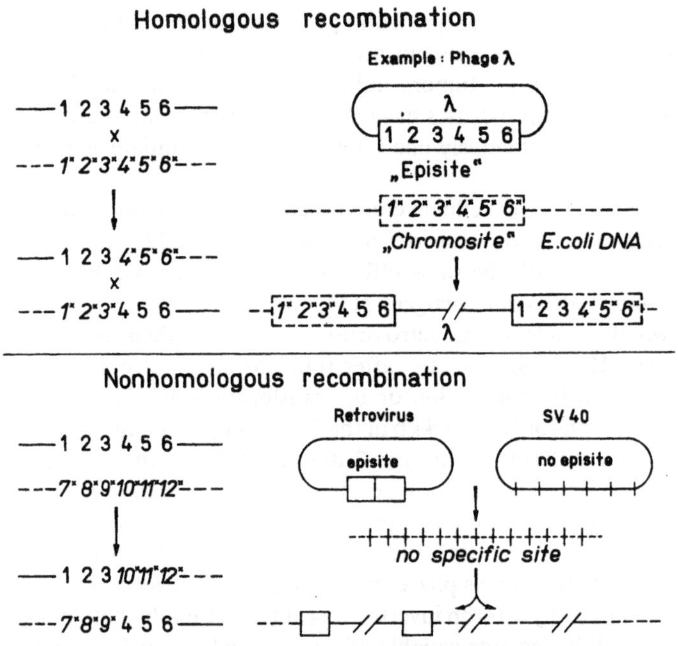

Fig. 7. While the λ prophage is integrated by homologous recombination, involving specific sites in bacterial (chromosite) and phage DNA (episite), DNA and RNA tumor viruses are integrated by nonhomologous recombination at no specific host sites (or not at all). Integration of retroviral DNA takes place at the closed ends, while no episite of oncogenic DNA viruses is known.

Nevertheless this mechanism is not typical for integration of movable elements. Instead, they are integrated by **nonhomologous recombination**, a process designated as "illegitimate" in the past. This term was first used to describe fusion between bacterial and phage DNA which leads to the recombinant chromosomes of special transducing phages [30, 51]. Nonhomologous recombination joins sequences that share no homology and are thus not joined by homologous recombination enzymes. Transposition of bacterial MEs, for instance, is independent of the different known **rec** enzymes.

The presence and widespread distribution of MEs among prokaryotic and eukaryotic organisms proves that such processes are by no means illegitimate but natural and even rather frequent. It is obvious that nonhomologous recombination greatly expands the possibilities for exchanging genetic material.

Transposons Can Make up a Substantial Fraction of the Genome

In *Drosophila melanogaster*, sequences consisting of mobile dispersed gentic elements can account for at least 5% of the total nuclear DNA, a figure which is not far from 13%, the total content of middle repetitive DNA in the *D. melanogaster* genome [168]. YOUNG and SCHWARTZ [189] estimate that "such repeated gene families compose nearly one sixth of the *D. melanogaster* chromosomal DNA".

SAPIENZA and DOOLITTLE [150] even "predict that most middle repetitive DNAs of most eukaryotes will ultimately be proven to be transposable elements, or the degenerate and immobilized descendants of such elements".

Furthermore, multiple copies of endogenous retroviruses coexist in the same cellular genome. They represent a significant fraction of the cellular DNA: "The total complement of endogenous retroviral sequences accounts for a minimum of 0.04—0.3% of the Mus species. Similar results have been obtained for other Mus species.

Finally, both the VL 30 sequences and those elements transcribed symmetrically should be mentioned in this context.

Summing up, a lot of different movable elements coexist in insect, avian and mammalian genomes. Even if single transposition events are extremely rare, the overall frequency of such processes should be rather high and lead to considerable plasticity of the genome.

Table 3
Endogenous viral gene frequency in Mus musculus (TODARO et al., 174)

Viral genome	copy number per haploid genome	complexity (bp)	percentage of genome
C-I	6	6×10^4	4×10^{-3}
C-II	10	1×10^5	$6{,}7 \times 10^{-3}$
M432 (IAP)	1,000	1×10^7	$6{,}7 \times 10^{-1}$
MMTV	10	1×10^5	$6{,}7 \times 10^{-3}$
total	1,026	$1{,}026 \times 10^7$	$6{,}8 \times 10^{-1}$

Sponataneous Mutations in Phages Are Mainly Caused by Movable Elements

ARBER et al. [6,7] demonstrated that most spontaneous mutants of phage P1 and its relative P1-15 are caused by IS transposition. Most P1 mutants (70%) were caused by deletions clustered arround the single IS1 element carried by wildtype P1. 27% of the P1 mutants and all mutants of P1-15, which does not harbor an IS1, carried insertions or partial duplications. Only two of the 165 P1 mutants and none of the P1-15 mutants were caused by nucleotide substitutions or very small insertions or deletions (fig. 8).

Furthermore, gene mutations induced by MEs have been described in retroviruses, yeast cells, *Drosophila*, and mouse as mentioned above.

It is therefore tempting to assume that numerous mutants in other prokaryotic and eukaryotic systems are caused by MEs.

MEs May Cause Macromutations

It has been suggested that macroevolution, "changes above the species level — involving the origin of new species and the establishment of higher taxonomic patterns" [98], is caused by "macromutations". The existence of macromutations or "systemic mutations" was proposed by R. GOLDSCHMIDT [57, p. 183], more than 40 years ago. He argued some years later, "that evolution requires such profound changes of genetic material as the origination of new genes, if we stay within the classical theory, and/or a type of mutant

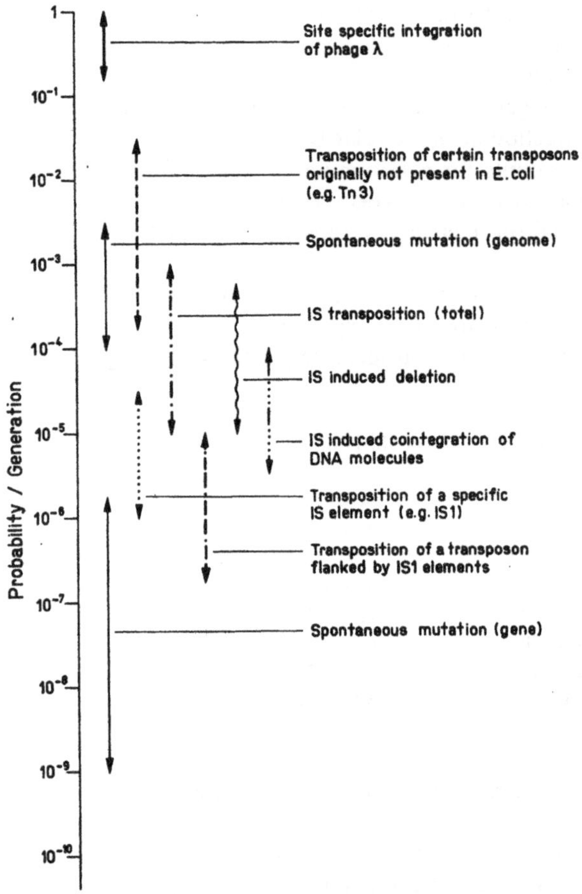

Fig. 8. Probabilities of genetic effects induced by MEs in *E. coli* (from ARBER, 1981, modified).

change which is different from the small variants of genetic effects with which Neo-DARWINian theory deals [58, p. 481]. "Considerations like these have led me to postulate 1940, as a hypothesis, that scrambling and repatterning of the polarized sequences of chromosomal sections may occur occasionally in a single event, which I called 'systemic mutation'. Such repatternings in all grades, from small inversions or transpositions in one chromosome to a complete repatterning of all chromosomes, may lead, if viable, to a large over-all effect changing major features of development and producing in one step (or a few successive ones) a major evolutionary deviation" [58, p. 488].

GOLDSCHMIDT was quite correct in saying that "the solution of the problem of new genes is clearly dependent upon insight into the nature of the genetic material" [58, p. 481]. Many such insights have now accumulated and lie — in contradiction to the opinion of Stephan Jay GOULD, Niles ELDREDGE and other "punctualists" [cf. 98] — completely within the frame of the Synthetic Theory. As STEBBINS and AYALA [161, p. 969], for instance, have already pointed out: "Single-gene or chromosome mutations may have large effects on the genotype because they act early in the embryo and their effects become magnified through development. Single-gene 'macromutations' have been carefully analyzed, for example, in *Drosophila melanogaster* mutations such as 'bithorax'." Interestingly enough it turned out only some weeks after publication of STEBBINS' and AYALA's paper that jumping genes are involved in the generation of bithorax "macromutations".

The **bithorax** complex of *Drosophila melanogaster* consists of some ten genes located in a 200 to 300 kb region of chromosome 3 and involves in a major part of the development of the segmented body of *Drosophila* [110]. Bithorax mutations interfere with the normal development of the fruit fly body. Two of them, **bithorax (bx)**, and **postbithorax (pbx)**, respectively, convert the haltere of the first thoracic segment (fig. 9) to a larger wing-like structure. Flies homozygous for both mutants develop a second pair of wings! Another mutation, **bithoraxoid (bxd)**, influences the development of the first abdominal segment and may lead to the development of a fourth pair of legs on the changed segment.

Recently David HOGNESS and Welcome BENDER, who are mapping the bithorax DNA, found that several mutants were caused by insertions of a few kb DNA that have the characteristics of movable elements [cf. 110]. Other **bithorax** mutations were caused by deletions. **Pbx**^{1}, for instance, was caused by deletion of a 17-kb fragment. The same fragment can be found in another mutant, **cbx**^{1} (first contra**bithorax**), reinserted about 10 kb from its previous position. These findings demonstrate unambigously that MEs can in fact induce macromutations thought to be important in macroevolution.

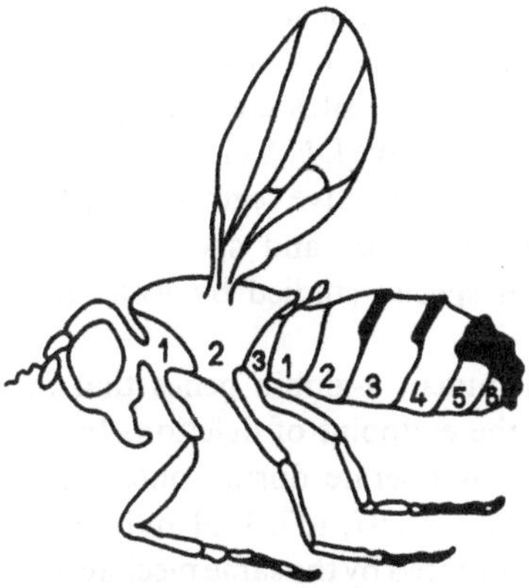

Fig. 9. Schematic presentation of *Drosophila melanogaster*.

Genome Rearrangements May Be Caused by MEs, too

Not only single-gene macromutations but also chromosomal aberrations lie within the conceptual frame of systemic mutations discussed by GOLDSCHMIDT. They seem to play a key role in the evolutionary process [cf. 90]. Such rearrangements, for example, may contribute to establishing gametic barriers, and thus play a role in speciation in both animals, including man [cf. 40], and plants. Thus, Gabriel DOVER proposed during a recent meeting [cf. 101] that repeated sequences contribute to the origin of new species by introducing reproductive incompatibility between one population of individuals and another. As the mutated members of a family might spread through the population independently of natural selection it might lead to "accidental speciation" and would even be in line with an evolution by punctuated equilibria as discussed by ELDREDGE and GOULD [98].

The importance of such processes in plants is indicated by that fact that more than 70% of the DNA in **Secale** chromosomes is highly repeated [141]. FLAVELL et al. [48] and BEDBROOK et al. [12] conclude from their studies on the distribution of repeated sequences in the genome of higher plants that rearrangements of these sequences, including permutations, amplifications and/or deletions, are responsible for major differences between the chromosomes of related species. They suggest that not the function of a certain family is important but the relative orientations of single members of a family or of complexes of sequences belonging to different families relative to certain genes controlling their activity [48]. Furthermore, single repeats could be directly involved in control if integrated into regulatory or structural genes.

Knowing the ability of movable elements to induce all kinds of chromosomal aberrations, we cannot exclude the possibility that MEs may have been involved in the rearrangement of these sequences. FLAVELL et al. [48] consider it unlikely, however, that these sequences themselves represent movable elements "because the resulting instability would lead to mutations at an unacceptable frequency".

WYMAN and WHITE [187] started looking for DNA sequence variations detectable as variants in the lengths of DNA fragments produced by restriction endonucleases ("restriction fragment length polymorphism", RFLP). By hybridizing with a cloned segment of human DNA (λCH4A-r^{Hs18}), which contains a 16-kb single-copy human DNA segment, they detected a human locus exhibiting a high degree of DNA sequence polymorphism. The structural basis for this polymorphism is due to a rearrangement series. WYMAN and WHITE comment that this observation "suggests, by analogy with sites containing transposable elements found in *E. coli*, yeast and *Drosophila*, that an active DNA element may reside at this locus". The authors are now looking for the presumed active segment [99].

Transpositions caused by MEs could also explain the existence and species-specific distribution of satellite DNAs in human beings and other speoies studied by JONES [86]. Its amplification seems to play a role in evolution [123].

Finally, LAUER et al. [96] have described deletions in human α-globin genes that are similar to deletions occurring in bacteria with respect to the endpoint of deletion. In this connection it should be mentioned that similar mechanisms operate during ontogenesis of higher animals. Both the development of antibody-diversity [cf. 97, 137, 175] and the development of (at least some types of) cancer [27, 93 a] seem to proceed by the same mechanisms as the evolution of new species. CAIRNS [27] therefore assumes that the evolution of cancer cells should usually be due to gene rearrangement rather than to local changes in base sequence. In this respect it is worth mentioning again that retrovirus DNAs are very similar to movable elements at least, and that (some) oncogenic DNA viruses cause chromosomal rearrangements besides gene mutations.

Numerous Experiments Demonstrate that Gene Transfer Is Possible

In prokaryotic systems gene transfer processes involving viral vestors have been known since 1952. By **generalized transduction** bacterial genes were transduced randomly by transducing particles, which can be designated "pseudovirions" containing host DNA instead of phage DNA. Specialized transduction transfers bacterial genes neighbouring the chromosite (prophage insertion site). Specialized transducing particles are produced by imprecise excision, i.e. by unhomologous recombination of the prophage and therefore contain hybrid DNA molecules consisting of covalently bound phage and host DNA sequences.

ANDERSON [2] proposed more than ten years ago "that viral transduction is a key mechanism for transporting segments of DNA across species and phylum barriers, and that evolution depends largely on this transfer" because "the evolution of one organism depends on contributions and new ideas from all". He argues that, among other things, viruses cross species barriers with ease and transfer host DNA into other cells, and that virus genomes may be incorporated into germ cells.

Moreover, ZHDANOV and TIKCHONENKO [191] discussed the possible capacity of animal pseudovirions, which are viral nucleocapsids filled with host DNA fragments, for generalized, i.e. nonspecific, transduction: although donor DNA is transferred into the nuclei of recipient cells by pseudovirions, several authors have failed to transfer thymidine kinase or inosine pyrophosphatase markers by means of pseudovirions of papovaviruses [10, 72]. In the light of successful DNA-mediated transformation experiments [cf. 132, 152] it is tempting to assume that these negative results were caused by technical reasons and not by a general inability of pseudovirions to transduce donor genes. Pseudovirions should therefore be considered as natural molecular vectors which are active at least on rare occasions.

In this connection it should be mentioned that V. APOSHIAN [4] has developed a system for producing "polyoma-like particles" by incubating empty capsides with DNA isolated from polyomaviruses or from other sources. Attempts to use these particles as molecular vehicles are in progress.

MERRILL et al. [120] claimed to have cured human galactosemic cells by infection with specialized transducing lambda **dg**-phages. Because they did not use established cell lines which could be cloned and studied after transduction, the fate of the transferred donor DNA could not be studied. Only one group has been able to reproduce MERRILL's results: HORST et al. [73]. HORST [74] furthermore reported that he had introduced the bacterial **lacZ** gene by transduction with lambda **plac** phage into human GM_1-gangliosidosis fibroblasts which respond by producing bacterial β-galactosidase.

On the other hand, in vitro recombination has been used successfully to construct hybrid DNA molecules which are able to transduce active prokaryotic and eukaryotic genes into prokaryotic or eukaryotic cells and to demonstrate the expression of these genes in their heterologous environment.

Furthermore it is also possible to introduce genes and viral genomes into the germ line of mammals artificially. JAENISCH [82] succeeded in introducting retrovirus-DNA into the germ line of mice by infection of preimplantation embryos with M-MuLV [83] or with cloned genomic retroviral DNA [68]. The virus genome is integrated at different sites into the host genome and behaves in crosses like a MENDELian gene. Interestingly enough two of the resulting substrains, **Mov**-3 and **Mov-13**, cannot be obtained homozygous (Rudolf JAENISCH 1981, personal communication). Possibly the integration of M-MuLV caused recessive lethal mutations in these strains. PELLICER et al. [131] introduced the HSV tk gene and the human β-globin gene into mouse teratocarcinoma cells. If these cells can be stimulated to participate in embryo formation by the methods developed by MINTZ [121] this work will provide a system for introducing genes not only into mice but also into their germ line.

Similar work is being done by GORDON et al. [61], who introduced a recombinant plasmid into fertilized mouse embryos by microinjection. The injected sequences could be detected in two of 78 newborn mice.

Furthermore, Karl ILMENSEE and Axel ULLRICH [109a], E. F. WAGNER et al. [180] and T. WAGNER and P. HOPPE [181] succeeded in transferring functional human and rabbit genes coding for insulin and β-globin respectively, as well as the HSV thymidine kinase gene into fertilized mouse eggs. Some animals which developed from these eggs expressed the foreign genes and synthesized the donor protein. They could even demonstrate that some of the functional donor genes they transferred were inherited when mice expressing them were mated.

Taken together these results demonstrate that it is possible to transfer genes across the barriers of all systematic groups and to get these genes expressed in their new host cells.

Retroviruses Pick up Host Genes and Transfer them

TEMIN [170] and SHIMOTOHNO and TEMIN [155] discuss the "origin of retroviruses from cellular movable genetic elements"[1], pointing to the remarkable similarity between integrated MEs and integrated proviruses of retroviruses discussed above. They propose that a small ME that induces 5 bp duplications of host DNA upon transposition may have been the ultimate ancestor of retroviruses.

This hypothesis is plausible because the last step in the evolution of strongly transforming (acute) retroviruses can be observed directly: type C sarcoma viruses are genetic recombinants containing sequences of replication-competent leukemia viruses linked to sarcoma virus-specific sequences designated **onc** or **src** genes. Today at least a dozen different **onc** genes are known, and a corresponding protooncogene has been identified in the cellular genome for each of them [19]. Such cellular **onc** genes are conserved during evolution, and the rate of their evolutionary divergence parallels that of globin sequences [145]. Therefore even in the human genome **onc** genes can be detected which are homologous to the transforming genes of monkey retroviruses. In this way sequences have been found which are related to the transforming genes of simian sarcoma virus (**c-sis**), to feline sarcoma virus (**c-fes**), and to avian myelocytomatosis virus (**c-myc**) [45].

Deletions in the oncogenes of highly transforming viruses such as Rous sarcoma virus (RSV) can be repaired by recombination with a proto-oncogene [182], and new oncogenic capacities can be recovered from cells by viral transduction [138]. Moloney murine sarcoma virus (M-MuSV), for instance, was isolated by passage of Moloney murine leukemia virus (M-MuLV) through BALB/c mice and was formed by recombination between M-MuLV and cellular genes [178]. Uninfected mouse DNA contains a 14.0 kb EcoRI fragment (C-**mos**) which hybridizes with cloned DNA encompassing the sarcoma specific region of the M-MuSV genome (V-**mos**^{M}). The 14.0 kb fragment contains an uninterrupted stretch of about 1,000 nucleotides that is homologous to the **src** specific region of the M-MuSV genome [87].

By contrast, the cellular counterparts of the oncogenes of Rous sarcoma virus (RSV), avian erythroblastosis virus, Abelson leukaemia virus, Harvey sarcoma virus, Simian sarcoma virus and Snyder-Theilen feline sarcoma virus are known to include intervening sequences (introns) which do not appear in the corresponding viral **src** genes [50].

Correspondingly, GREEN [64] wonders whether the viral RNA (of insects) could serve as a template for reverse transcription resulting in DNA which serves as an insertion sequence. A similar new "somatic selection hypothesis" was forwarded recently by STEELE [179].

1 On the other hand David BALTIMORE has proposed [56] that mdg elements might be related to the genomes of proretroviruses integrated into the genomes of ancestors of *D. melanogaster* or of other species.

This hypothesis is described as follows [100]: "During an organism's early life . . . mutations occur in the somatic cells, some of which may be better suited to prevailing environmental conditions. These cells will proliferate relative to their nonmutant relatives and will eventually dominate the organ. Because of the relative abundance of the mutant cells in the organism, there is a great likelihood, STEELE contends, that retroviruses will pick up some of the new genetic material, transport it to the germ line cells, and then insert it into the genome, whose role is to make the next generation."

STEELE's ideas base on experiments done with R. GORCZYNSKI, which, they claim, demonstrated LAMARCKian inheritance of acquired immunological tolerance [59, 60]. But their results could not be reproduced (cf. PASTERNAK, this volume).

But even if they could have been, and if it had been demonstrated that retroviruses or other MEs are capable of transposing alleles from mutant cells into the germ line, one could not classify such processes as LAMARCKian because the rules of the central dogma would not be violated.

As HOWARD [75] points out, however, "if viral vectors are engaged in the transmission of DNA between cells, and, as implied by GORCZYNSKI and STEELE's recent paper . . ., are no respecters of the linkage group, we should have no loci and no alleles. The chromosomes of the germ cells would be a randomized dump of DNA sequences, gathered from cells all over the body . . . Every time alleles behave like alleles and linkage groups behave like linkage groups it is another conformation of the generality of MENDELian genetics and of the integrity of the germ line, and a further exception to the generality of STEELE's hypothesis."

Of course, this argument is not convincing if, as discussed above, such processes happen only on very rare occasions. And there are in fact indications that such rare events took place.

Besides, participants at the previously mentioned meeting on "Genome evolution and phenotypic variation" agreed that mobility of DNA sequences in the eukaryotic genome "is mostly but not exclusively the prerogative of the repeated sequences. The genome can therefore be viewed as a relatively stable island of structural genes immersed in a steadily shifting tide of changing repeated sequences" [101].

In this context it should be mentioned that A. LEDER et al. [96a] observed the dispersion of two mouse α pseudogenes from their primary locus on chromosome 11 to chromosome 15 (α-ψ^3) and 17 (α-ψ^4). They even obtained preliminary evidence that a retroviral genome might have been involved in transposition of the α-ψ^4 sequence.

Horizontal Transmission of Movable Elements Between Distantly Related Species

The possibility of MEs being involved in intraspecies transfer of genetic information is strengthened especially by the demonstration of a number of such transfers by retroviral genes and/or oncogenes.

BENVENISTE and TODARO [13] found that baboon retrovirus RD114, which seems to have existed in the primate genome for about 30 mio years, was transferred to the ancestors of the domestic cat some three to five mio years ago.

Moreover, FeLV was transmitted from an ancestor of the rat to an ancestor of the domestic cat [14]. Another endogenous virus was acquired by an ancestor of the domestic pig from an ancestor of the mouse [15].

Furthermore, studies on the origin of the woolly monkey-gibbon ape group of infectious primate type C viruses suggested that these retroviruses were of murine origin. In fact in Asian species of **Mus** at least two additional previously unrecognized classes of endogenous retroviruses have been identified. One of them, designated type C-I, is related to the woolly monkey-gibbon ape group of retroviruses [174].

Also, R-MuLV, Rauscher murine leukemia virus, and gibbon ape leukemia virus (GaLV) are all closely related [103].

Viral gene sequences of two Old World monkey retroviruses, CPC-1, from *Colobus polykomos*, and MAC-1, the endogenous stumptailed monkey virus, are closely related to one another and to cellular DNAs of various carnivores [20], in which they seem to have been present for at least 30 mio years.

OROSZLAN et al. [129] and RICE et al. [140] detected homologies between both the major structural protein and the genome of the horizontally transmitted avian reticuloendotheliosis virus type A (REV-A) and members of the MAC-1, MMC-1, CPC-1 group of retroviruses which are endogenous in maçaques and colobus monkeys. Both groups of virus are therefore suggested to be descendants of a common ancestral virus. It is impossible to decide whether REV represents an escaped primate virus, whether the maçaque-colobus viruses are derivatives of an exogenous bird virus, or whether both groups derived from a common ancestor residing in another, hitherto unknown species.

Furthermore, SNV, avian spleen necrosis virus, another member of the avian reticuloendotheliosis viruses, is related to Old World monkey retroviruses CPC-1 and MAC-1 but not to the avian leukosis-sarcoma group of viruses [104]. Since SNV and similar viruses have no detectable homology with the DNAs of their natural hosts, water fowl and turkeys, LOVINGER et al. [104] conclude that "an ancestor of avian reticuloendotheliosis virus was propably transmitted from a monkey to an avian host".

Furthermore a relationship has been reported between avian reticuloendotheliosis viruses and two viruses of New World species, the endogenous owl monkey virus OMC-1, and deer kidney virus [11].

ANDERSEN et al. [2] even demonstrated immunologic relatedness of the viper retrovirus VRV isolated from spleen tissue of a tumor-bearing RUSSELL's viper (*Vipera russellii*) to endogenous type D retroviruses of Old World primates. If it could be established that both viruses are evolutionarily related "this would represent the first known example of evolutionarily linked retroviruses that are genetically transmitted within more than one class of vertebrate" [2].

But MEs are transposed even between still more distantly related systematic entities. PIRUZIAN et al. [134] found that DNA of phage Mu is incorporated into fragments of the genome of *Pisum sativum* if the DNA of this plant is cloned in *E. coli*. Similarly, the bacterial transposon Tn5 is inserted into adenoviral genes cloned with plasmid pXC1 (Randy MCKINNON 1981, personal communication).

Horizontal gene transfer is the basic mechanism of crown gall plant tumor formation. Crown galls were induced by infection of dicotyledonous plants with *Agrobacterium tumefaciens*. The ability of that bacterium to induce tumors depends on the presence of the large Ti plasmid [190]. Part of the Ti plasmid, the "T-region", can be shown to be bound covalently to nuclear DNA at multiple sites in multiple copies in crown-gall tumor cells [35, 119, 173]. The bacterial DNA in the plant tumor cells is transcribed into poly(A)-containing RNA [38, 67] which can be translated at least in vitro [118]. Proteins controlled by T-DNA are involved in cell transformation and/or in the synthesis of "opine compounds" (octopine, nopaline, agropine) which are unique to plant tumor cells [68].

The insertion of movable elements into the T-region of Ti plasmids result in mutated strains which may induce morphologically distinguishable tumors [127]. The MEs were transferred into the plant cell DNA as part of the T-DNA [128]. Their insertion impairs the function of part(s) of the T-DNA which control the activity of phytohormones.

"The demonstration that *Agrobacterium tumefaciens* can move some of its plasmid genes into the host chromosome and have them properly expressed in this new location" is discussed by BORST and GRIVELL [21] as "a present day analogy" of their hypothesis "that mitochondria arrived in the eukaryotic cell as bacterial endosymbionts and that these symbionts transferred most of their genes to the nucleus, but that this process stopped before it was complete".

In fact, the results of nucleic sequence analysis indicate that the eukaryotic cell is a chimera at a very basic level. While still nothing is known regarding the origin of the nucleus, the chloroplast and perhaps the mitochondrion both seem to be of eubacterial origin, while at least one ribosomal protein seems to be archaebacterial. A third line of descent is represented in the 18S ribosomal RNA, which is unrelated to both archaebacterial and eubacterial rRNA [49]. On the other hand the 5.8S RNA present in the large ribosomal subunits seems to be related to the 5′ end of the eukaryotic large subunit 23S rRNA [81a].

Participation of MEs in Exon Transfer Cannot Be Excluded

We have already mentioned (GEISSLER and HÖRZ, this volume) that people now speculate that one way of creating new eukaryotic genes might be to reshuffle exons. It is difficult, however, to think of such processes in terms of homologous recombination. Therefore the idea forwarded by CALOS and MILLER [28], i.e., that "the penomenon of discontinuous coding sequences in eukaryotes may also be related to transposable elements. The intervening sequences are generally flanked by short imperfect direct repeats and may have arrived in the gene as insertion sequences", is suggestive.

There are first indications that the gene coding for copper-zinc superoxide dismutase of teleost fishes has been transfered to the bioluminescent bacterium *Photobacter leiognathi* which is a symbiont of the teleost ponyfish *Leiognathus splendens* [109]. Moreover, the ability of legumes to produce the oxygen binding protein leghemoglobin (Lb) might possibly be a consequence of some ancient gene transfer from animal to plant, because the corresponding legume gene looks like a primitive globin gene: there are homologies in the amino acid sequences of Lb and other globins, and the two introns found in α- and β-globins can be located at the corresponding positions in a Lb gene of soybean [85]. The Lb gene does include a third exon, however. This is in line with the suggestion that the vertebrate globin precursor gene lost an intron during its history, as already mentioned (GEISSLER and HÖRZ, this volume).

Furthermore, a lot of evidence suggests that the lysozymes of hen egg-white and of bacteriophage T4 evolved from a common precursor (8, 111, 112). Finally KOIDE and MARVO [94] and LEROITH et al. [97a] describe the presence of mammalian hormonelike proteins in bacteria. This might be the consequence of integration of mammalian DNA by bacteria sometimes in the past.

Such and similar observations can be explained only if some mechanisms of exon and gene transfer between different organisms really do exist, regardless of whether the organisms belong to the same species or not.

Summary

In conclusion, movable elements are evolutionarily important for the following reasons:

1. They are obviously not only widespread among prokaryotes and eukaryotes but are also often found in relatively high copy numbers.
2. They cause mutations, thus contributing to genetic variability.
3. They cause not only duplications and other types of gene and chromosome mutations but are themselves duplicated during transposition, thereby increasing the amount of genetic material available for further variation.
4. They cause gross chromosomal aberrations which can be regarded as "macromutations" responsible for macroevolutionary processes.
5. Some of them are sometimes transfered between distantly related species, thus reshuffling different genetic material among different genomic backgrounds.

References

[1] ANANIEV, E. V., V. A. GVOZDEV, YU. V. ILYIN, N. A. TCHURIKOV, and G. P. GEORGIEV: Reiterated genes with varying location in intercalary heterochromatin of *Drosophila melanogaster* polytene chromosomes, Chromosoma **70** (1978), 1—17.

[2] ANDERSEN, P. R., M. BARBACID, S. R. TRONICK, H. F. CLARK, and S. A. AARONSON: Evolutionary relatedness of viper and primate endogenous retroviruses, Science **204** (1981), 318—321.

[3] ANDETSON, N. G.: Evolutionary significance of virus infection, Nature **227** (1970), 1346—1347.

[4] APOSHIAN, V. and S. SLILATY: Assembly of a polyoma-like particle in a cell free system, 5th Intern. Congr. Virology, Abstracts, (1981), p. 309.

[5] ARBER, W.: Das Bakterium *E. coli* unter der Lupe der Molekulargenetiker, Mannheimer Forum 81/82, Boehringer Mannheim (1981), 9—11.

[6] ARBER, W., M. KÜMBELIN, P. CASPERS, H. J. REIF, S. IIDA, and J. MEYER: Spontaneous mutations in the *Escherichia coli* prophage P1 and IS-mediated processes. Cold Spring Harbor Symposium Quant. Biol. **45** (1981), 38—40.

[7] ARBER, W., S. IIDA, H. JÜTTE, P. CASPERS, J. MEYER, and C. HÄNNI: Rearrangements of genetic material in *Escherichia coli* as observed on the bacteriophage P1 plasmid, Cold Spring Harbor Symp. Quant. Biol. **43** (1979), 1197—1208

[8] ARTYMIUK, P. J., C. C. F. BLAKE, and A. E. SIPPEL: Genes pieced together — exons delineate homologous structures of diverged lysozymes, Nature **290** (1981), 287—288

[9] ASTRIN, S. M.: Endogenous viral genes of the White Leghorn chicken: Common site of residence and sites associated with specific phenotypes of viral gene expression, Proc. Natl. Acad. Sci. USA **75** (1978), 5941—5945.

[10] AXELROD, D. and D. TRILLING: Interaction of pseudovirions with infectious virions of papovaviruses in susceptible host cells, Adv. Biosci. **8** (1972), 175—186.

[11] BARBACID, M., M. D. DANIEL, and S. A. AARONSON: Immunological relationships of OMC-1, an endogenous virus of owl monkeys, with mammalian and avian type C viruses, J. Virol. **33** (1980), 561—566.

[12] BEDBROOK, J., J. JONES, and R. FLAVELL: Evidence for the involvement of recombination and amplification events in the evolution of *Secale* chromosomes, Cold Spring Harbor Symp. Quant. Biol. **45** (1981), 755—760.

[13] BENVENISTE, R. and G. J. TODARO: Evolution of C-type viral genes: inheritance of exogenously acquired viral genes, Nature **252** (1974), 456—459.

[14] BENVENISTE, R. E., C. J. SHERR, and G. J. TODARO: Evolution of type C viral genes: Origin of feline leukemia virus, Science **190** (1975), 886—888.

[15] BENVENISTE, R. E. and G. J. TODARO: Evolution of type C viral genes: Preservation of ancestral murine type C viral sequences in pig cellular DNA, Proc. Natl. Acad. Sci. USA **72** (1975), 4090—4094.

[16] BERTANI, L. E.: Split-operon control of a prophage gene, Proc. Natl. Acad. Sci. USA **65** (1970), 331—336.

[17] BINGHAM, P. M.: A novel dominant mutant allele at the white locus of *Drosophila melanogaster* is mutable, Cold Spring Harbor Symp. Quant. Biol. **45** (1981), 519—525.

[18] BINGHAM, P. M. and B. H. JUDD: A copy of the copia transposable element is very tightly linked to the w^a allele at the white locus of *D. melanogaster*, Cell **25** (1981), 705—711.

[19] BISHOP, J. M.: Enemies within: the genesis of retrovirus oncogenes, Cell **23** (1981), 5—6.

[20] BONNER, T. I. and G. J. TODARO: Carnivores have sequences in their cellular DNA distantly related to the primate endogenous virus, MAC-1, Virol. **94** (1979), 224—227.

[21] BORST, P. and L. A. GRIVELL: The mitochondrial genome of yeast, Cell **15** (1978), 705—723.

[22] BOTCHAN, M. R., W. C. TOPP, and J. SAMBROOK: Studies on Simian virus 40 excision from cellular chromosomes, Cold Spring Harbor Symp. Quant. Biol. **43** (1979), 709.

[23] BOTCHAN, M. R., J. STRINGER, T. MITCHISON and J. SAMBROOK: Integration and excision of SV40 DNA from the chromosome of a transformed cell, Cell **20** (1980), 143—152.

[24] BREGLIANO, J. C., G. PICARD, A. BUCHETON, A. PELISSON, J. M. LAVIGE and P. L'HERITIER: Hybrid dysgenesis in *Drosophila melanogaster*, Science **207** (1980), 606—611.

[25] BUKHARI, A. I., J. A. SHAPIRO, and S. L. ADHYA (eds.): DNA Insertion Elements, Plasmids, and Episomes, Cold Spring Harbor Laboratory, Cold Spring Harbor, N.Y. (1977).

[26] BURR, B. and F. A. BURR: Detection of changes in maize DNA at the shrunken locus due to the intervention of Ds elements, Cold Spring Harbor Symp. Quant. Biol. **45** (1981), 463—465.

[27] CAIRNS, J.: The origin of human cancers, Nature **289** (1981), 353—357.

[28] CALOS, M. P. and J. H. MILLER: Transposable elements, Cell **20** (1980), 579—595.

[29] CAMERON, J. R., E. Y. LOH, and R. W. DAVIS: Evidence for transposition of dispersed repetitive DNA families in yeast, Cell **16** (1979), 739—751.

[30] CAMPBELL, A.: Episomes, Adv. Genet. **11** (1962), 101—145.

[31] CAMPBELL, A.: Some general questions about movable elements and their implications, Cold Spring Harbor Symp. Quant. Biol. **45** (1981), 1—9.

[31a] CAMPBELL, A.: Evolutionary significance of accessory DNA elements in bacteria, Ann. Rev. Microbiol. **35** (1981), 55—83.

[32] CAMPBELL, A., D. E. BERG, D. BOTSTEIN, E. M. LEDERBERG, R. P. NOVICK, P. STARLINGER, and W. SZYBALSKI: Nomenclature of transposable elements in prokaryotes, Gene **5** (1979), 197—206.

[33] CHALEFF, D. and G. FINK: Genetic events associated with an insertion mutation in yeast, Cell **21** (1980), 227—237.

[34] CHAIN, H. W., C. F. GARON, E. H. CHANG, D. R. LOWY, G. HAGER, E. M. SCOLNICK, R. REPASKE, and M. A. MARTIN: Molecular cloning of the Harvey sarcoma virus circular DNA intermediates. II. Further structural analysis, J. Virol. **33** (1980), 845—855.

[35] CHILTON, M.-D., R. K. SAIKI, N. YADAV, M. P. GORDON, and F. QUETIER: T-DNA from *Agrobacterium* Ti plasmid is in the nuclear DNA fraction of crown gall tumor cells, Proc. Natl. Acad. Sci. USA **77** (1980), 4060—4064.

[36] CHOW, L. T. and A. I. BUKHARI: The invertible DNA segments of coliphages Mu and P1 are identical, Virol. **74** (1976), 242—248.

[37] DHAR, R., W. L. MCCLEMENTS, L. W. ENQUIST, and G. F. VAN DE WOUDE: Nucleotide sequences of integrated Moloney sarcoma provirus long terminal repeats and their host and viral junctions, Proc. Natl. Acad. Sci. USA **77** (1980), 3937—3941.

[38] DRUMMOND, M. H., M. P. GORDON, E. W. NESTER, and M.-D. CHILTON: Foreign DNA of bacterial plasmid origin is transcribed in crown gall tumors, Nature **269** (1977), 535—536.

[39] DUNSMUIR, P., W. J. BROREIN jr., M. A. SIMON and G. M. RUBIN: Insertion of the *Drosophila* transposable element copia generates a 5 base pair duplication, Cell **21** (1980), 575—579.

[40] DUTRILLAUX, B.: Chromosomal evolution in primates: Tentative phylogeny from *Microcebus murinus* (prosimian) to man, Hum. Genet. **48** (1979), 251—314.

[41] EGEL, R.: "Flip-flop" control and transposition of mating-type genes in fission yeast, in: "DNA Insertion Elements, Plasmids and Episomes" (eds. A. BUKHARI, J. SHAPIRO, and S. ADHYA). Cold Spring Harbor Laboratory, Cold Spring Harbor, N.Y. (1977), 447—455.

[42] ENGELS, W. R.: Hybrid dysgenesis in *Drosophila* and the stochastic loss hypothesis, Cold Spring Harbor Symp. Quant. Biol. **45** (1981), 561—565.

[43] ERREDE, B., T. S. CARDILLO, G. WEVER, and F. SHERMAN: Studies on transposable elements in yeast. I. ROAM mutations causing increased expression of yeast genes: Their activation by signals directed toward conjugation functions and their formation by insertion of Ty1 repetitive elements, Cold Spring Harbor Symp. Quant. Biol. **45** (1981), 593—602.

[44] FARABAUGH, P. and G. FINK: Insertion of the eukaryotic transposable element Ty1 creates a 5 bp duplication, Nature **286** (1980), 352—356.

[45] FAVERA, R. D., E. P. GELMANN, R. C. GALLO, and F. WONG-STAAL: A human **onc** gene homologous to the transforming gene (v-sis) of simian sarcoma virus, Nature **292** (1981), 31—35.

[46] FINK, G., P. FARABAUGH, G. ROEDER, and D. CHALEFF: Transposable elements (Ty) in yeast, Cold Spring Harbor Symp. Quant. Biol. **45** (1981), 575—580.

[47] FLAVELL, A. J. and D. ISH-HOROWICZ: Extrachromosomal circular copies of the eukaryotic transposable element copia in cultured *Drosophila* cells, Nature **292** (1981), 591—595.

[48] FLAVELL, R. B., M. O'DELL, and J. HUTCHINSON: Nucleotide sequence organization in plant chromosomes and evidence for sequence translocation during evolution, Cold Spring Harbor Symp. Quant. Biol. **45** (1981), 501—508.

[49] FOX, G. E., E. STACKEBRANDT, R. B. HESPELL, J. GIBSON, J. MANILOFF, T. A. DYER, R. S. WOLFE, W. E. BALCH, R. S. TANNER, L. J. MAGRUM, L. B. ZABLEN, R. BLAKEMORE, R. GUPTA, L. BONEN, B. J. LEWIS, D. A. STAHL, K. R. LUEHRSEN, K. N. CHEN, and C. R. WOESE: The phylogeny of prokaryotes, Science **209** (1980), 457—463.

[50] FRANCHINI, G., J. EVEN, C. J. SHERR, and F. WONG-STAAL: **Onc** sequences (v-**fes**) of Snyder-Theilen feline sarcoma virus are derived from noncontiguous regions of a cat cellular gene (c-**fes**), Nature **290** (1981), 154—157.

[51] FRANKLIN, N. C.: Illegitimate recombination in: "The Bacteriophage Lambda" (ed. A. D. HERSHEY). Cold Spring Harbor Laboratory, Cold Spring Harbor, N.Y. (1971), 175—194.

[52] GAFNER, J. and P. PHILIPPSEN: The yeast transposon Ty1 generates duplications of target DNA upon insertion, Nature **286** (1980), 414—418.

[53] GEHRING, W. J. and R. PARO: Isolation of a hybrid plasmid with homologous sequences to a transposing element of *Drosophila melanogaster*, Cell **19** (1980), 897—904.

[54] GEISSLER, E.: Die Gentheorie ist mit der materialistischen Dialektik vereinbar, Dtsch. Z. Phil. **26** (1978), 765—770.

[55] GEISSLER, E., S. SCHERNECK, H. WAEHLTE, W. ZIMMERMANN, L. LUEBBE, H. KRAUSE, M. THEILE, H.-J. HEROLD, M. RUDOLPH, F. WEICKMANN, and G. NISCH: Further studies on the relationship of SV40-like viruses to human tumors in: "Viruses in Naturally Occurring Cancers, Cold Spring Harbor Harbor Conf. on Cell Proliferation" **7** (1980), 343—356.

[56] GEORGIEV, G. P., Y. V. ILYIN, V. G. CHMELIAUSKAITE, A. P. RYSKOV, D. A. KRAMEROV, K. G. SKRYABIN, A. S. KRAYEV, E. M. LUKANIDIN, and M. S. GRIGORYAN: Mobile dispersed genetic elements and other middle repetitive DNA sequences in the genomes of *Drosophila* and mouse: Transcription and biological significance, Cold Spring Harbor Symp. Quant. Biol. **45** (1981), 641—654.

[57] GOLDSCHMIDT, R.: The Material Basis of Evolution, Yale University Press, New Haven, CO (1940).

[58] GOLDSCHMIDT, R.: Theoretical Genetics, University of California Press, Berkeley and Los Angeles (1955).

[59] GORCZYNSKI, R. M. and E. J. STEELE: Inheritance of acquired immunological tolerance to foreign histocompatibility antigens in mice, Proc. Natl. Acad. Sci. USA **77** (1980), 2871—2875.

[60] GORCZYNSKI, R. M. and E. J. STEELE: Simultaneous yet independent inheritance of somatically acquired tolerance to two distinct H-2 antigenic haplotype determinants in mice, Nature **289** (1981), 678—681.

[61] GORDON, J. W., G. A. SCANGOS, D. J. PLOTKIN, J. A. BARBOSA, and F. H. RUDDLE: Genetic transformation of mouse embryos by microinjection of purified DNA, Proc. Natl. Acad. Sci. **77** (1980), 1031—1044.

[62] GREEN, M. M.: Genetic instability in *Drosophila melanogaster*: De novo induction of putative insertion mutants, Proc. Natl. Acad. Sci. USA **74** (1977), 3490—3493.

[63] GREEN, M. M.: The genetic control of mutation in *Drosophila*, Stadler Symp. **10** (1978), 95—104.

[64] GREEN, M. M.: Transposable elements in *Drosophila* and other diptera, Ann. Rev. Genet. **14** (1980), 109—120.

[65] GREER, H., M. IGO, and F. de BRUIJN: Transposable elements involving the his4 region of yeast, Cold Spring Harbor Symp. Quant. Biol. **45** (1981), 567—574.

[66] GURLEY, W. B., J. D. KEMP, M. J. ALBERT, D. W. SUTTON, and J. CALLIS: Transcription of Ti plasmid derived sequences in three octopine-type crown gall tumor lines, Proc. Natl. Acad. Sci. USA **76** (1979), 2828—2832.

[67] GUYON, P., M.-D. CHILTON, A. PETIT, and J. TEMPÉ: Agropine in "null-type" crown gall tumors: Evidence for generality of the opine concept, Proc. Natl. Acad. Sci. USA **77** (1980), 2693—2697.

[68] HARBERS, K., D. JÄHNER, and R. JAENISCH: Microinjection of cloned retroviral genomes into mouse zygotes: integration and expression in the animal, Nature **293** (1981), 540—542.

[69] HAYNES, S. R., T. P. TOOMEY, L. LEINWAND, and W. R. JELINEK: The Chinese hamster **Alu**-equivalent sequence: a conserved, highly repetitious, interspersed deoxyribonucleic acid sequence in mammals has a structure suggestive of a transposable element, Mol. Cell. Biol. **1** (1981), 573—583.

[70] HAYWARD, W. S., B. G. NEEL and S. M. ASTRIN: Activation of a cellular **onc** gene by promoter insertion In ALV-induced lymphoid leukosis, Nature **290** (1981), 475—480.

[71] HICKS, J. B., J. N. STRATHERN, and I. HERSKOWITZ: The cassette model of mating-type interconversion, In: "DNA Insertion Elements, Plasmids, and Episomes", eds. A. BUKHARI, J. SHAPIRO and S. ADHYA, Cold Spring Harbor Laboratory, Cold Spring Harbor, N.Y., (1977), pp. 457—462.

[72] HIRT, B. and C. WIDMER-FAVRE: Expression and integration of viral DNA in animal cells, Advan. Biosci. **8** (1972), 167—174.

[73] HORST, J., H. STANBRO and C. R. MERRIL: On procaryotic gene expression in eucaryotic systems, Hum. Genet. **54** (1980), 289—302.

[74] HORST, J.: Möglichkeiten, Probleme und Grenzen einer Gentherapie, In: „Genetic engineering und der Mensch." VII. Kühlungsborner Kolloquium. Hrsg. E. GEISSLER und W. SCHELER, Berlin (1981), S. 55—63.

[75] HOWARD, J. C.: A tropical Volute shell and the Icarus syndrome, Nature **290** (1981), 441—442.

[76] HUGHES, S. H., F. PAYVAR, D. H. SPECTOR, R. T. SCHIMKE, H. L. ROBINSON, J. M. BISHOP, and H. E. VARMUS: Heterogeneity of genetic loci in chickens: Analysis of endogenous viral and non-viral genes by cleavage of DNA with restriction endonucleases, Cell **18** (1979), 347—359.

[77] HUGHES, S. H., K. TOYOSHIMA, J. M. BISHOP, and H. E. VARMUS: Organization of the endogenous proviruses of chickens: Implications for origin and expression, Virol. **108** (1981), 189—207.

[78] IINO, T. and K. KUTSUKAKE: Trans-acting genes of bacteriophage Pl and Mu mediate inversion of a specific DNA segment involved in flagellar phase variation in *Salmonella*, Cold Spring Harbor Symp. Quant. Biol. **45** (1981), 11—16.

[79] ILYIN, Yu. V., N. A. TCHURIKOV, E. V. ANANIEV, A. P. RYSKOV, G. N. YENIKOPOLOV, S. A. LIMBORSKA, N. E. MALEEVA, V. A. GVOZDEV, and G. P. GEORGIEV: Studies on the DNA fragments of mammals and *Drosophila* containing structural genes and adjacent sequences. Cold Spring Harbor Symp. Quant. Biol. **42** (1978), 959—969.

[80] ILYIN, V. G. CHMELIAUSKAITE and G. P. GEORGIEV: Double stranded sequences in RNA of *Drosophila melanogaster*. Relation to mobile dispersed genes, Nucleic Acids Res. **8** (1980), 3439—3457.

[81] ISING, G. and K. BLOCK: Derivation-dependent distribution of insertion sites for a *Drosophila* transposon, Cold Spring Harbor Symp. Quant. Biol. 45 (1981), 527—544

[81a] JACQ, B.: Sequence homologies between 5.8S rRNA and the 5'end of prokaryotic 23S rRNA: Evidences for a common evolutionary origin, Nucleic Acids Res. **9** (1981), 2913—2932

[82] JAENISCH, R.: Germ line integration and Mendelian transmission of the exogenous Moloney leukemia virus, Proc. Natl. Acad. Sci. USA **73** (1976), 1260—1264

[83] JAENISCH, R., P. NOBIS, D. JÄHNER and H. STUHLMANN: Molecular parameters and immunotherapy of leukemia in BALB/Mo mice, In: "Viruses in Naturally Occurring Cancers, Book B". eds. M. ESSEX, G. TODARO and H. ZUR HAUSEN, Cold Spring Harbor, N.Y. (1980), pp. 1021—1032.

[84] JENKINS, N. A., N. G. COPELAND, B. A. TAYLOR and B. K. LEE: Dilute (d) coat colour mutation of DBA/2J mice is associated with the site of integration of an ecotropic MuLV genome, Nature **293** (1981), 370—374.

[85] JENSEN, E. Ø., K. PALUDAN, J. J. HYLDIG-NIELSEN, P. JØRGENSEN and K. A. MARCKER: The structure of a chromosomal leghaemoglobin gene from soybean, Nature **291** (1981), 677—679.

[86] JONES, K. W.: Speculations on the functions of satellite DNA in evolution, Z. Morphol. Anthropol. **69** (1978), 143—171.

[87] JONES, M., R. A. BOSSELMAN, F. A. v. d. HOORN, A. BERNS, H. FAN, and I. M. VERMA: Identification and molecular cloning of Moloney mouse sarcoma virus-specific sequences from uninfected mouse cells, Proc. Natl. Acad. Sci. USA **77** (1980), 2651—2655.

[88] JU, G., L. BOONE and A. SKALKA: Isolation and characterization of recombinant DNA clones of avian retroviruses: Size heterogeneity and instability of the direct repeat, J. Virol. **33** (1980), 1026—1033.

[89] JU, G. and A. M. SKALKA: Nucleotide sequence analysis of the long terminal repeat (LTR) of avian retroviruses: Structural similarities with transposable elements, Cell **22** (1980), 379—386.

[90] KAINA, B. and R. RIEGER: Chromosomenmutationen, Karyotyp-Polymorphismus and Speciation, Biol Zbl. **98** (1979), 661—697.

[91] KESHET, E., Y. SHAUL, J. KAMINCHIK and H. AVIV: Heterogeneity of "virus-like" genes encoding retrovirus-associated 30s RNA and their organization within the mouse genome, Cell **20** (1980), 431—439.

[92] KESHET, E. and Y. SHAUL: Terminal direct repeats in a retrovirus-like repeated mouse gene family, Nature **289** (1981) 83—85.

[93] KLAPWIJK, P. M., J. van BREUKELEN, K. KOREVAAR, G. OOMS, and R. A. SCHILPEROOT: Transposition of Tn904 encoding streptomycin resistance into the octopine Ti plasmid of *Agrobacterium tumefaciens*, J. Bact. **141** (1980), 129—136.

[93a] KLEIN, G.: The role of gene dosage and genetic transpositions in carcinogenesis, Nature **294** (1981), 313—318.

[94] KOIDE, S. S. and T. MARÑO: Bacterial factor with gonadotropin activity, Recombinant DNA Technical Bulletin **4** (1981), 5—10.

[94a] KOPECKO, D. J.: Specialized genetic recombination systems in bacteria: Their involvement in gene expression and evolution, Progr. Mol Subcell. Biol. **7** (1980), 135—234.

[95] LANDY, A. and W. ROSS: Viral integration and excision: Structure of the lambda att sites, Science **197** (1977), 1147—1160.

[96] LAUER, J., C.-K. J. SHEN, and T. MANIATIS: The chromosomal arrangement of human α-like globin genes: Sequence homology and α-globin gene deletions, Cell (1980), 119—130.

[96a] LEDER, A., D. SWAN, F. RUDDLE, P. D'EUSTACHIO, and P. LEDER: Dispersion of α-like globin genes of the mouse to three different chromosomes, Nature **293** (1981), 196—200.

[97] LEDER, P., E. E. MAX, J. G. SEIDMAN, S.-P. KWAN, M. SCHARFF, M. NAU, and B. NORMAN: Recombination events that activate, diversify, and delete immunoglobulin genes, Cold Spring Harbor Symp. Quant. Biol. **45** (1981), 859—865.

[97a] Le ROITH, D., J. SHILOACH, J. ROTH, and M. A. LESNIAK: Insulin or a closely related molecule is native to *Escherichia coli*, J. Biol. Chem. **256** (1981), 6533—6536.

[98] LEWIN, R.: Evolutionary theory under fire, Science **210** (1980), 883—887.

[99] LEWIN, R.: Jumping genes help trace inherited diseases, Science **24** (1981a), 690—692.

[100] LEWIN, R.: LAMARCK will not lie down, Science **213** (1981b), 316—321.

[101] LEWIN, R.: Do jumping genes make evolutionary leaps? Science **213** (1981c), 634—636.

[102] LIM, J. K.: Site-specific intrachromosomal rearrangements in *Drosophila melanogaster*: Cytogenetic evidence for transposable elements, Cold Spring Harbor Symp. Quant. Biol. **45** (1981), 553—560.

[103] LOVINGER, G. G. and G. SCHOCHETMAN: 5'-terminal nucleotide sequences of the Rauscher leukemia virus and gibbon ape leukemia virus genomes exhibit a high degree of correspondence, J. Virol. **32** (1979), 803—811.

[104] LOVINGER, G. G., G. MARK, G. J. TODARO, and G. SCHOCHETMAN: 5'-terminal nucleotide noncoding sequences of retroviruses: Relatedness of two Old World primate type C viruses and avian spleen necrosis virus, J. Virol. **39** (1981), 238—245.

[105] LUKASH, L. L., T. I. BUZHIEVSKAYA, N. B. VARSHAVER, and N. I. SHAPIRO: Oncogenic adenovirus as mutagen for Chinese hamster cells in vitro, Somatic Cell Genet. **7** (1981), 133—146.

[106] MAJORS, J. E., R. SWANSTROM, W. J. DE LORBE, G. S. PAYNE, S. H. HUGHES, S. ORTIZ, N. QUINTRELL, J. M. BISHOP, and H. E. VARMUS: DNA intermediates in the replication of retroviruses are structurally (and perhaps functionally) related to transposable elements, Cold Spring Harbor Symp. Quant. Biol. **45** (1981), 731—738.

[107] MARENGO, C., M. MBIKAY, J. WEBER, and J.-P. THIRION: Adenovirus-induced mutations at the hypoxanthine phosphoribosyltransferase locus of Chinese hamster cells, J. Virol. **38** (1981), 184—190.

[108] MARSHAK, M., I. VARSHAVER, and N. B. SHAPIRO: Induction of gene mutations and chromosomal aberrations by Simian virus 40 in cultured mammalian cells, Mutat. Res. **30** (1980), 383—395.

[109] MARTIN, jr., J. P. and I. FRIDOVICH: Evidence for a natural gene transfer from the ponyfish to its bioluminescent bacterial symbiont *Photobacter leiognathi*, J. Biol. Chem. **256** (1981), 6080—6089.

[109a] MARX, J. L.: More progress on gene transfer, Science **213** (1981), 996—997.

[110] MARX, J. L.: Genes that control development, Science **213** (1981), 1485—1488.

[111] MATTHEWS, B. W., M. G. GRÜTTNER, W. F. ANDERSON, and S. J. REMINGTON: Common precursor of lysozymes of hen egg-white and bacteriophage T4, Nature **290** (1981a), 334—335.

[112] MATTHEWS, B. W., S. J. REMINGTON, M. G. GRÜTTER, and W. F. ANDERSON: Relation between hen egg-white lysozyme and bacteriophage T4 lysozyme: Evolutionary implications, J. Mol. Biol. **147** (1981b), 545—558.

[113] MCCLEMENTS, W. L., L. W. ENQUIST, M. OSKARSSON, M. SULLIVAN, and G. F. VANDE WOUDE: Frequent site specific deletion of coliphage λ murine sarcoma virus recombinants and its use in the identification of a retrovirus integration site, J. Virol. **35** (1980), 488—497.

[114] MCCLEMENTS, W. L., R. DHAR, D. G. BLAIR, L. ENQUIST, M. OSKARSSON, and G. F. VANDE WOUDE: The long terminal repeat of Moloney sarcoma provirus, Cold Spring Harbor Symp. Quant. Biol. **45** (1981), 699—705.

[115] MCCLINTOCK, B.: Chromosome organization and genic expression, Cold Spring Harbor Symp. Quant. Biol. **16** (1951), 13—47.

[116] MCCLINTOCK, B.: Controlling elements and the gene, Cold Spring Harbor Symp. Quant. Biol. **21** (1956), 197—216.

[117] MCCLINTOCK, B.: Mechanisms that rapidly reorganize the genome, Stadler Symp. **10** (1978), 25—48.

[117a] MCKINNON, R. D., S. BACCHETTI, and F. L. GRAHAM: Tn5 mutagenesis of the transforming genes of human adenovirus type 5, Gene **19** (1982), 33—42.

[118] MCPHERSON, J. C., E. W. NESTER, and M. P. GORDON: Proteins encoded by *Agrobacterium tumefaciens* Ti plasmid DNA (T-DNA) in crown gall tumors, Proc. Natl. Acad. Sci. USA **77** (1980), 2666—2670.

[119] MERLO, D. J., R. C. NUTTER, A. L. MONTOYA, D. J. GARFINKEL, M. H. Drummond, M.-D. CHILTON, M. P. GORDON, and E. W. NESTER: The boundaries and copy numbers of Ti plasmid T-DNA vary in crown gall tumors, Mol. Gen. Genet. **177** (1980), 637—645.

[120] MERRIL, C. R., M. R. GEIER, and J. C. PETRICIANNI: Bacterial virus gene expression in human cells, Nature **233** (1971), 398—400.

[121] MINTZ, B.: Mutant mice from mutagenized teratocarcinoma cells, In: "Differentiation and Development". Miami Winter Symposia, Vol. 15. eds. F. AHMAD, J. SCHULTZ, T. R. RUSSEL, and R. WERNER. Academic Press, New York, San Francisco, London (1978), S. 441—452.

[122] MIRZABEKOV, A. D.: Mobile genetic elements: new Soviet studies, Nature **291** (1981), 12.

[123] MITCHELL, A. R., J. R. GOSDEN, and O. A. Ryder: Satellite DNA relationships in man and the primates, Nucleic Acids Res. **9** (1981), 3235—3249.

[124] NEEL, B. G., W. S. HAYWARD, H. L. ROBINSON, J. FANG, and S. M. ASTRIN: Avian leukosis virus-induced tumors have common proviral integration sites and synthesize discrete new RNAs: oncogenesis by promoter insertion, Cell **23** (1981), 323—334.

[125] NEIMAN, P., K. BEEMON, and J. A. LUCE: Independent recombination between avian leukosis virus terminal sequences and host DNA in virus-induced proliferative disease, Proc. Natl. Acad. Sci. USA **78** (1981), 1896—1900

[126] NEVERS, P. and H. SAEDLER: Transposable genetic elements as agents of gene instability and chromosomal rearrangements, Nature **268** (1977), 109—115.

[127] OOMS, G., P. M. KLAPWIJK, J. A. POULIS, and R. A. SCHILPEROORT: Characterization of Tn904 insertions in octopine Ti plasmid mutants of *Agrobacterium tumefaciens*, J. Bacteriol. **144** (1980), 82—91.

[128] OOMS, G., P. J. J. HOOYKAS, G. MOOLENAAR, and R. A. SCHILPEROORT: Crown gall plant tumors of abnormal morphology, induced by *Agrobacterium tumefaciens* carrying mutated octopine Ti plasmids; analysis of T-DNA functions, Gene **14** (1981), 33—50.

[129] OROSZLAN, S., M. BARBACID, T. D. COPELAND, S. A. AARONSON, and R. V. GILDEN: Chemical and immunological characterization of the major structural protein (p28) of MMC-1, a rhesus monkey endogenous type C virus: Homology with the major structural protein of avian reticuloendotheliosis virus, J. Virol. **39** (1981), 845—854.

[130] OSKARSSON, M. W. L. MCCLEMENTS, D. G. BLAIR, J. V. MAIZEL, and G. F. VANDE WOUDE: Properties of a normal mouse cell DNA sequence (sarc) homologous to the Src sequence of Moloney sarcoma virus, Science **207** (1980), 1222—1224.

[131] PELLICER, A. J., E. F. WAGNER, A. H. KAREN, M. J. DEWEY, A. J. REUSER, S. SILVERSTEIN, R. AXEL, and B. MINTZ: Introduction of a viral thymidine kinase gene and the human β-globin gene into developmentally multipotential mouse teratocarcinoma cells, Proc. Natl. Acad. Sci. USA **77** (1980), 2078—2102.

[132] PERUCHO, M. and M. WIGLER: Linkage and expression of foreign DNA in cultured animal cells, Cold Spring Harbor Symp. Quant. Biol. **45** (1981), 829—838.

[133] PETERSON, P. A.: Instability among the components of a regulatory element transposon in maize, Cold Spring Harbor Symp. Quant. Biol. **45** (1981), 447—455.

[134] PIRUZIAN, Ė., V. ANDRIANOV, M. MOGUTOV, E. KRIVTSOVA, V. YUZEEVA, A. VETOSHKIN, and N. KOBETS: Specificity of bacteriophage Mu integration into DNAs of different origins, Cold Spring Harbor Symp. Quant. Biol. **45** (1981), 365—369.

[135] POTTER, S. S., W. J. BROREIN jr., P. DUNSMUIR, and G. M. RUBIN: Transposition of elements of the 412, copia and 297 dispersed repeated gene families in *Drosophila*, Cell **17** (1979), 415—427.

[136] POTTER, S., M. TRUETT, M. PHILLIPS, and A. MAHER: Eucaryotic transposable genetic elements with inverted terminal repeats, Cell **20** (1980), 639—647.

[137] RABBITTS, T. H., D. L. BENTLEY, W. DUNNICK, A. FORSTER, G. E. A. R. MATTHYSSENS and C. MILSTEIN: Immunoglobulin genes undergo multiple sequence rearrangements during differentiation, Cold Spring Harbor Symp. Quant. Biol. **45** (1981), 867—878.

[138] RAPP, U. R. and G. TODARO: Generation of oncogenic mouse type C viruses: In vitro selection of carcinoma-inducing variants, Proc. Natl. Acad. Sci. USA **77** (1980), 624—628.

[139] RASMUSON, B., B. M. WESTERBERG, Å. RASMUSON, V. A. GVOZDEY, E. S. BELYAEVA, and Y. V. ILYIN: Transpositions, mutable genes, and the dispersed gene family Dm225 in *Drosophila melanogaster*, Cold Spring Harbor Symp. Quant. Biol. **45** (1981), 545—551.

[140] RICE, N. R., T. I. BONNER, and R. Y. GILDEN: Nucleic acid homology between avian and mammalian type C viruses: Relatedness of reticuloendotheliosis virus cDNA to cloned proviral DNA of the endogenous colobus virus CPC-1, Virol. **114** (1981), 286—290.

[141] RIMPAU, J., D. SMITH, and R. FLAVELL: Sequence organisation analysis of the wheat and rye genomes by interspecies DNA/DNA hybridisation, J. Biol. **123** (1978), 327—359.

[142] ROBERTSON, M.: Gene families, hopeful monsters, and the selfish genetics of DNA, Nature **293** (1981), 333—334.

[143] ROEDER, G. S. and G. FINK: DNA rearrangements associated with a transposable element in yeast, Cell **21** (1980), 239—249.

[144] ROTWEIN, P., R. CHYN, J. CHIRGWIN, B. CORDELL, H. M. GOODMAN, and M. A. PERMUTT: Polymorphism in the 5'-flanking region of the human insulin gene and its possible relation to type 2 diabetes, Science **213** (1981), 1117—1120.

[145] ROUSSEL, M., S. SAULE, C. LAGROU, C. ROMMENS, H. BEUG, T. GRAF, and D. STEHELIN: Three new types of viral oncogene of cellular origin specific for haematopoietic cell transformation, Nature **281** (1979), 452—456.

[146] RUBIN, G. M., W. J. BROREIN jr., P. DUNSMUIR, A. J. FLAVELL, R. LEVIS, E. STROBEL, J. J. TOOLE, and E. YOUNG: copia-like transposable elements in the *Drosophila* genome, Cold Spring Harbor Symp. Quant. Biol. **45** (1981), 619—628.

[147] SAGER, R., A. ANISOWICZ, and N. HOWELL: Genomic rearrangements and tumor-forming potential in an SV40-transformed mouse cell line and its hybrid and cybrid progeny, Cold Spring Harbor Symp. Quant. Biol. **45** (1981a), 747—754.

[148] SAGER, R., A. ANISOWICZ, and N. HOWELL: Genomic rearrangements in a mouse cell line containing integrated SV40 DNA, Cell **23** (1981b), 41—50.

[149] SALAMINI, F.: Controlling elements at the opaque-2 locus of maize: Their involvement in the origin of spontaneous mutation, Cold Spring Harbor Symp. Quant. Biol. **45** (1981), 467—476.

[150] SAPIENZA, C. and W. F. DOOLITLE: Genes are things you have whether you want them or not, Cold Spring Harbor Symp. Quant. Biol. **45** (1981), 177—182.

[151] SASTRY, G. R. K., K. M. ASLAM, and V. JEFFRIES: The role of controlling elements in the instability of flower color in *Antirrhinum majus* and *Impatiens balsamina*, Cold Spring Harbor Symp. Quant. Biol. **45** (1981), 477—486.

[152] SCANGOS, G. and F. H. RUDDLE: Mechanisms and applications of DNA-mediated gene transfer in mammalian cells — a review, Gene **14** (1981), 1—10.

[153] SCOLNICK, E. M., W. C. VAAS, R. S. HAWK, and P. H. DUESBERG: Defective retrovirus-like 30S RNA species of rat and mouse cells are infectious if packaged by type C helper virus, J. Virol. **29** (1979), 964 to 972.

[154] SHERMAN, F., T. S. CARDILLO, B. ERREDE, L. FRIEDMAN, G. MCKNIGHT, and J. I. STILES: Yeast mutants overproducing iso-cytochromes c, In: "Molecular Genetics in Yeast." eds. D. von WETTSTEIN, J. FRIIS, M. KIELLAND-BRANDT and A. STENDERUP. Copenhagen (1981), pp. 154—174.

[155] SHIMOTOHNO, K. and H. M. TEMIN: Evolution of retroviruses from cellular movable genetic elements, Cold Spring Harbor Symp. Quant. Biol. **45** (1981), 719—730.

[156] SHOEMAKER, C., S. GOFF, E. GILBOA, M. PASKIND, S. W. MITRA, and D. BALTIMORE: Structure of a cloned circular Moloney murine leukemia virus DNA molecule containing an inverted segment: Implication for retrovirus integration, Proc. Natl. Acad. Sci. USA **77** (1980), 3932—3936

[157] SHOEMAKER, C., S. GOFF, E. GILBOA, M. PASKIND, S. W. MITRA, and D. BALTIMORE: Structure of cloned retroviral circular DNAs: Implications for virus integration, Cold Spring Harbor Symp. Quant. Biol. **45** (1981), 711—717.

[158] SKALKA, A., G. JU, F. HISHINUMA, P. J. DEBONA, and S. ASTRIN: Structural analogies among avian retroviral DNAs and transposable elements, Cold Spring Harbor Symp. Quant. Biol. **45** (1981), 739—746.

[159] SMITH, G. R.: DNA supercoiling: Another level for regulating gene expression, Cell **24** (1981), 599—600.

[160] STARLINGER, P.: IS elements and transposons, Plasmid **3** (1980), 241—259.

[161] STEBBINS, G. L. and F. J. AYALA: Is a new evolutionary synthesis necessary?, Science **213** (1981), 967—971.

[162] STEELE, E.: Somatic Selection and Adaptive Evolution, Williams-Wallace International and Croom Helm, Toronto and London (1979).

[163] Stiles, J. I., L. R. Friedman, C. Helms, S. Consaul, and F. Sherman: Transposition of the gene cluster CYC1-OSM1-RAD7 in yeast, J. Mol. Biol. **148** (1981), 331—346.

[164] Strathern, J. N., E. Spatola, C. McGill and J. B. Hicks: Structure and organization of transposable mating type cassettes in *Saccharomyces* yeasts, Proc. Natl. Acad. Sci. USA **77** (1980), 2839—2843.

[165] Strobel, E., P. Dunsmuir, and G. M. Rubin: Polymorphisms in the chromosomal locations of elements of the 412, copia and 297 dispersed repeated gene families in *Drosophila*, Cell **17** (1979), 429—439.

[166] Swanstrom, R., W. J. DeLorbe, J. M. Bishop, and H. E. Varmus: Nucleotide sequence of cloned unintegrated avian sarcoma virus DNA: Viral DNA contains direct and inverted repeats similar to those in transposable elements, Proc. Natl. Acad. Sci. USA **78** (1981), 124—128.

[167] Taylor, A. L.: Bacteriophage-induced mutation in *Escherichia coli*, Proc. Natl. Acad. Sci. USA **50** (1963), 1043—1051.

[168] Tchurikov, N. A., E. S. Zelentsova, and G. P. Georgiev: Clusters containing different mobile dispersed genes in the genome of *Drosophila melanogaster*, Nucleic Acids Res. **8** (1980), 1243—1258.

[169] Tchurikov, N. A., Y. V. Ilyin, K. G. Skryabin, E. V. Ananiev, A. A. Bayev jr., A. S. Krayev, E. S. Zelentsova, V. V. Kulguskin, N. V. Lyubomirskaya, and G. P. Georgiev: General properties of mobile dispersed genetic elements in *Drosophila melanogaster*, Cold Spring Harbor Symp. Quant. Biol. **45** (1981), 655—665.

[170] Temin, H. M.: Origin of retroviruses from cellular movable genetic elements, Cell **21** (1980), 599—600.

[171] Theile, M., S. Scherneck, and E. Geissler: Mutagenesis by Simian virus 40. I. Detection of mutations in Chinese hamster cell lines using different resistance markers, Mutat. Res. **37** (1976), 111 to 124.

[172] Theile, M., M. Strauss, L. Luebbe, S. Scherneck, H. Krause and E. Geissler: SV40-induced somatic mutations: Possible relevance to viral transformation, Cold Spring Harbor Symp. Quant. Biol. **44** (1980), 377—382.

[173] Thomashow, M. F., R. Nutter, K. Postle, M.-D. Chilton, F. R. Blattner, A. Powell, M. P. Gordon, and E. W. Nester: Recombination between higher plant DNA and the Ti plasmid of *Agrobacterium tumefaciens*, Proc. Natl. Acad. Sci. USA **77** (1980), 6448—6452.

[174] Todaro, G. J., R. Callahan, U. R. Rapp, and J. E. DeLarco: Genetic transmission of retroviral genes and cellular oncogenes, Proc. R. Soc. Lond. B **210** (1980), 367—385.

[175] Tonegawa, S., H. Sakano, R. Maki, A. Traunecker, G. Heinrich, W. Roeder, and Y. Kurosawa: Somatic reorganization of immunoglobulin genes during lymphocyte differentiation, Cold Spring Harbor Symp. Quant. Biol. **45** (1981), 839—858.

[176] Tooze, J. (ed.): DNA Tumor Viruses, Cold Spring Harbor Laboratory, Cold Spring Harbor, N. Y. (1981).

[177] Truett, M. A., R. S. Jones, and S. S. Potter: Unusual structure of the FB family of transposable elements in *Drosophila*, Cell **24** (1981), 753—763.

[178] van Zaane, D. and H. P. J. Bloemers: The genome of the mammalian sarcoma viruses, Biochim. Biophys. Acta **516** (1978), 249—268.

[179] Varmus, H. E., N. Quintrell, and S. Ortiz: Retroviruses as mutagens: Insertion and excision of a nontransforming provirus alter expression of a resident transforming provirus, Cell **25** (1981), 23—36.

[180] Wagner, E. F., T. A. Stewart, and B. Mintz: The human beta-globin gene and a functional viral thymidine kinase gene in developing mice, Proc. Natl. Acad. Sci. USA **78** (1981), 5016—5020.

[181] Wagner, T. E., P. C. Hoppe, J. D. Jollick, D. R. Scholl, R. L. Hodinka, and J. B. Gault: Microinjection of a rabbit β-globin gene into zygotes and its subsequent expression in adult mice and their offspring, Proc. Natl. Acad. Sci. USA **78** (1981), 6376—6380.

[182] Wang, L.-H., P. Snyder, T. Hanafusa, and H. Hanafusa: Evidence for the common origin of viral and cellular sequences involved in sarcomagenic transformation, J. Virol. **35** (1980), 52—64.

[183] Weiss, R. A., N. M. Teich, H. Varmus, and J. M. Coffin (eds.): RNA Tumor Viruses, Cold Spring Harbor Laboratory, Cold Spring Harbor, N.Y. (1982).

[184] WENSINK, P. C., S. TABATA, and C. PACHL: The clustered and scrambled arrangement of moderately repetitive elements in *Drosophila* DNA, Cell **18** (1979), 1231—1246.

[185] WILLIAMSON, B.: The cloning revolution meets human genetics, Nature **293** (1981), 10—11.

[186] WILLIAMSON, V. M., E. T. YOUNG, and M. CIRIACY: Transposable elements associated with constitutive expression of yeast alcohol dehydrogenase II, Cell **23** (1981), 605—614.

[187] WYMAN, A. R. and R. WHITE: A highly polymorphic locus in human DNA, Proc. Natl. Acad. Sci. USA **77** (1980), 6754—6758.

[188] YAMAMOTO, K. R., V. L. CHANDLER, S. R. ROSS, D. S. UCKER, J. C. RING, and S. C. FEINSTEIN: Integration and activity of mammary tumor virus genes: Regulation by hormone receptors and chromosomal position, Cold Spring Harbor Symp. Quant. Biol. **45** (1981), 687—697.

[189] YOUNG, M. W. and H. E. SCHWARZT: Nomadic gene families in *Drosophila*, Cold Spring Harbor Symp. Quant. Biol. **45** (1981), 629—640.

[190] ZAENEN, I. N., van LAREBEKE, H. TEUCHY, M. van MONTAGU, and J. SCHELL: Supercoiled circular DNA in crown gall inducing *Agrobacterium* strains, J. Mol. Biol. **86** (1974), 109—127.

[191] ZHDANOV, V. M. and T. I. TIKCHONENKO: Viruses as a factor of evolution: Exchange of genetic information in the biosphere, Adv. Virus Res. **19** (1974), 361—394.

[192] ZIEG, J., M. SILVERMAN, M. HILMEN, and M. SIMON: Recombinational switch for gene expression, Science **196** (1977), 170—172.

[193] ZIEG, J. and M. SIMON: Analysis of the nucleotide sequence of an invertible controlling element, Proc. Natl. Acad. Sci. USA **77** (1980), 4196—4200.

Forschungszentrum für Molekularbiologie und Medizin der Akademie der Wissenschaften der DDR, Berlin-Buch, DDR

Inheritance of Immunological Tolerance: Report on a Controversial Hypothesis

G. Pasternak

"Lamarck will not lie down" was the headline of the "Research News" in a July issue of the journal Science [1]. What has happened? Certain immunological experiments on mice have been claimed as evidence of Lamarckian inheritance.

The basic idea of the so-called somatic selection hypothesis as created by the young Australian scientist, Dr. Edward Steele, is that Darwinism may account only to some extent for the diversity and abundance of cells or organisms. There remains a suspicion, says Steele, that it provides no satisfactory explanation for our intuitive belief that there appears to be an element of "directional" progress in the complexity and sophistication of adapted living things.

Steele's aim is not to replace Darwinism by his form of Lamarckism, rather he seeks "to enrich evolutionary theory" by adding a directional element to it, explains Roger Lewin in the Science article.

According to the hypothesis certain mutations occurring in the somatic cells will proliferate and dominate the organ. Therefore, there is a great likelihood that retroviruses will pick up some of the new genetic material and transfer it to the germ line cells. The genetic information is then inserted into the genome and will be expressed in the descendants. Thus, retroviruses are important factors which speed the evolution of their hosts.

Steele and his colleague Dr. Reg Gorczynski at the Ontario Cancer Institute developed an experimental device to proof this hypothesis. The experiments included the induction of immunological tolerance to foreign histocompatibility antigens in one strain of mice and its inheritance to the first and second generation progeny [2].

Immunological tolerance to foreign histocompatibility antigens in mice has been shown first about 30 years ago by Medawar, Brent and Billingham in Great Britain [3].

The characteristic feature of this phenomenon is that mice given lymphoid cells from a foreign strain neonatally accept and tolerate skin grafts from this particular strain later in life. Normally, the immune system of the organism recognizes skin grafts bearing alien histocompatibility antigens as foreign. In the process of an effective immune response to these antigens the graft is rejected. Graft rejection would be the result also in clinical organ transplantation if the patient were not subjected to immunosuppressive therapy. The reaction is immunological in nature. Antibodies and particularly immune lymphoid cells develop to the foreign histocompatibility antigens of the graft. Immunization is detectable in vivo by enhanced graft rejection and in vitro by cellular cytotoxicity against donor cells. Gorczynski and Steele employed a cytotoxicity test by which spleen cells to be investigated were stimulated in culture with X-ray killed foreign cells. Stimulation to cytotoxicity is detectable only if the spleen cells are sensitized to the corresponding histocompatibility antigen. Spleen cells from tolerant animals do not respond to the antigenic stimulus.

In the crucial experiment of a second paper [4] B10 male mice were rendered tolerant simultaneously to both H-2^k and H-2^d haplotypes by neonatal inoculation of 100×10^6 lymphoid cells and by repeated injections every 2 weeks. Tolerant male mice were chosen for mating with normal B10 females. This experimental system was used because transmission of the tolerance state through the male is assumed to be more probable. Male germ cells in contrast to female ones are formed continuously throughout life. Animals of the first and second generation progeny were tested for tolerance to H-2^k and H-2^d determinants by cytotoxicity. Cytotoxicity was diminished in a high proportion of animals. The results appeared to demonstrate that somatically acquired immunological tolerance to foreign histocompatibility antigens induced in male inbred mice can be transmitted at a high frequency to the offspring without exposure of these animals to the tolerizing antigens. Individual male mice made tolerant to two different H-2 haplotypes seem to transmit the tolerance state to each haplotype independently or even simultaneously.

There was much critizism on the hypothesis itself and a number of reports disproved the results of GORCZYNSKI and STEELE. TEMIN, who himself argued that a product of protovirus evolution might have a chance to become integrated into the germ line, doubted the frequency with which transmission might occur between somatic and germ cells [1]. Furthermore, a major problem is that there is apparently no particular specificity in where the viruses insert their passenger DNA into the genome.

Meanwhile, BRENT et al. [5] were unable to confirm the inheritance of immunological tolerance when they used the model of GORCZYNSKI and STEELE. They had included skin grafting in their experiments which did not show any tolerance in the progeny of tolerant fathers.

Manipulation of experimental data, incorrect statistical analysis, insufficient scientific knowledge, and suppression of revolutionary ideas are the arguments in a contest that is no longer scientifically in this way. I enjoyed the final sentence in the Science article by Roger LEWIN when he says: "If the results cannot be repeated independently, then the phenomenon they represent must be said to be not generalizable and therefore not very interesting."

References

[1] LEWIN, R.: LAMARCK will not lie down. Science **213** (1981), 316—321.

[2] GORCZYNSKI, R. M. and E. J. STEELE: Inheritance of acquired immunological tolerance to foreign histocompatibility antigens in mice. Proc. Nat. Acad. Sci. **77** (1980), 2871—2875.

[3] BILLINGHAM, R. E., L. BRENT, and P. B. MEDAWAR: "Actively acquired tolerance" of foreign cells. Nature **172** (1953), 603—606.

[4] GORCZYNSKI, R. M., and E. J. STEELE: Simultaneous yet independent inheritance of somatically acquired tolerance to two distinct H-2 antigenic hypotype determinants in mice. Nature **289** (1981), 678—681.

[5] BRENT, L., L. S. RAYFIELD, P. CHANDLER, W. FIERZ, P. B. MEDAWAR, and E. SIMPSON: Supposed LAMARCKian inheritance of immunological tolerance. Nature **290** (1981), 508—512.

Discussion

H. A. ROSENTHAL: I would like to mention that the so-called acute RNA tumor viruses can also integrate at sites adjacent to cellular onc genes and thereby activate the latter. In this respect I see no true difference between acute and chronic viruses. There is another point which makes me wonder whether integration of RNA tumour viruses into host genome always influences the transcription of adjacent cellular genes. What I am thinking about is the experiment by JAENISCH who got the Moloney virus integrated at different sites of the murine genome early in embryogenesis and succeeded in breeding 12 or so different mouse clones that differed in the time at which the virus was expressed during ontogenesis. In this particular case it is not the virus which influences the expression of the host gene but the host gene which, when active during differentiation, renders the virus active.

GEISSLER: I fully agree with Dr. ROSENTHAL, but the difference between acute and chronic retrovirusses was not my point. I mentioned the latter only in the context of activating genes by integration of movable elements. I do agree also with ROSENTHAL's reference to the interesting experiments done by JAENISCH. In fact I have mentioned this work in the written version of my paper, the more so as JAENISCH has provisional evidence that the integration of the MuLV provirus might have been mutagenic in at least one of the strains he obtained.

F. VOGEL: Some human germ cell mutations might be caused by ME., especially those with a small paternal age effect (tumor diseases: bilateral retinoblastoma; neurofibromatosis; tuberous sclerosis).

ANTONOV: Movable elements comprise a large portion of eucaryotic genomes, they are multifunctional. Then why are species so evolutionary stable?

GEISSLER: You have raised a very important question. Of course, I can only speculate. In my mind MEs either are not "selfish" in the respects discussed by CRICK, ORGEL and SAPIENZA or only species that developed mechanisms to protect their essential genes from integration of MEs, or to cut them out as soon as they become integrated, survived during phylogeny. Besides most ME induced transpositions in the germ line might be lethal to the organism affected. It could be interesting to study, however, whether rearrangements caused by MEs can be found in somatic cells with a measurable frequency.
At any rate we must consider that MEs as well as ME caused mutations can be found in at least some higher organisms.

F. JUNG: What is the mutation rate by these transposable elements compared with the knwon mutation rates for bacteria or even human beings?

GEISSLER: I have no exact figures here but I can refer you to the experiments done by my coworker Michael THEILE and his group on the mutagenic activities of oncogenic DNA viruses. Depending on the marker studied SV40 enhances the frequency of mutants five, ten or even one hundredfold. Of course, as I have already pointed out, SV40 is no ME.

PFEIFFER: Irrespective of whether STEELE's and GORCZYNSKI's experiments are valid or not, I would like to make three remarks. Firstly, I remind you of DARWIN's opinion from experience that more or less intensive training of an individual organ might influence this organ in the offspring of the individual. Expressed in modern terms, this means that there could be channels of information leading from different parts of the body to the germ cells. Likewise, DARWIN was convinced that a first father could influence the offspring of a female generated by following fathers. This would mean the existence of analogous channels from the developing germ in the womb of the female. Indeed, as I pointed out at the annual meeting of the Leopoldina Academy in 1967 (Nova Acta Leopoldina **33**, 266—267, 1968), there are certain reasons to presume the existence of such information to the germ cells, and I then suggested that this hypothesis should be tested seriously. One of the reasons I mentioned was the opinion of V. BERTALANFFY, expressed in a paper read at the same meeting, who judged random mutations not to be sufficient to explain the relatively quick development of higher organisms. Another more general reason is the following: Any form of information to the germ cells would be a kind of feedback. Now we know the role numerous feedback mechanisms play in individual life, enabling the organism quickly to adapt to changing conditions. Therefore, from a general viewpoint, the absence feedback in the life of a species should not be regarded as probable.

Secondly: Adopting DARWIN's original concept, one might deduce from the very nature of the assumed effect that the transition of acquired characters to the following generations might be only temporary, as, for instance, the hereditary transmission of various so-called modifications is known to be.

Thirdly: STEELE's hypothesis of the role of the RNA as a mobile messenger between the cells of the soma and the germ cells seems not to be the only one possible. Perhaps chemical influences upon the RNA in the germ by humoral paths should also be regarded as not being impossible, the more so as chemical influences upon the RNA may be presumed also in other cases. For instance, the wellknown putting out of action of specific parts of the RNA in the various parts of the body may be caused by the continuously developing differentiation of chemisms in the various parts of the germ.

Ruse: 1. It is clear that Ted STEELE hopes that his neo-LAMARCKian mechanism will prove to have a very broad application going beyond the special effects he claims to have found and explained.

2. In fact "Lamarckism", that is to say, the inheritance of acquired characteristics, is only a minor part of LAMARCK's theory, and is also used by Charles DARWIN in the "Origin of Species."

3. Darwinism must threaten many people (e. g. Arthur KOESTLER, Karl POPPER, and Peter MEDAWAR). I forecast that although STEELE's work may fail (I suspect that it has already failed), it will not be long before another neo-LAMARCKian starts to proclaim this position. In other words, it will be felt that one can counter the blind materialism of Darwinism.

GEISSLER: I can't agree with Prof. PFEIFFER. Firstly, as already mentioned by Prof. PASTERNAK, there is no reproducible experiment known which violates the "Central Dogma". Even STEELE discusses his observations, which is obviously not reproducible, in terms of some type of transduction, e. g. within the scope of the "Central Dogma". Secondly it is of course not possible to prove the existence of some somatic induction by referring to DARWIN's belief in the inheritance of acquired characters. DARWIN and Friedrich ENGELS simply had to believe in this very hypothesis because they were neither aware of DNA and genes nor of the mechanisms of gene expression.

Of course I agree with Prof. RUSE that Lamarickism cannot be reduced to inheritance of acquired characters whereas it makes no sense and is even incorrect to denounce Darwinism as "blind materialism".

Institut für Theoretische Chemie und Strahlenchemie der Universität Wien, Währingerstraße 17, A-1090 Wien, Austria

Evolution of Polynucleotides — an Analytical Access to DARWIN's Theory from the View of Molecular Biology

P. SCHUSTER

Polynucleotides optimize their properties with respect to high rates of replication provided the environmental conditions keep the system apart from equilibrium. The two fundamental concepts in DARWIN's theory, variation and natural selection can be traced down to molecular origins in ensembles of primitive replicating entities.

1. Introduction

About 130 years ago Charles DARWIN was working on his fundamental theory of biological evolution. He had in mind highly developed organisms, plants and animals which he had studied for a lifetime. He built his universal concept on two basic principles the consequences of which he saw in nature. These principles are 1. the variability of multiplying organisms and 2. natural selection as a permanently operating optimization process. The quantity to be optimized is reproductive success under conditions dictated by the environment and the internal constraints of the organism.

The variability of nature has never been seriously called in question. It is so obvious to everybody who performs studies in the fields. With natural selection, however, DARWIN had some troubles. The concept of "survival of the fittest" degenerates to a mere tautology when the degree of fitness cannot be measured independently of the probability of survival. It is interesting to see how DARWIN discussed this problem. He directed the readers' attention towards selection through man and refered to the results achieved by animal breeders and farmers working in nursery gardens. In particular, he writes in the "Origin of species" [1]: ". . . As man can produce, and certainly has produced, a great result by his methodical and unconsious means of selection, what may not natural selection effect? Man can only act on external and visible characters: nature . . . cares nothing for appearances, except so far as they are useful to any being. She can act on every internal organ, on every shade of constitutional difference, on the whole machinery of life. Man selects only for his own good: nature only for that of the being which she tends . . .".

More than one hundred years later an enormous and impressive abundance of detailed information has become available. We know a lot on the mechanism of inheritance and gene distribution through the marvelous and enlightening works in population genetics and molecular biology. Nevertheless the discussion on natural selection is still going on, sometimes carried by exceedingly superficial arguments, sometimes conducted on more profound grounds [2]. Despite all the progress made, there seems to be still a need for deeper understanding of the mechanisms through which natural selection works. It is not unnecessary, therefore, to search for simple enough systems which allow to study how biological evolution works. In this contribution we shall consider a particularly simple example. The entire system will contain a piece of viral RNA and a specific RNA replicase together with the necessary low molecular weight materials. This simple system, nevertheless, will provide some insight into the molecular mechanisms of variation and selection.

2. Some Data on Simple RNA Bacteriophages

One of the simplest viruses known is the bacteriophage $Q\beta$. It belongs to a class of single stranded RNA phages with very similar properties. The phage MS2 belongs to this family and is a close relative of $Q\beta$. We shall mention data of both organisms since not all the information we need is available for a single virus. MS2 forms a virus particle consisting of one piece of plus strand RNA about 3500 bases long. In the native virus particle the RNA is condensed onto a protein molecule called A-protein. This complex is surrounded by a capsula built up from 200 coat protein molecules [3].

The nucleotide sequence of MS2 has been determined (figure 1). The genome consists of four genes which are read in different frames and which overlap partially [4, 5].

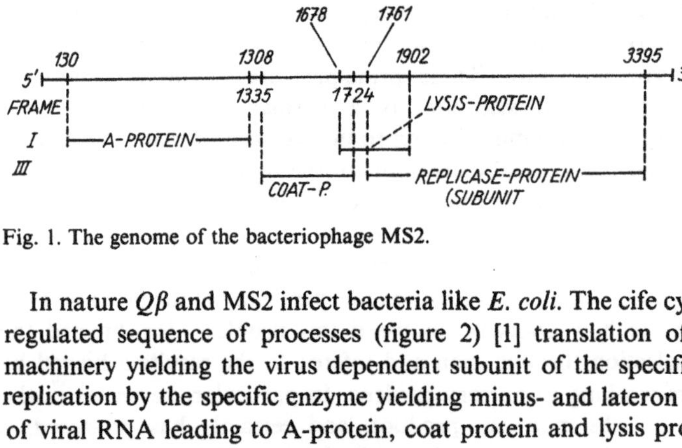

Fig. 1. The genome of the bacteriophage MS2.

In nature $Q\beta$ and MS2 infect bacteria like *E. coli*. The cife cycle of the virus consists of a regulated sequence of processes (figure 2) [1] translation of viral RNA by the host's machinery yielding the virus dependent subunit of the specific RNA replicase, [2] RNA replication by the specific enzyme yielding minus- and lateron plus-strands, [3] translation of viral RNA leading to A-protein, coat protein and lysis protein, [4] formation of virus particles and [5] lysis of the bacterial cell.

The basic properties of viral RNA which are used in the life cycle of the bacteriophage are the capabilities to act as template for replication as well as to be recognized as messenger RNA by the host translation system. In this contribution we shall be concerned mainly with replication because this process can be regarded as a simple version of Darwinian evolution. The mechanism of RNA replication is shown schematically in figure 3. The enzyme processes template and newly synthesized strand in such a way that the formation of (plus-minus) double strands is avoided. The experimentally determined rate of RNA chain elongation is highly variable [6]. The replicase stops replication temporarily at several "pause sites" in order to allow secondary structure, in particular hairpin formation in the product strand and/or reformation of hairpins in the template strand. The avoidance of double strand formation is extremly important for

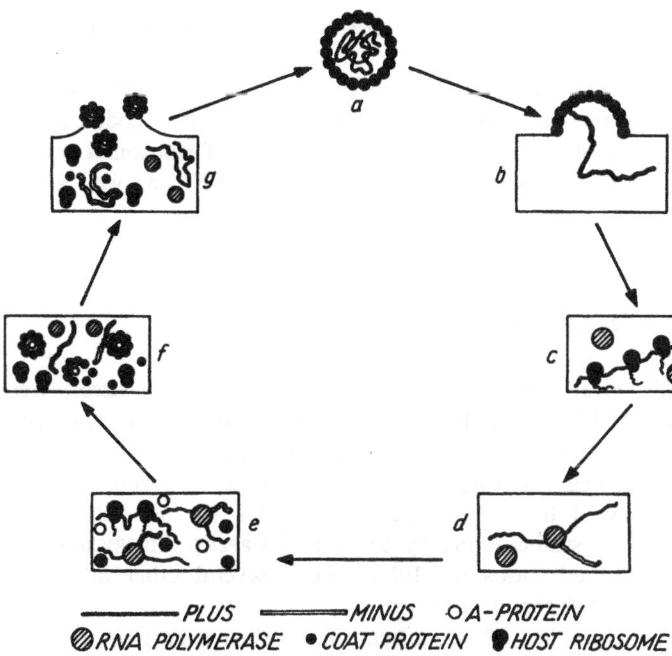

Fig. 2. The life cycle of simple bacteriophages (Qβ or MS2) in a bacterial cell (*E. coli*).

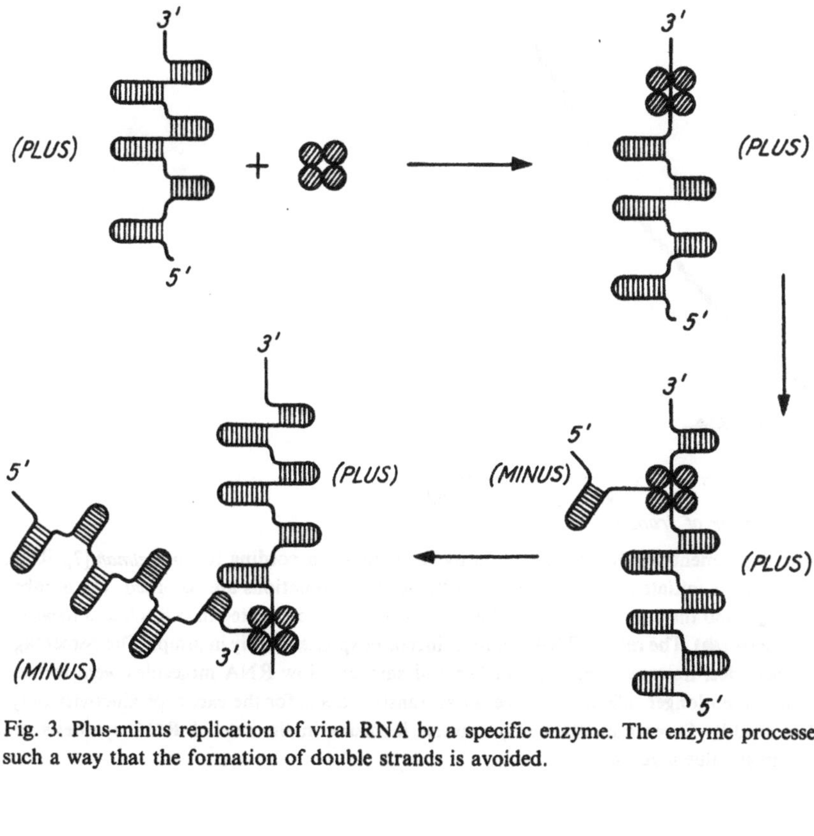

Fig. 3. Plus-minus replication of viral RNA by a specific enzyme. The enzyme processes the template in such a way that the formation of double strands is avoided.

efficient RNA synthesis. Double strands have to melt before they are recognized as templates and replicated by enzyme.

The replicating enzymes, e.g. $Q\beta$ replicase, are characterized well biochemically. $Q\beta$ replicase is a tetramer. One subunit is dependent on viral RNA, i.e. it is coded on the RNA of the bacteriophage. The other three subunits are factors of ribosomal protein synthesis of the host, namely the translation factors EFT_u and EFT_s as well as the ribosomal protein S_1.

3. Experimental Studies on RNA Replication

The first attempts to study viral RNA replication systematically "in vitro" were undertaken by Sol SPIEGELMAN and his group. In particular, they did the first serial transfer experiments on $Q\beta$-RNA replication by its specific replicase (figure 4) [7]. The most exciting result they got was a stepwise increase in the rate of replication accompanied by changes in the other properties of the RNA, infectiousness for *E. coli* cells or the molecular weight. Careful kinetic studies done by SUMPER, KÜPPERS, BIEBRICHER and LUCE in the laboratory of Manfred EIGEN [8—10] relevated several other interesting details on the mechanism of $Q\beta$-RNA replication:

[1] The specific enzyme recognizes an RNA molecule only when it carriers a certain recognition site which consists of at least two CCC-tripletts, [2] template-free enzyme

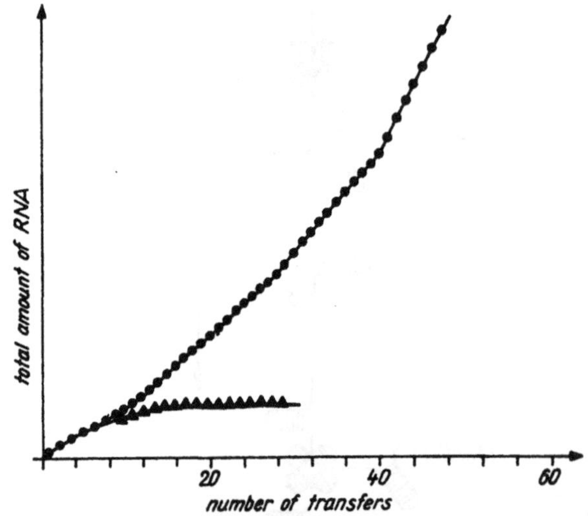

Fig. 4. Serial transfer experiments on RNA replication by Qβ replicase according to *Spiegelman* [7]. RNA replication in the test tube is initiated by addition of Qβ RNA. Two quotations are recorded: the number of infectious particles (▲) and the amount of nucleoside trihphosphate incorporated into RNA as a measure of polynucleotide synthesis (●). The rate of RNA synthesis increases spontaneously in jumps. The percentage of infectious polynucleotides, however, becomes smaller and smaller. Now RNA molecules were formed which grow faster but are no longer infectious. Since serial transfer selects for the excess productivity only, infectiousness is lost readily. Several spontaneous, stepwise increases in the rate of RNA synthesis are observed until an optimal value is reached.

synthesizes RNA "de novo" when nucleoside triphosphates are available and [3] RNA-synthesis at excess nucleoside triphosphate concentrations and constant amount of enzyme shows three phases of growth in the sequence exponential, linear and parabolic. Exponential growth is observed at excess enzyme concentration, i.e. at low concentration of polynucleotides. It changes to linear growth when the enzyme is saturated by template. The parabolic phase at high concentrations of polynucleotides is consistent with product inhibition. All data can be interpreted by means of the many-step mechanism shown in figure 5.

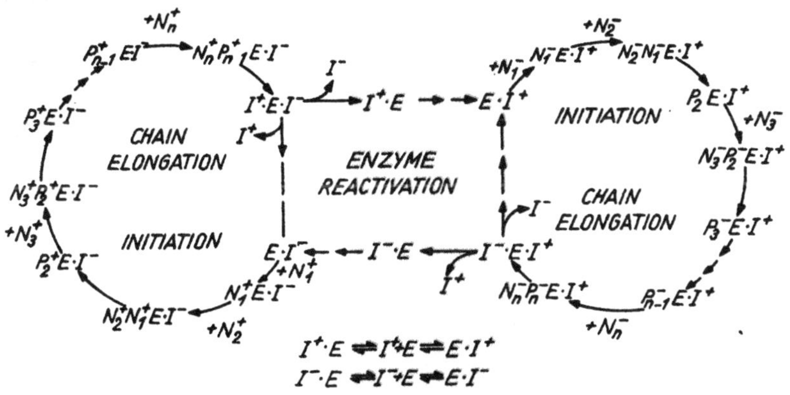

Fig. 5. A many step mechanism of RNA replication by Qβ replicase according to BIEBRICHER, EIGEN and LUCE [9]. E stands for the enzyme I^+ and I^- are plus and minus strands of the RNA; $N_1^{(\pm)}, N_2^{(\pm)}, \ldots, N_n^{(\pm)}$ are the appropriate nucleoside triphosphates for the corresponding position and $P_2^{(\pm)}, P_3^{(\pm)}, \ldots, P_{n-1}^{(\pm)}$, finally, represents the growing polynucleotide chain bound to the enzyme.

4. Autocatalysis, Exponential Growth and Selection

The fact that replication or multiplication leads to exponential growth of populations has been predicted long ago by the well-known British economist T. R. MALTHUS who lived in the 18th and 19th century. MALTHUS' ideas had a certain influence on DARWIN when he worked on his theory of evolution about fifty years later. What is unique about self-replication and exponential growth? In order to be able to see the underlying principle let us consider an exceedingly simple example.

A simple, single step autocatalytic process leads to exponential growth when it is run under conditions far off equilibrium. For the reaction.

$$I + A \underset{f_2}{\overset{f_1}{\rightleftharpoons}} 2I \tag{1}$$

we obtain the kinetic equation (concentrations are denoted by $[I] = x$ and $[A] = a$):

$$\frac{dx}{dt} = \dot{x} = (f_1 a - f_2 x)\, x\,. \tag{2}$$

Equation (2) can be integrated without difficulties. We obtain

$$x(t) = \frac{f_1(a_0 + x_0)\, x_0 \exp\{f_1(a_0 + x_0)\, t\}}{f_1(a_0 + x_0) - (f_1 + f_2)\, x_0(1 - \exp\{f_1(a_0 + x_0)\, t\})} \tag{3}$$

wherein a_0 and x_0 stand for the concentrations of A and I at time $t = 0$. In figure 6 we show a concrete numerical example. We observe an initial phase of exponential growth but then when the reaction is left alone it approaches equilibrium through a common relaxation process. In order to keep the autocatalytic reaction within the interesting range of exponential growth we have to couple the system to an external reservoir or to other reactions which can be driven from outside. An easy to handle three-step mechanism contains a degradation and a recycling reaction together with the autocatalytic process:

$$I + A \underset{f_2}{\overset{f_1}{\rightleftarrows}} 2I \tag{4a}$$

$$I \underset{d_2}{\overset{d_1}{\rightleftarrows}} B \tag{4b}$$

$$B \xrightarrow{g(E)} A \tag{4c}$$

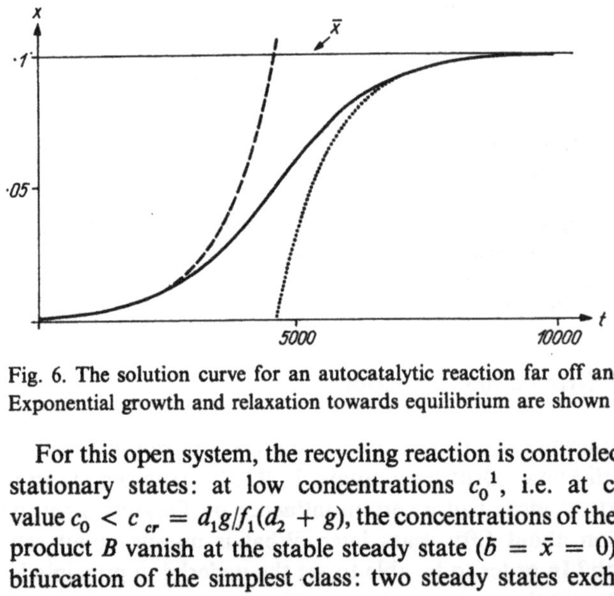

Fig. 6. The solution curve for an autocatalytic reaction far off and close to thermodynamic equilibrium. Exponential growth and relaxation towards equilibrium are shown as broken lines.

For this open system, the recycling reaction is controled by the environment, we find two stationary states: at low concentrations c_0[1], i.e. at concentrations below the critical value $c_0 < c_{cr} = d_1 g / f_1(d_2 + g)$, the concentrations of the autocatalyst I and its degradation product B vanish at the stable steady state ($\bar{b} = \bar{x} = 0$). The point $c_0 = c_{gr}$ represents a bifurcation of the simplest class: two steady states exchange their properties. Above the critical concentration the stable steady state is unstable and another state is stable (figure 7).

This second steady state is characterized by non-vanishing concentrations of all three compounds. At the bifurcation point the linear functions $\bar{a}(c_0)$, $\bar{b}(c_0)$ and $\bar{x}(c_0)$ show a characteristic tilt. Another important feature of bifurcations is the extremely slow approach towards the steady state in the surrounding of the critical point. This phenomenon is commonly called "critical slowing down".

The assumption that reactions (4a) and (4b) are run under conditions of practical irreversibility ($f_2 = 0$, $d_2 = 0$) does not change the qualitative features of the system. The only major change concerns the stationary concentration of A above the bifurcation point: $\bar{a}$ is independent of c_0 in case of irreversibility: $\bar{a} = d_1/f_1$ for $c_0 > c_{cr} = d_1/f_1$. The autocatalytic

1 The concentration $c_0 = a + b + x$ is a conserved quantity in reaction (4). Hence, it can be considered as a parameter which is under external control.

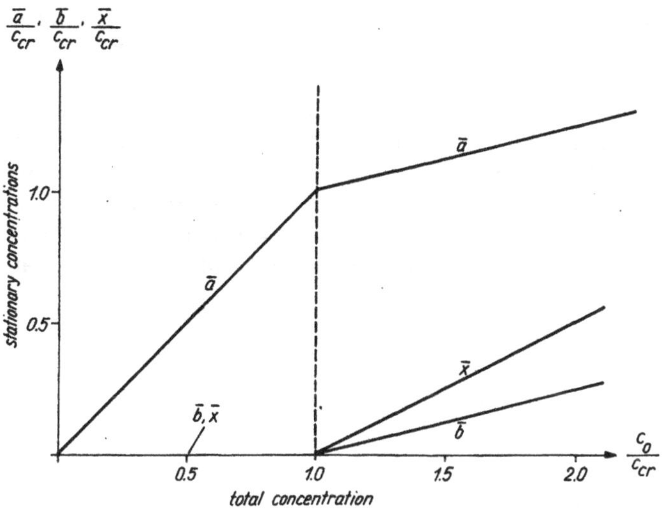

Fig. 7. Stable stationary states and the bifurcation point in a concrete numerical example of the three-step mechanism shown in equation (4). The numerical values of the rate constants are: $f_1 = 4$, $f_2 = 2$, $d_1 = 3$, $d_2 = 1$ and $g = 5$ in arbitrary time and concentration units. The critical concentration then is: $C_{cr} = 5/8$.

reaction step thus consumes all A accessible except the minimum quantity which is necessary to keep the reaction going.

Replication of synthetic homopolyribonucleotides has been studied by Friedemann SCHNEIDER and coworkers [11, 12]. They studied RNA-synthesis at poly(a)-poly(U) templates by the specific RNA polymerase from *E. coli* in a stirred flow reactor. In general this reaction follows an overall autokatalytic kinetics. In the flow reactor the recycling process is replaced by an influx of A and an outflux of the material in the reactor. It is worth noticing that critical slowing down has been observed experimentally in this system.

Next, we consider two autocatalysts growing on the same resource. We use the same recycling reaction as before in order to keep the system off equilibrium. For the sake of simplicity the reactions are assumed to be irreversible for practical purposes:

$$A + I_1 \xrightarrow{f_1} 2I_1 \tag{5a}$$

$$A + I_2 \xrightarrow{f_2} 2I_2 \tag{5b}$$

$$I_1 \xrightarrow{d_1} B \tag{5c}$$

$$I_2 \xrightarrow{d_2} B \tag{5d}$$

$$B \xrightarrow{g(E)} A \tag{5e}$$

The reaction mechanisms (5) can be analysed in straigth forward manner. We find three steady states and two bifurcation points (figure 8). Only one bifurcation is relevant since one steady state remains unstable for all values of the total concentration $c_0 = a + b + x_1 + x_2$. As we see from figure 8 the two autocatalysts compete and selection occurs. The autocatalyst which leads to the bifurcation at a lower critical concentration, $c_{cr}(I_i) = d_i/f_i$,

$i = 1$ or 2, is the one that will be selected. Systems which have the potentiality to grow exponentially lead to selection. Thus, we observe one principle of DARWIN's theory already with fairly simple chemical reactions.

Selection can be studied appropriately and in great detail under conditions which allow a straight forward mathematical analysis [13, 14]. A convenient system, more elaborate than serial transfer experiments or stirred flow reactors, consists of an evolution reactor (figure 9) which allows to control the concentrations of all important reactants or products. In the simplest case we keep these concentrations constant as well as the sum of all concentrations of autocatalysts. Then, the relative concentrations of autocatalysts are the only remaining variables.

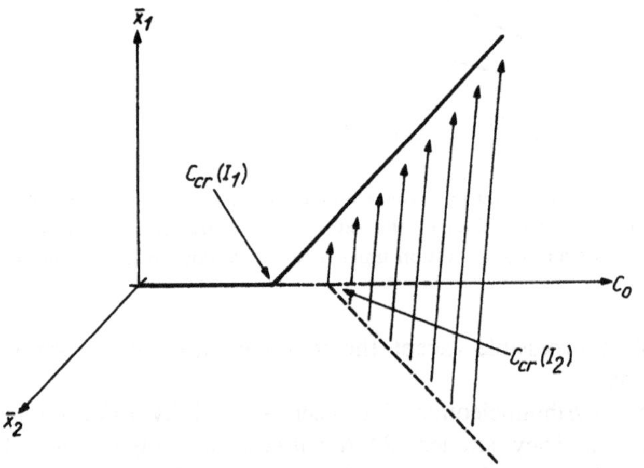

Fig. 8. Stationary states and bifurcation points in the five step reaction (5). Note, that the stationary state which branches off the zero state is always stable whereas the other stationary state is always unstable. Autocatalysts, hence, compete and selection takes place.

Under these conditions a dynamical analysis of the processes taking place in the reactor is straigth forward. Chemical kinetics leads to a system of differential equations:

$$\frac{dx_i}{dt} = \dot{x}_i = \Gamma_i(x_1 \ldots x_n) - \frac{x_i}{C}\psi; \qquad i = 1, 2 \ldots n \tag{6}$$

Here we use C for the total concentration: $C = \Sigma x_i$. The growth function Γ_i describes the kinetics of replication for the compound I_i, Φ is the constraint imposed on the system, in particular, the dilution flux in the evolution reactor. In general, the growth function determines the macroscopic phenomena we observe, like co-existence, selection or co-operation. When mass action kinetics is applied the mathematical analysis is straight forward [14—16]. The results are summarized in table 1.

Linear growth functions, the cases which are of particular interest here, inevitably lead to competition and selection and thus to DARWINian behaviour as it is shown schematically in figure 10. The appearance of a more efficient mutant leads to selection of this variant. During this evolutionary process the excess production, $E_i = f_i - d_i$, turns out to be the

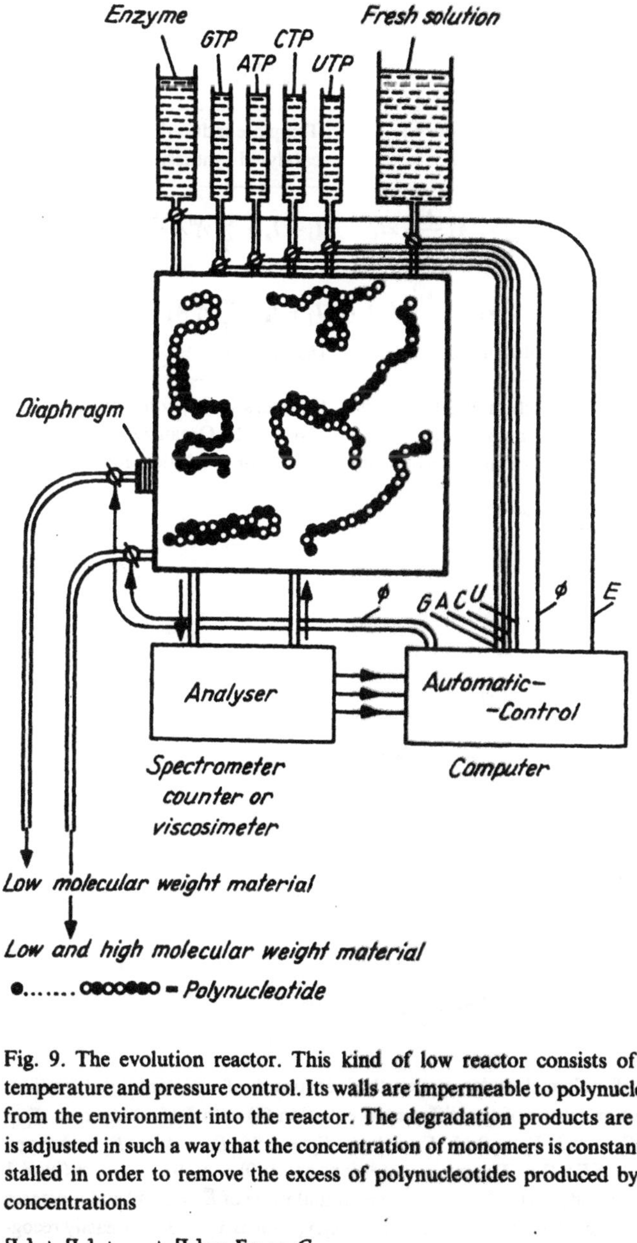

Fig. 9. The evolution reactor. This kind of low reactor consists of a reaction vessels which allows for temperature and pressure control. Its walls are impermeable to polynucleotides. Energy rich material is poured from the environment into the reactor. The degradation products are removed steadily. Material transport is adjusted in such a way that the concentration of monomers is constant in the reactor. A dilution flux Φ is installed in order to remove the excess of polynucleotides produced by multiplication. Thus the sum of the concentrations

$$[I_1] + [I_2] + \dots + [I_n] = \sum_i x_i = C$$

maybe controlled by the flux Φ. Under "constant organization" Φ is adjusted such that the concentration $C\ C\ C_0$ is constant. The regulation of Φ requires internal control which may be achieved by analysis of the solution and data processing by a computer as indicated above.

Table 1: Polynominal expansion of the growth function Γ_i

Γ_i	$= k^{(i)}$	$+ \sum_{j,k} k^{(i)}_{jk} x_j x_k$	$+ \sum_j k^{(i)}_j x_j$
Reaction	Spontaneous formation of I_i	Template induced synthesis if I_i	Template induced **and** catalysed synthesis of I_i
Mass action kinetics	$\sum_\lambda \nu^{(i)}_\lambda A_\lambda \xrightarrow{k^{(i)}} I_i$	$I_i + \sum_\lambda \nu^{(i)}_\lambda A_\lambda \xrightarrow{k^{(i)}_i} 2I_i$	$I_i + I_k + \sum_\lambda \nu^{(i)}_\lambda A \xrightarrow{k^{(i)}_{ik}} 2I_i + I_k$
		$I_j + \sum_\lambda \nu^{(i)}_\lambda A_\lambda \xrightarrow{k^{(i)}_j} I_i + I_j$ (Mutation)	$I_j + I_k + \sum_\lambda \nu^{(i)}_\lambda A_\lambda \xrightarrow{k^{(i)}_{jk}} I_i + I_j + I_k$ (Mutation)
Phenomenon	Co-existence	Selection and formation of a quasispecies "Darwinian" evolution	Co-operation of competitors Hypercyclic Organization "Once for ever" decisions

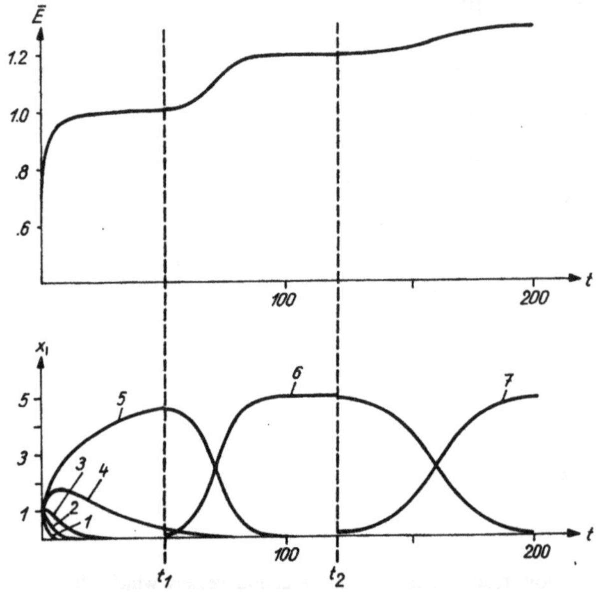

Fig. 10. Selection in an ensemble of polynucleotides under the constraint of constant total concentration (C = const.). We present solution curves of the differential equation (6) with $\Gamma_i(f_i - d_i)\, x_i = E_i x_i$; $\dot{x}_i = x_i(E_i - \bar{E})$, $i = 1,2, ..., 7$ with $E_1 = 0.3$, $E_2 = 0.65$, $E_3 = 0.82$, $E_4 = 0.96$, $E_5 = 1.01$, $E_6 = 1.2$ and $E_7 = 1.3$ in arbitrary time units and $\bar{E} = \sum E_i x_i / \sum x_i$. The initial conditions are $x_1(0) = x_2(0) = ... = x_5(0) = 1$ and $x_6(0) = x_7(0) = 0$. The mean excess productivity $\bar{E}$ starts from an initial value of $\bar{E} = 0.748$ and increases steadily as the population becomes homogeneous: $\bar{E} \to 1.01$ and $x_1, x_2, x_3, x_4 \to 0$ as $x_5 \to 5$. One easily recognizes the cause for the increase of $\bar{E}$: The less efficiently growing polynucleotides are eliminated. They disappear in the same sequence as their excess productivity E_i increases, namely 1,2 ... 5. At $t = t_1 = 50$ we observe a fluctuation $x_6 = 0 \to x_6 = \delta$ the appearence of I_6 as a favourable mutant leads to further increase in $\bar{E}$ which now approaches the valqe $\bar{E} = 1.2$ as I_5 disappears when the population becomes homogeneous again: $x_1 = ... = x_5 = 0$ $x_6 = 1$. The same story happens when a new more efficient mutant appears at $t = t_2 = 120$. Now I_6 is replaced by I_7 and $\bar{E}$ approaches the temporary optimum $\bar{E} = 1.3$. This example illustrates well the nature of the selection process and the role of the mean excess productivity $\bar{E}$ as the quantity to be optimized.

crucial quantity[2]. This excess production simply is the number of templates I_i produced in excess over degradation.

The mean excess production $\bar{E} = \sum_i E_i x_i / C$ plays an important role: it grows monotonously during the evolution in the reactor. In mathematical terms $\bar{E}$ is an Lyapounov function, in physical terms it represents the quantity which is optimized during the evolutionary process.

5. Mutations and Its Consequences for Replicating Systems

Variability of organisms caused by the imperfection of reproduction represents the other leg upon which DARWIN's theory stands. Variations are introduced already at the level of viral RNA replication. Point mutations are one major source of variability. They are caused by mismatching of bases during replication. Systematic studies concerning the nature and the frequencies of point mutations were performed on the bacteriophage $Q\beta$ by Charles WEISSMANN and his coworkers [17—19]. We shall consider the consequences of point mutations on replicating ensembles of polynucleotides later on.

Another source of variations observed with $Q\beta$ replicase consists in major changes in sequence and length of the RNA. Most probably deletions or duplications of whole segments occur at certain sites on the polynucleotide. The most complete studies on the replication process performed so far [9, 10] were undertaken in order to clarify an interesting catalytic property of the replication enzyme first reported by SUMPER and LUCE [20]: in absence of any detectable amount of longer RNA templates[3], $Q\beta$ replicase synthesizes RNA "de novo". For given external conditions, these are the conditions under which the serial transfer experiments are conducted (temperature, concentrations of reactants and enzyme, ionic strength etc.) the "de novo" products have defined chain lengths and replication properties. A change in the environment, e.g. an increase in ionic strength or the addition of ethidium bromide, leads to increased resistance of the variants against this agents. In case of ionic strength the increase in salt resistence is accompanied by a stepwise increase of the RNA's molecular weight. Three well defined variants, 86, 113 and 220 bp long have been isolated and characterized [10]. A hint on the nature of the mechanism of this stepwise change in the chain lengths can be seen in the discontinuous dynamics of the polymerization process mentioned in section. 2. The replicating enzyme stops at several "pause sites" and idles around for some while before polymerization is continued. Insertions and deletions, eventually are the results of irregular continuation of the polymerization process at these "pause sites".

SPIEGELMAN's serial transfer experiments and the optimization of "de novo" products have never been carried out under identical conditions. Nevertheless, both series of experiments led to similar adaption products coming either from larger or from smaller chain length. What we observe with this replicating polynucleotides is DARWINian evolution in its most simple form: an interplay between variation through mutations and selection through competition of self-reproducing molecules.

2 The rate constant d_i is defined as in reaction (5d, d), f_i, however, has a somewhat different meaning. The constant concentration of A, a_0, and eventually other constant concentrations are included as factors: $f_i = f_i^0 \cdot a_0$.

3 The experiments were carried out under cloning conditions which allow to find single molecules of RNA.

Replication with errors can be studied by means of equation (6). We put

$$\Gamma_i = (f_i Q_i - d_i)\, x_i + \sum_{j \neq i} w_{ij} x_i\,; \qquad i, j = 1, \ldots, n \tag{7}$$

and analyze the corresponding differential equation. Herein, f_i and d_i are the rate constants of nucleotide synthesis and degradation as before, Q_i is the quality factor of the replication process, i.e. the probability to obtain an error-free copy w_{ij} finally is the probability to obtain the sequence I_i as an error copy of I_j multiplied by the corresponding rate constant. The problem has been formulated and studied extensively in the literature [13, 14, 21, 22]. Depending on the accuracy of replication we distinguish two cases: (1) The replication process is accurate enough to sustain a stable ensemble of polynucleotide sequences. This ensemble consists of a master sequence and a distribution of frequently occurring mutants. Such an ensemble has been called "quasispecies" [14] in order to indicate some analogy to the notion of a species in biology. Apart from cases of kinetic degeneracy[4] the master sequence selected, I_m, is that with the maximum selective value w_{mm}:

$$\begin{aligned} I_m\colon w_{mm} &= \max\,\{w_{ii}\,; i = 1, \ldots, n\}\,; \\ w_{ii} &= f_i Q_i - d_i \end{aligned} \tag{8}$$

Case (1), therefore, is in complete analogy to the previously treated dynamics of error-free replication; the only important difference is the fact that a quasispecies instead of a single sequence. (2) There is no stable distribution of sequences in case the accuracy of replication is too low. No sequences appear whereas those present at the moment will disappear after a few generations.

The two cases discussed above are separated by a precisely defined error threshold which corresponds to a critical accuracy of the replication process ($Q_{\min}$). Selection of a quasispecies with the polynucleotide I_m as master sequence occurs if

$$Q_m > Q_{\min}^{(m)} = \sigma_m^{-1} = \frac{d_m + \bar{E}_{-m}}{f_m}\,. \tag{9}$$

Herein we define a superiority parameter σ_m for the master sequence. The "mean but the best" excess production $\bar{E}_{-m}$ reads in explicit form ($E_i = f_i - d_i$)

$$\bar{E}_{-m} = \frac{\sum_{i \neq m} x_m E_m}{\sum_{i \neq m} x_m}\,. \tag{10}$$

Let us consider now polynucleotide replication in more detail. We can define a mean single digit accuracy $\bar{q}$ which is a measure of the mean probability for the correct incorporation of a single base into the growing chain. The accuracy of replication for the whole sequence is given by

$$Q = \bar{q}^{\nu} \tag{11}$$

when ν is the degree of polymerization. Here we neglect deletions and insertions. In contrast to this results in serial transfer experiments these variations play a minor role only in the "in vivo" reproduction of RNA viruses. In serial transfer experiments the

4 The situation is somewhat more complicated in case we are dealing with two or more different sequences which have the same or almost the same selective values.

rate of replication is the exclusive target of selection. The RNA in "in vivo" systems has to fulfil other requirements as well like to be translatable into functional proteins, and almost all deletions and insertions will be lethal.

Equation (9) has been derived by zero-order perturbation theory. Nevertheless, it leads to excellent agreement with the exact results when ν exceeds a certain limit ($\nu > 10$). In figure 11 we compare the exact solution with the expression from perturbation theory for $\nu = 50$ [23]. This figure shows the percentage of master sequence present in the quasispecies. This percentage decreases with decreasing $\bar{q}$ and becomes very small near the critical accuracy, $Q \simeq Q_{min}$.

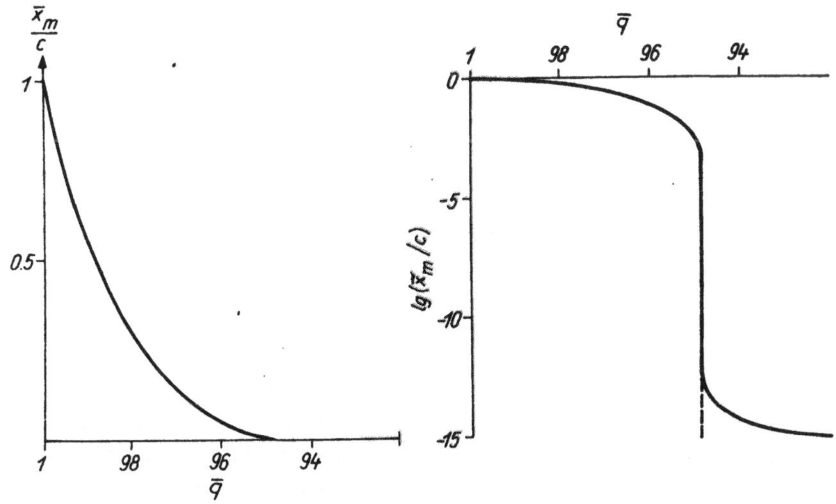

Fig. 11. Percentage of the master copy in a stable quasispecies. We present the fraction x_m/c as a function of $\bar{q}$ for polynucleotides with a chain length of $\nu = 50$. Note that the results of perturbation theory (broken line) coincide almost perfectly with the exact solution [23]. The only exception being the range below critical accuracy (Q_{min}). Here, perturbation theory predicts zero concentration whereas a very small fraction of the master copy is still present due to the statistics of erroneous replication.

The minimum accuracy for the replication of the whole polynucleotide chain (Q_{min}) can be converted into a maximum chain length (ν_{max}) for a given mechanism of replication. Combination of equations (9) and (11) leads to the inequality

$$\nu < \nu_{max} = -\frac{ln\sigma}{ln\bar{q}} \simeq \frac{ln\sigma}{1-\bar{q}} \tag{12}$$

The mean single digit accuracy of replication has been determined experimentally for three cases: enzyme-free template induced replication [24], enzyme-catalysed RNA replication in simple bacteriophages [17] and DNA replication in procaryotes [25]. We can use equation (12) to estimate the maximum chain length of polynucleotides which are compatible with the fidelities of given mechanisms of replication (table 2).

Without the help of enzymes replicating RNA sequences seem to be limited to about the size of *t*-RNA like molecules. The fidelity of viral RNA replication with specific replicases allows to reproduce molecules up to a few thousands bases long. These are the actual sizes

Table 2: Single digit accuracy and error threshold in some chemical and biological systems

Single digit accuracy $\bar{q}_m$	Error rate per digit $1-\bar{q}_m$	Superiority σ_m	Maximum digit content ν_{max}	Biological examples
0,95	$5 \cdot 10^{-2}$	2	14	Enzyme-free RNA replication
		20	60	
		200	106	t-RNA precursor, $\nu = 80$
0,9995	$5 \cdot 10^{-4}$	2	1386	Single stranded RNA replication via specific
		20	5991	replicases
		200	10597	Phage Qβ, $\nu = 4500$
0,999999	$1 \cdot 10^{-6}$	2	$0,7 \cdot 10^6$	DNA replication via
		20	$3,0 \cdot 10^6$	polymerases and proofreading
			$5,3 \cdot 10^6$	*E. coli*, $\nu = 4 \cdot 10^6$

of RNA molecules found with simple bacteriophages and with most animal viruses. Nature, thus, approaches the error-limit rather closely. Two experimental findings support this interpretation: (1) The virus populations studied in detail so far resemble a quasispecies growing closely to the limit. Cloning experiments have revealed substantial nucleotide sequence heterogeneity of the RNA in natural populations of the bacteriophage *Qβ* [19], of foot-and mouth disease virus [26] and of influenza virus [27, 28]. (2) Simple bacteriophages, the complete nucleotide sequences of which are known, have overlapping genes in their genomes. This, certainly, leads to a restricted adaptability of the proteins encoded. The existence of a maximum chain length close to that observed provides an explanation: restricted adaptability is still better than the loss of stability in replication. DNA replication involves a highly developed enzymatic machinery which uses a proof reading mechanism in order to increase the single digit accuracy. Thereby, genomes as long as some million base pairs become possible. In procaryots nature again seems to operate close to the error-limit. Bacteria, nevertheless, have retained a certain flexibility to adjust mutation rates to the evolutionary needs of the moment. They do not seem to replicate as closely to the limit as viruses do.

Biological evolution, at least in the early stages, appears to be governed by limits set by the accessible accuracy of polynucleotide replication.

6. Concluding Remarks

In this review we tried to show that the basic concepts of DARWIN's theory have their molecular roots in the properties of polynucleotides. Variation and selection occur in systems which sustain replication of viral RNA. These systems are simple enough to allow experimental studies and their interpretation at the molecular level. DARWIN's principles seem to have played an important role even earlier at some of the stages of prebiotic evolution (for recent reviews see [29]). They guide through the whole evolution of the biosphere from the first appearence of replicating molecules until present time. Molecular biology has provided and is still providing new and profound insights into the mechanisms operating in the evolving organisms. This knowledge often places the emphasis differently from the central problems of last century but DARWIN's ingenious insight has not at all lost his relevance [30].

Acknowledgements

Financial support of the work done in Vienna by the Austrian „Fonds zur Förderung der wissenschaftlichen Forschung" as well as generous supply with computer time by the „Interuniversitäres Rechenzentrum" is gratefully acknowledged. The author is indebted to Mrs. J. Jakubetz and Mr. J. König for typing the manuscript and drawing the figures.

References:

[1] Darwin, C.: The Origin of Species. Reprinted by J. M. Dent & Sons, Everyman's Library, London (1967), p. 83.

[2] Gould, S. J.: Ever since Darwin, Burnett Books, London (1978), pp. 39—55.

[3] Eoyang, L. and J. T. August: Reproduction of RNA Bacteriophages, in H. Fraenkel-Conrat and R. R. Wagner, eds. Comprehensive Virology, Vol. 2.

[4] Fiers, W., R. Contreras, F. Duerinck, G. Haegman, D. Iserentant, J. Merregaert, W. Min Jou, F. Molemans, A. Raemaekers, A. Van der Berghe, G. Volckaert and M. Ysebaert: Nature **260** (1976), 500—507.

[5] Atkins, J. F., J. A. Steitz, C. W. Anderson, P. Model: Cell **18** (1979) 247—256 and M. W. Beremand and T. Blumenthal: Cell **18** (1979), 257—266.

[6] Mills, D. R., C. Dobkin and F. R. Kramer: Cell **15** (1978), 541—550.

[7] Spiegelman, S.: Quart. Rev. of Biophysics **4** (1971), 213—253.

[8] Review: B. O. Küppers, Naturwissenschaften **66** (1979), 228—243.

[9] Biebricher, C. K., M. Eigen and R. Luce; J. Mol. Biol. **148** (1981), 369—390 and 391—410.

[10] Biebricher, C. K., S. Diekmann and R. Luce: Structural analysis of self-replicating RNA synthesized by Qβ-replicase, J. Mol. Biol., **154** (1982), 629—648.

[11] Schneider, F. W., D. Neuser and M. Heinrichs: in M. Balaban, ed., Molecular Mechanisms of Biological Recognition, Elsevier (1979), pp. 241—252.

[12] Heinrichs, M. and F. W. Schneider: Ber. Bunsenges. Phys. Chem. **84** (1980), 857—865.

[13] Eigen, M.: Naturwissenschaften **58** (1971), 465—526.

[14] Eigen, M. and P. Schuster: The hypercycle — a principle of natural self-organization, Springer-Verlag, Berlin (1979).

[15] Schuster, P., K. Sigmund and R. Wolff: J. Math. Anal. Appl. **78** (1980), 88—112.

[16] Hofbauer, J., P. Schuster and K. Sigmund: J. Math. Biol. **11** (1981) 115—168.

[17] Domingo, E., R. A. Flavell and C. Weissmann: Gene **1** (1976), 3—25.

[18] Batschelet, E., E. Domingo and C. Weissmann: Gene **1** (1976), 27—32.

[19] Domingo, E., D. Sabo, T. Taniguchi and C. Weissmann: Cell 13 (1978), 735—744.

[20] Sumper, M. and R. Luce: Proc. Natl. Acad. Sci. USA **72** (1975), 162—166.

[21] Thompson, C. J. and J. L. Mc Bride: Math. Bioscience **21** (1974), 127—142.

[22] Jones, B. L., R. H. Enns and S. S. Rangnekar: Bull. Math. Biol. **38** (1976), 15—28.

[23] Swetina J. and P. Schuster: Self-replication with errors — a model for polynucleotide replication, Biophys. Chem. **16** (1982), 329—345.

[24] Lohrmann, R., P. K. Bridson and L. E. Orgel: Science **208** (1980), 1464—1465.

[25] Kunkel, T. A. and L. A. Loeb: J. Biol. Chem. **254** (1979), 5718—5725 and T. A. Kunkel, R. R. Meyer and L. A. Loeb, Proc. Natl. Acad. Sci. USA **76** (1979), 6331—6335.

[26] Domingo, E., M. Davila and J. Ortin: Gene **11** (1980), 333—346.

[27] Ortin, J., R. Najera, C. Lopez, M. Davila and E. Domingo: Gene **11** (1980), 319—331.

[28] Fields, S., G. Winter: Gene **15** (1981), 207—214.

[29] Schuster, P., Prebiotic Evolution, in H. Gutfreund, ed., Biochemical Evolution, Cambridge University Press Cambridge U.K. (1981), pp. 15—87 and Eigen, M., W. Gardiner, P. Schuster and R. Winkler-Oswatitsch, Sci. Am. **244** (4) (1981), 88—118.

[30] Eigen, M.: Angew. Chem. Internat. Ed. **20** (1981), 233—241.

Appendix:

Comments on "Non-DARWINian"-Phase of Evolution and Hypercycles

P. SCHUSTER

This is a symposium on "DARWIN today". So, I thought I was supposed to talk about DARWIN's theory and its roots in molecular biology. My lecture, therefore, was restricted to these questions. The issues raised by Prof. KUHN give me a chance to continue and to digress into another important principle of the evolution of macromolecules. In order to make my point clear I would like to start with the fossil record (fig. 1). There seems to be an interesting structure in the time course of global evolution. Long periods of rather quiescent development and gradual changes are interrupted by fast and burst-like phases, during which new levels of complexity are attained.

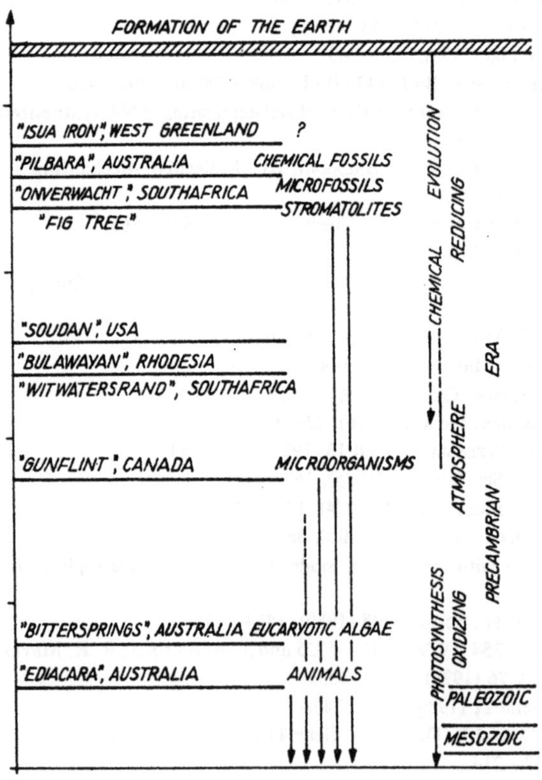

Fig. 1. The fossile record.

The most striking and well documented event of this kind is the origin of eucaryotic organisms on the Earth. Primitive life consisting in ancient microorganisms seems to have originated very early, about 3.6 billion years ago. The fossils of the first eucaryots are rather"young", less than 1 billion years old. We observe a long more or less quiet period preceeding the burst in the development of higher plants and animals. These higher organisms differ from procaryots by a rich intracellular structure and organization as well as a highly sophisticated mechanism of cell division. This great innovation, interestingly, is very similar not to say uniform in most cells of higher organisms. Such uniformity appears also at other occasions. In particular, the machinery of translation and the genetic code are universal apart from a few marginal differences.

In order to search for the reasons of such discontinuities in development it seems appropriate to go back to the more primitive molecular systems where we can realize the principles much easier. Eventually, there exists also a common basic mechanism which allows to pass over critical barriers in evolution. Polynucleotides, as I tried to show in my lecture, can be replicated faithfully up to a certain length only. Above such a treshold an error catastrophy takes place: the sequences and together with them secondary structures and other molecular properties are no longer conserved. How to increase the content of information that can be stored in polynucleotides? We encounter a vitious cycle: for more information to be transmitted we need a more sophisticated and more precisely operating machinery of replication. To build this machinery we need more information. Evolution faces an information crisis. The trivial answer to use more different sequences, each present in many copies, does not immediately solve the problem. The different sequences will compete through replication and only one, eventually a whole quasispecies, will survive. We proposed a solution of different nature [1]. Competition can be suppressed by higher order catalytic interaction in case these interactions form a positive feed-back loop. Then, mutual catalytic help leads to an internally controled system. The relative concentrations of all members of such a system are strictly regulated. We called such a feed-back loop a hypercycle. A simple but not necessarily the most realistic and probable

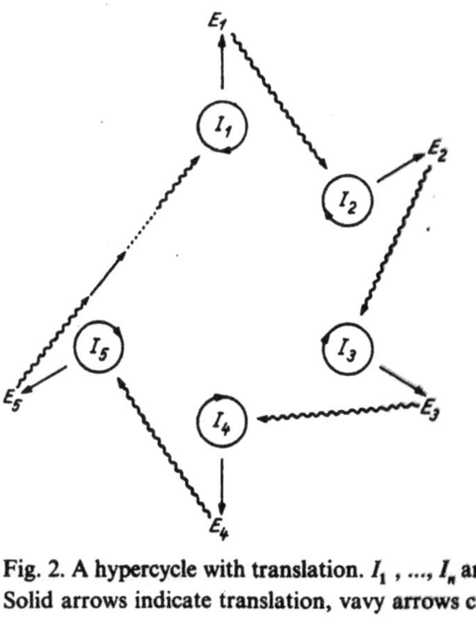

Fig. 2. A hypercycle with translation. $I_1, \ldots, I_n$ are polynucleotides and $E_1, \ldots, E_n$ their translation products. Solid arrows indicate translation, vavy arrows catalytic help during replication.

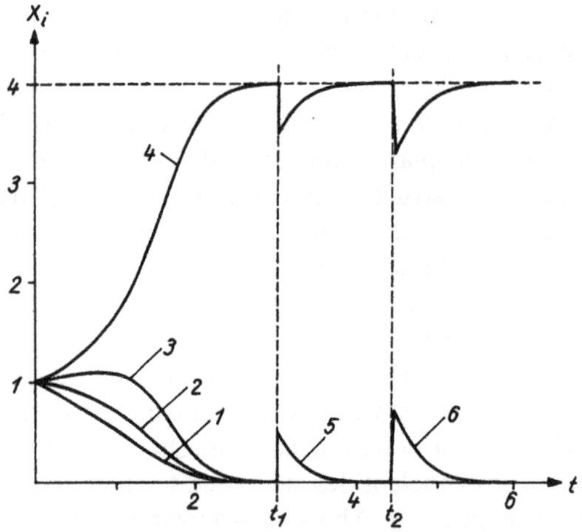

Fig. 3. Selection in an ensemble of hypercycles. We present solution curves of the differential equation describing competition between hypercycles $\dot{x}_i = f_i x_i^2 - x_i \Phi_0 / c_0$, $i = 1,2 \ldots, n$ with $f_1 = 0.3[t^{-1}c^{-1}]$, $f_2 = 0.6[t^{-1}c^{-1}]$, $f_3 = 0.9[t^{-1}c^{-1}]$, $f_4 = 1.1[t^{-1}c^{-1}]$, $f_5 = 1.2[t^{-1}c^{-1}]$ and $f_6 = 1.3[t^{-1}c^{-1}]$, $c_0 = \sum_i x_i$ and $\Phi_0 = \sum_i f_i x_i^2$. [c] and [t] are arbitrary time and concentration units. The initial conditions are $x_1(0) = x_2(0) = x_3(0) = x_4(0) = 1$, $x_5(0) = x_6(0) = 0$. In the first phase of the development the system with the largest value of f_i is selected. After one species reached a stage of dominance it is almost impossible for a favourable mutant to replace this selected species. Even extremely large fluctuations ($t_1 = 3$ [t], $\Delta x_5 = 0.5$ [c]; $t_2 = 4.5$ [t], $\Delta x_6 = 0.7$ [c]) die out in rather short time. Thus, no optimization takes place after a first process of selection has been completed. "once for ever" decisions occur.

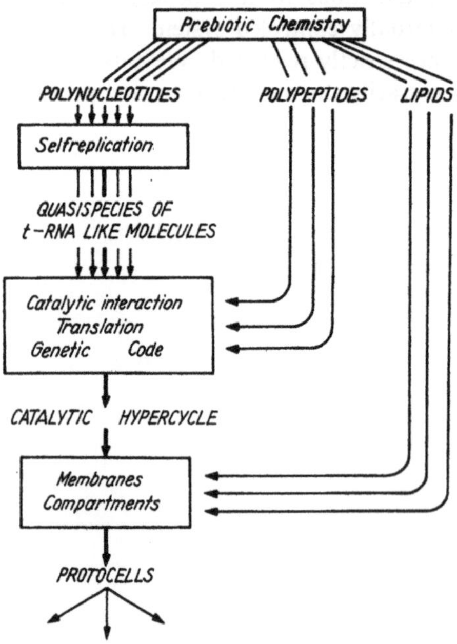

Fig. 4. A scheme for the logical sequence of steps in prebiotic evolution leading from unorganized macromolecules to protocells.

way to introduce such coupling terms is via translation of polynucleotides into proteins which have the desired catalytic properties (figure 2). A biochemically more acceptable version of a hypercycle with many intermediate steps has recently been stimulated numerically by BIEBRICHER and GARDINER: it shows the same properties of co-operation and internal control.

Competition between hypercycles has an interesting feature (figure 3). After a first initial phase optimization is suppressed. New hypercycles have an exceedingly small chance to be selected even when they appear as very favourable mutants. Due to the non-linearities in the growth functions fluctuations have no chance to grow. The only way hypercycles can evolve further is through incorporation of new members which appear in the mutant distributions existing around each of the constituents of the cycle. Thereby we observe strict conservation of what has been selected before. "Once for ever" decisions occur which lead to uniformity. According to this model bursts in developments are characterized by new achievements which are conserved in all successful descendants. A model showing where hypercycles may have played a role in prebiotic evolution is shown in figure 4.

Reference:

[1] EIGEN, M., P. SCHUSTER: The Hypercycle — a principle of natural self-organization. Springer, Berlin (1979).

Max-Planck-Institut für Biophysikalische Chemie, Molekularer Systemaufbau, D-3400 Göttingen-Nikolausberg

Selforganization in Molecular Aggregates and Origin of Life

Hans Kuhn

Darwin's basic ideas are well supported by modern molecular biology. The genetic machinery is essentially the same in all living species, as to be expected from their common origin. The question arises of how first systems with such machinery could develop by selforganization of matter.

How could simple systems arise that learn by evolution? How could machinery for the translation of genetic information that contains proteins as essential translation products arise? Can the laws of physics be applied towards its understanding?

Darwin had the idea that simplest living systems appeared in "some warm little pond with all sorts of ammonia and phosphoric salts, light, heat, electricity etc. present, that a protein compound was chemically formed, ready to undergo still more complex changes" [1]. He did not speak about the driving force of the process and expressed the opinion that the initial step was due to a breath of the creator [2].

Basic Concepts of the Approach

In the view presented here the first simple translation machinery is already the result of a long evolutionary process, and under particular environmental influences this process is inevitable on the basis of physical laws.

The process begins in an appropriate system that contains energy-rich nucleotides and strands formed somewhere by spontaneous condensation of nucleotides. All of a sudden a strand arises that can replicate thereby creating a system that adapts itself again and again to a changing environment, in many steps of replication and selection. Such progressive adaption to the environment represents a process of "learning" for the system that requires many generations.

It seems difficult to find possibilities of how a machinery with the capability of learning could have originated in a spontaneous process without confrontation with a complex spatial and temporal environment. A cardinal point in the present approach is that a particular pre-existing external spatial and temporal structure is essential for the formation of the living machinery. A machine is constructed by fitting its interacting parts together via external directed action. On the primordial planet the role of the external designer is replaced, in a way, by the enormous variety of environmental influence offering appropriate conditions somewhere by chance.

In this view then the appearance of first systems that reproduce themselves and that eventually change by error during the copying process represents a jump in quality, in which a fundamental property of matter suddenly manifests itself. Systems begin to be carriers of information, of a meaningful message, with a content capable of growing as the process advances. Prior to this not even the faintest trace of this property existed, but once the breakthrough has occurred, the process of learning by DARWIN's mechanism goes on inexorably via the continued confrontation of evolving systems with their surroundings and their adjustment to environmental changes by multiplication, mutation and selection (Fig. 1).

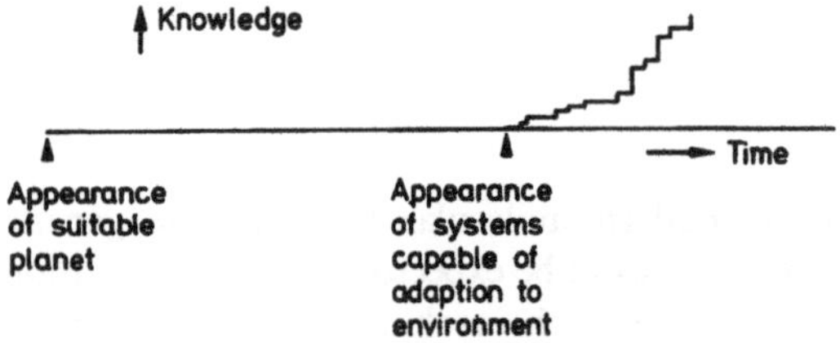

Fig. 1. Emergence of life. Learning systems suddenly appear, without prior traces of this new quality. Knowledge of evolving systems increases discontinuously.

The Method of Model Paths

The methodological program followed here for recognizing and understanding the grand connections in the process of selforganization is to consider a specific path consisting of many small physically and chemically plausible model steps that lead to a selforganization of matter. It is driven by periodic temperature changes and a multifaceted spatial environment, both of which were available in suitable locations on the primordial planet.

This detailed model path was described elsewhere [3—6] and shall not be considered here. The procedure shall be illustrated by indicating some model steps and by describing the overall organizational structure.

Limitation in Rate of Copying Errors — A General Feature of a Learning-device

For studying the important question of how a learning device can originate, it is useful to discuss first some general features of the learning process. The zig-zag-lines in Fig. 2a were obtained by proceeding in equal steps from left to right, each time deciding by tossing a coin whether the step should go up or down. Afterwards zig-zag-lines are cancelled according to a rule, until half of the original number is present (selection phase). According to this rule the zig-zag-lines to be cancelled are determined by the dice in such a way, that each line that fits better into a given form than another has the better chance to survive. (In the case of Fig. 2a, b zig-zag-line 4 has the best survival chance). The surviving zig-zag-lines are copied. Eventually copying errors are introduced by drawing lots. A step may go down instead of up (Fig. 2c) or the chain may add a step or loose a step. This multiplication phase is followed by a selection phase defined by the rule described above, and the process is repeated again and again. Zig-zag lines are obtained that are adjusted to the given form, since such lines have the best chance to survive. If the given form is slowly changed in shape, the zig-zag lines will adjust to the new shape in a sufficient number of generations (Fig. 2d).

This example demonstrates that the adaption process is only possible if the frequency of copying errors is neither too large nor too small. In the case of too large a copying error rate zig-zag lines cannot adjust to the form, since a zig-zag line of appropriate shape is likely to suffer a change during the time needed for its selection. In the case of too small error rate adjustment to a form that is slowly changing in shape is not possible. The system of evolving zig-zag lines is not sufficiently adaptable. For not loosing the message given by a zig-zag line this line must be copied error-free with sufficient probability, and therefore a longer zig-zag line (an increased amount of genetic information) requires a smaller copying error rate per step[1]. Most barriers that must be surmounted in the evolution of life are related to the difficulty of suppressing copying errors.

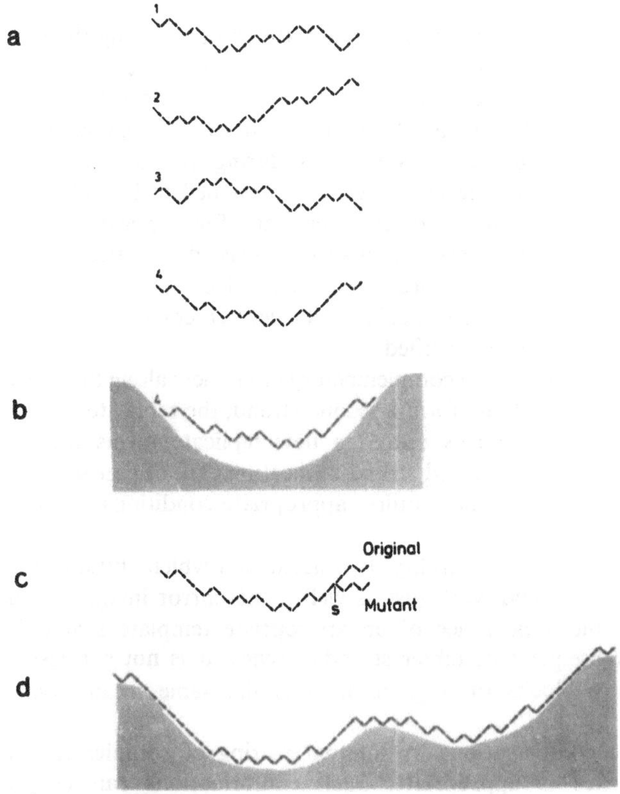

Fig. 2. Arrangement to demonstrate process of learning
a) Arbitrary zig-zag lines 1—4
b) selection of zig-zag line 4
c) copying-error in step s
d) zig-zag line adapted after many generations.

1 The proposed relation between length of the genome and error rate in the copying process is actually observed in biosystems, [3, 14] and this is not surprising, since it is based on this general property of learning systems. The agreement is expected for any reasonable model of the origin of life and cannot be used for discriminating between models.

First Steps in the Origin of Life

A simple learning system with nucleotides can well be imagined to have evolved on a planet such as the primeval earth. It has been demonstrated that energy-rich nucleotides and oligo-nucleotides can be obtained by simulating conditions thought to have existed on prebiotic earth. The prebiotic synthesis required solid state reactions as well as reactions in aqueous solution and in the gas phase. It is plausible that such substances would have accumulated on primeval earth in particular locations where a multitude of special conditions were fulfilled that allowed a succession of very different reactions to occur, requiring highly diversified and structural regions. ORGEL and LOHRMANN [7] have succeeded in the enzyme-free polymerization of nucleotides on nucleic acid templates, more than 90% of the nucleotides of the replicate strand being complementary to those on the template strand.

We may thus imagine a short nucleic acid strand on primordial earth containing the two complementary bases guanine and cytosine that are anchored together in the strand by ribose and phosphate groups and are capable of interlinking by three hydrogen bonds. It is further supposed that this short strand has diffused into a special region, after having been formed elsewhere by accidental condensation during the drying of a solution of monomers. In this strand, all of the monomers are presumed to be linked in such a way that the strand can serve as a precise template for replication. By considering a solution of the different monomers that may have existed on earth at the location considered, it is possible to estimate the probability with which the very special original template strand can be formed through accidentally correct condensation. In this way, the plausibility of such a step can be justified.

Monomers in the solution attach themselves to complementary monomers along the given strand. The monomers become inter-linked and form a second strand, the replicate or (—) strand. The two strands separate; on the new strand a new replicate forms that is identical with the original or (+) strand. Repeated strand replication can, of course, take place only with suitable kinds of monomers and requires appropriate conditions, such as specific periodic temperature changes.

All of this requires monomers of the same chirality. It is accidental which chirality the original strand has, since the process would work just as well if the mirror images of all monomers were used. However, the appearance of an appropriate template strand is decisive for the chirality of all subsequent daughter strands. Hence, it is not surprising that the chirality of the building blocks of a given kind is the same in all living systems.

Strands may fold into specific conformations by internal pairing of complementary bases, provided the base sequence is appropriate. Such conformations can confer selective advantages upon existing forms. Some conformations make strands resistant against chemical attacks, for example by mutual protection of adjoining regions. However, any given conformation is tied to an exact sequence of nucleotides and is therefore lost by almost any new error. The situation is similar to that in Fig. 2. Longer strands have selective advantage under appropriate conditions (they do not easily diffuse from the minute region in which the required regime of critical temperature changes exists), but there is a limit to the number of monomers a replicable strand may contain, because of the unavoidability of copying errors. That is, the probability that replicate strands be error free must be large enough to assure their survival.

Faulty Replicates Rejected during Aggregate Formation

In the model the resulting impasse is broken by a mechanism in which the strands convolute into a conformation that allows their interlocking into aggregates. Not only is diffusion of such aggregates slowed by their size, but it is of the greatest importance that practically all erroneous copies are rejected during aggregation, because of the precision required for interlocking. Aggregation therefore serves the function of an error filter.

The survival changes of strands in an aggregate are increased by their mutual protection against chemical attack, or because it may be less easy for aggregated strands to leave the favorable region. Aggregates may undergo selfreproduction. Suitable temperature and other environmental changes may cause their disassembly into component strands. The individual strands may replicate and again assume folded conformations. They move about at random and again aggregate through favorable accidental collisions, increasing the number of aggregates. All of this requires a very specific and highly detailed program of changes of temperature and other parameters (e.g. ionic composition of solution) that must be periodically repeated for repetition of the process described. Convolution of strands and aggregation are favored by cooling, disassemblage of aggregates and deconvolution of strands by warming.

It is of importance that the component strands do not diffuse too far from each other during the multiplication phase, because they would be lost and reaggregation could no longer occur. The phenomena discussed must therefore take place in a confined space, such as in small pores of a rock formation. The pore walls keep the strands together during their diffusion, so they are able to locate each other again and can form new aggregates. The rock formation is presumed to be inundated by a solution of suitable energy-rich monomers that can easily diffuse through the pore channels, while the strands formed are largely retained by them.

A great selective premium is associated with the evolution of machinery that facilitates the aggregation of interlocking subunits B_1, B_2, B_3, ... A model of such machinery in its simplest version can be an unfolded strand *C*. The function of *C* is that of a collector strand for the subunits B_1, B_2, B_3, ... that are guided by this strand to the growth region of the aggregate.

Consider strands B_1, B_2 that are capable of assuming the conformation of hairpins (Fig. 3a), with legs twisted into a double helix. Such hairpin strands would be capable of interlocking into a picket-fence like aggregate (Fig. 3b), stabilized perhaps by bivalent cations, such as Mg^{+2} or Ca^{+2} ions, that form links between negatively charged phosphatidyl groups on the outside of neighboring hairpin strands.

A detailed model with GC RNA strands in the conformation of a left-handed double helix analogous to the RICH-ARNOTT-DICKERSON Z conformation [8] exhibits an astonishingly good fit between neighboring hairpin subunits (Fig. 3c). Equally excellent is the steric fit between "anticodon" loops that form the hairpin bends and "codons" along the collector strand. The bases in these complementary triplets are stacked, providing stabilization, and the 3′5′ directions of the strands connected by complementary base triplets are opposite.[2]

2 Aggregation can also be achieved in the case of right handed double helices of nucleic acids when adding multivalent cations (formation of bundle-like aggregates of about 7 molecules) [9].

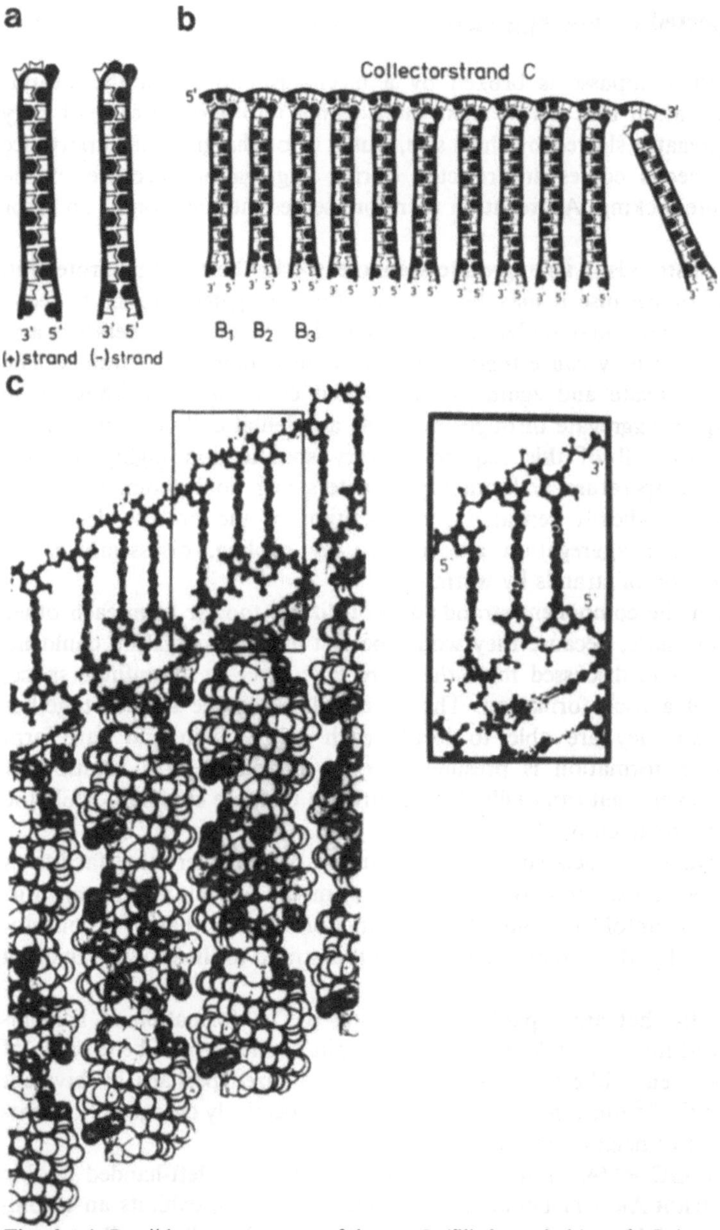

Fig. 3. a) Possible arrangements of bases *G* (filled symbols) and *C* (open symbols) that permits a hairpin conformation. (+) and (—) strand.

b) Schematic view of collector strand *C*, and subunits $B_1 B_2 B_3$ ... forming picket-fence like aggregate. Attachment of subunits to collector strand by base triplets. In the present case B_1 is a (+) strand, B_2 and B_3 are (—) strands.

c) Section of collector strand and attached subunits. For clarity nucleotides located along the collector strand and in the "anticodon" loops are shown as ball-and-strick models, the remaining portions of the hairpin strands are shown as room-filling models. The ends of the hairpins are shown bonded to each other in this figure. The inset depicts details of the bonding between "codon" and "anticodon".

Comparative studies by EIGEN [10] of t-RNA sequences indicate that all t-RNA's developed from a common ancestor which may have consisted mainly of G and C nucleotides. It is therefore not implausible that at first there were G and C nucleotides only. Morevover, the ancestral t-RNA sequence derived by EIGEN [10] from comparisons among the known nucleotide sequences of many present-day t-RNA's is such that only a few changes are needed to allow a perfect hairpin conformation.

An attractive feature of the hairpin conformation is that (+) and (—) strands are identical except for the nucleotide in the middle and the nucleotides at the open ends (Fig. 3a), a feature that permits great economy in the usage of strand material and the assemblage of aggregates. The nucleotide in the first position of the triplet at the hairpin bend can be C in (+) as well as (—) strands; in the third position it is G in both strand types. At corresponding first positions of the collector strand there must therefore be G and at third positions C. The midpositions can randomly be occupied by G or C.

Catalytic Activity of Aggregates: Polypeptides Serving as "Cellular" Envelopes

There exists selective pressure towards the invasion of pores that contain sufficient monomers, but are too large effectively to curb strand diffusion. It is proposed that such pores can be colonized by the aid of polypeptides formed, by catalytic action of the aggregates, from amino acids presumed to be present on prebiotic earth. Such polypeptides could form impediments in pore channels that slow diffusion; later they could coalesce into confining "cellular" envelopes that keep strands together as aggregates disassemble into strands which replicate and again aggregate. A strong evolutionary gradient would exist in this manner.

In our model the aggregates described develop catalytic properties in the following way. The open ends of the hairpin strands possess or develop affinity towards activated amino acids that are thereby brought into favorable proximity of each other and enabled to form polypeptide bonds; that is, the hairpin strands become adapters of amino acids. Open ends of the hairpin strands would be of help in strand replication as well, because replication could automatically begin at an end of a strand only and offer selective advantages in this way.

Both (+) and (—) strands would be capable of serving as adapters. To fix ideas, assume that the midpositions of a (+) strand are occupied by bases C, the ends by bases G; then there must be bases G at the midpositions and bases C at the ends of a (—) strand (Fig. 3a). The two kinds of strand ends could then have different affinities for two kinds of activated amino acids. Amino acids bonded to nucleotides could have existed prebiotically, for example, a_1 combined with C and a_2 with G. In this way (+) strands would serve automatically as adapters C_{a1} for a_1 and (—) strands as adapters G_{a2} for a_2.

Specifity of Amino Acid-Nucleotide Interaction: Origin of Stereospecificity of Amino Acids

A possible explanation of how a specific linkage of an amino acid to a nucleotide could have arisen is illustrated for (+) strands in Fig. 4a. The amino acid a_1 is activated by a nucleotide (G) (cf. Fig. 4a, step 1). Intercalation and complementary base pairing would mediate specific linkage and place the amino acid in such a position that reaction with the 2'-OH group of the ribose of the terminal nucleotide of the strand could occur. Our molecular model shows that the steric condition for reaction is only given for intercalation of the strand with the 3' end, and that the whole sequence of steps described would be sterically quite plausible, permitting reaction with the 2'-OH group of the ribose in question. This would correspond to the same linkage that is found in today's charged tRNA. The steric fit and the possibility of hydrogen bond formation between the amino group of the amino acid and the 3' OH-group of the ribose can only be realized with L-amino acids (Fig. 4b). This may indicate the origin of the stereospecifity of amino acids in biological systems and explain the astonishing fact that each amino acid selected in nature for the genetic machinery is L (not D). Amino acid a_2 could be activated in a corresponding way by a pyrimidine nucleotide (C) and then linked to the terminus of a (—) strand. The open ends with the amino acids have sufficient freedom of motion so peptide bonds can be formed in the aggregate since the two bases at the 3' end are not paired with

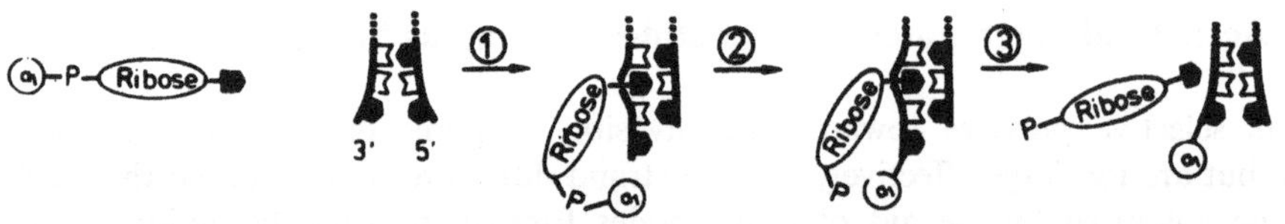

Fig. 4a. Possible mode of attachment of amino acid a_1 that has been activated by a purine nucleotide to the end of a (+) strand.

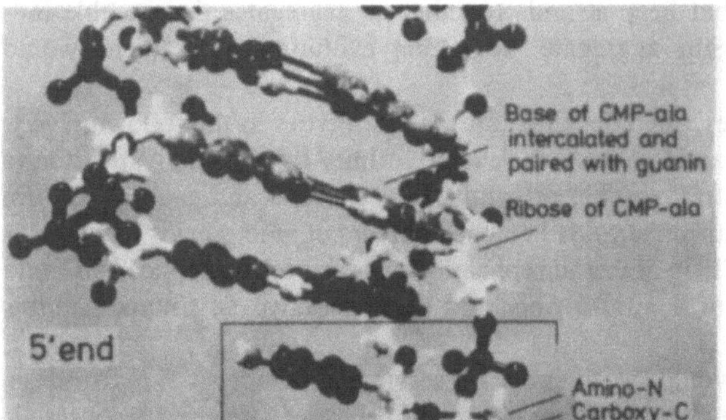

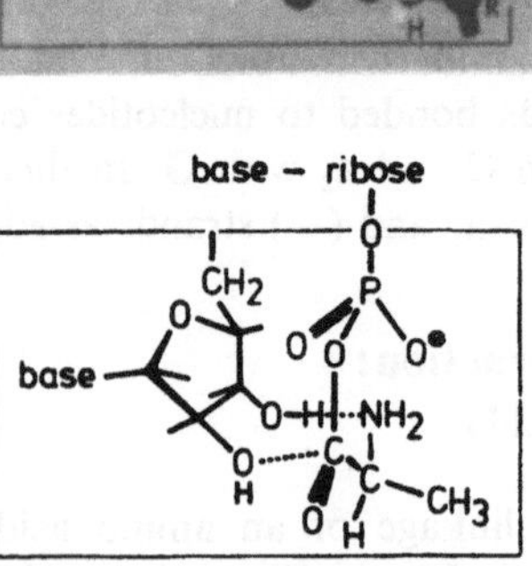

Fig. 4b. Model showing situations (2) in more detail.

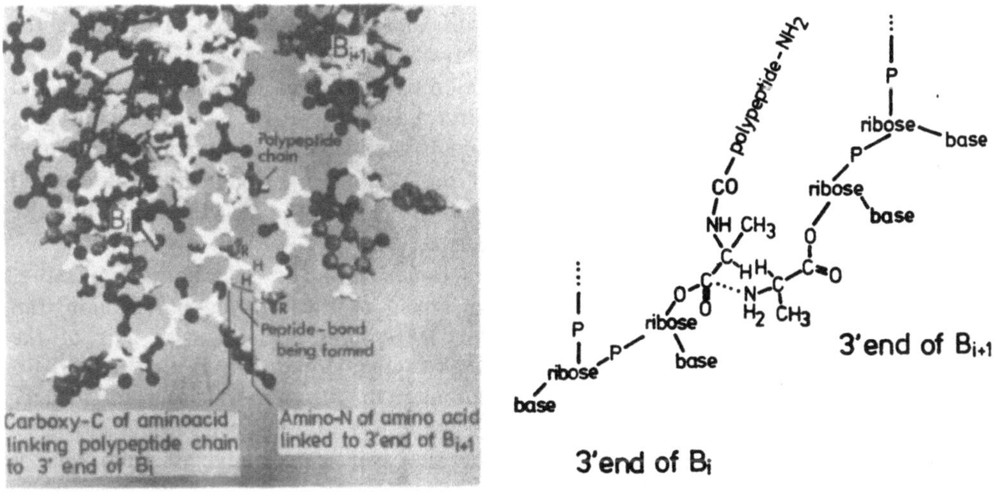

Fig. 4c. Model showing peptide bond formation in more detail. Polypeptide linked to subunit B_i and amino acid linked to subunit B_{i+1} at 3′ end by ester formation with the ribose 2′ OH group. The *C*—0 bond linking the polypeptide to B_i is opened while a peptide bond is formed between carboxy *C* of polypeptide and *N* of amino acid. By this process the polypeptide chain moving from subunit B_i to subunit B_{i+1} adds one chain member. The steric fit in Figs. 3b and 3c is no more present in the case of *D* amino acid (exchange of *R* and *H*).

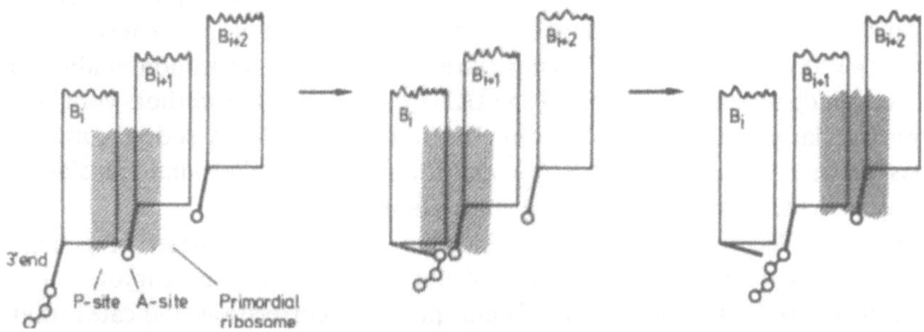

Fig. 4d. Possible role of primeval ribosome. "Ribosome" at line-up elements B_i and B_{i+1} charged with peptide and amino acid respectively (left). *P* site and *A* site occupied, ribosome fixed, allowing formation of additional peptide bond (middle). "Ribosome" (free to diffuse on aggregate-ribbon after peptide bond has been formed) reaches line-up elements B_{i+1} and B_{i+2} (right). "Ribosome" fixed to line-up elements catalyses again peptide bond formation, and in this manner the polypeptide is formed.
In our molecular model of the stabilized situation in the middle of Fig. 3d the base at the 3′ end of the "ribosome" is paired with the base at the 3′ end of B_{i+1}, the base at the 5′ end of the "ribosome" is paired with the base at the 3′ end of B_i and the arrangement is stabilized by stacking interaction with the "ribosome".

corresponding bases at the 5′ end. Our molecular model shows a steric fit allowing peptide bond formation (Fig. 4c). One can imagine possibilities for a specific activation of amino acids under prebiotic conditions. Dr. LEHMANN in our laboratory has shown by chromatographic separation of amino acids and nucleotides using salt solution as eluent, that glycine and the guanine nucleotide are eluted at the same speed, while alanine and the cytosine nucleotide eluted at another significantly different speed. This offers the possibility of in situ formation of the proper activated forms.

There is a striking similarity between the functioning of the aggregates here described and the operation of present-day genetic machinery, a similarity that is most encouraging if not altogether surprising. The collector strand **C** would be the precursor of messenger RNA, the subunits $\mathbf{B}_i$ those of tRNA's and the "cellular" envelope that of the cell membrane. As is the case today, codons along **C** are read in the 5′3′ direction while the strand direction is opposite for attached subunits $\mathbf{B}_i$.

Such an aggregate may be stabilized by interaction with one or several other molecules. In the region in which new amino acids are linked to the growing polypeptide, the aggregate "ribbon" might partially enfold an RNA double helix, with axis of the adapter hairpins parallel to the axis of the RNA double strand. The RNA double helix might glide along the aggregate "ribbon", remaining in the region in which the amino acids are linked to the growing polypeptide. In this way the RNA double helix would serve as a primordial ribosome, a precursor of the device that translocates adapters from what is called the A site to the P site (Fig. 4d).

It should be emphasized that the study of all possible conformations of nucleic double helixes is still a very new field, and further stable conformations may well be formed (see e.g. the DNA conformations recently proposed by HOPKINS [11]) and may be important in the search for better models. The essential point is the cooperative stabilization of a complex of collector strand and of at least two adjacent adaptors, and the close fit of codon and anticodon, regardless of whether this can be achieved by left-handed or right-handed double helixes or by other conformational details. All that is needed is a primitive translation device based on a mechanism for rejecting faulty replicates.

PÖRSCHKE and LABUDA [12] have recently demonstrated that codon-anticodon binding of $tRNA^{Phe}$ from yeast induces an association of the tRNA molecules. The authors conclude from their observed large free energy for dimerisation of the tRNA-codon complex that interactions between adjacent tRNA's are important for the ribosomal translation process. This may indicate vestiges of the early translation device.

The view that the collector strand is the primordial form of the carrier of genetic information is supported by a sequence analysis of the DNA of virus, procaryotes, and eucaryotes by SHEPHERD [13]. The clearly found periodic correlation indicates that originally there existed the reading frame PuNPy (Pu-purine, as G; Py-pyrimidine, as C; N-purin or pyrimidine), vestiges of which are still in evidence.

Experiments that would test the proposed molecular model should be very fruitful and are highly desirable.

The example of the emergence of a translation apparatus illustrates the method of of procedure in our model. Further breakthrough phases appear which are again concerned with the difficulties in increasing the amount of genetic information. These difficulties are due to the logic requirements in the different stages of the evolutionary process.

Organizational Framework of the Evolution Process Logical Requirements and their Realizations

Under appropriate circumstances a system appears that is capable of learning (Fig. 5, step 1). The complexity increases but stagnation sets in, because of the accumulation of

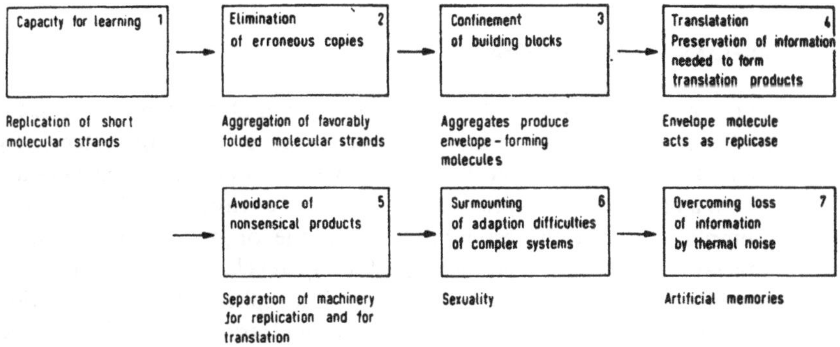

Fig. 5. Organizational framework of the evolution process. Logical requirements (boxes) and their realizations.

replication errors. To overcome it, machinery for the discarding of erroneous copies is needed and feasible (step 2). The complexity of the systems can again increase, until stagnation occurs, because the systems are tied to pre-existing compartmentation. To overcome this barrier machinery is needed and achievable that permits a confinement of building blocks that is independent of external structure (step 3). Given appropriate conditions this machinery develops by necessity into machinery for code translation and machinery that permits the preservation of the information needed for the forming of translation products (step 4). This development in turn allows an increase in complexity until stagnation sets in at a certain level, because of the accumulation of nonsense products. The surmounting of this hurdle (step 5) requires machinery that permits the bypassing of meaningless production and can be achieved by a reorganization of the systems. Further evolution leads to increasingly complex and thus ever more delicate systems, ending again in stagnation because of inadequate adaptability of the systems. The barrier can again be hurdled by a fundamental reorganization of the systems (step 6). This change permits the achievement of a level of complexity that cannot be raised further because of thermal noise, until another fundamental restructuring of the organization permits a conquest of this barrier (step 7).

Knowledge as a Measure of the Usefulness of Information

The process of evolution can be described as a phenomenon that starts with a bang, somewhere and sometimes in the development of any suitable planet (Fig. 1). All of a sudden a strand arises that can replicate, thereby creating a system that adapts itself again and again to a changing environment, in many steps of replication and selection. This evolution also proceeds in jumps. Nothing of importance happens for long periods, until a reorganization of the system ushers in a rapid new development. In a corresponding way there are discontinuous increases in the contents of the message of which the systems are carriers. An important function increases that is a measure of the evolution of the systems, the knowledge of the systems. We roughly define this function as the information (measured in bits or yes-no decisions) contained in the totality of the blueprints that has to be rejected by necessity until the stage of evolution considered was reached. Knowledge is a measure of the usefulness of the information accumulated during the course of evolution.

Origin of Life — DARWINian or non-DARWINian Process?

The basic idea of the present approach is that the origin of life is a DARWINian process, the sudden appearance of a new quality of matter, the emergence of learning systems. By this mechanism more and more sophisticated forms of cooperation between molecules are developed resulting from a continuous adaption of the system to environmental conditions. The contrasting assumption is that the essential event in the origin of life is a non-DARWINian self-organizational process: The buildup of a hypercycle in a homogenous or quasi homogeneous solution [14]. Hypercycles are formed if a replication and translation machinery is present, in other words a device by which nucleic acid strands are replicated and by which the information given by the sequence of nucleotides in these strands is translated into proteins of corresponding sequence of amino acids. Devices of this kind cooperate in such a way that the translation product of a first device catalyses the action of a second device, the second device acts similarly on a third and so on, and the last acts on the first. This complex reaction network dominates in an appropriate homogeneous solution and suppresses competition, and therefore no DARWINian evolution can occur. In this contrasting view then a new principle of selforganization, different from DARWIN's principle, is required to explain the origin of life.

The hypercycle is a principle of selforganization by which the catastrophy due to accumulation of copying errors can be avoided: The errors are suppressed by the functional integration of replicating units.

Difficulties in that approach are:

1 The translation machinery is presupposed as the result of some unspecified inherent selforganization process. In our opinion, the fundamental enigma is the mechanism that caused the emergence of a translation apparatus. The degree of sophistication of such an apparatus could only have developed after evolution of an effective mechanism for the removal of replication errors. The chicken-and-egg problem is unsolved: a device to suppress replication errors is needed to build up a mechanism — the hypercycle — whose purpose is to suppress replication errors.

2 The non-DARWINian nature of the hypercycle is due to the fact that the translation product of each strand is spread over the solution; there is no binding of gene and translation product. A mutated strand which has (by chance) a favorable translation product is not specially favored; the translation product is equally available to everybody in the molecular society; the originator of a favorable product has no particular advantage. The difficulty would be avoided by compartmentation. The gene would then be bound to its own translation product. The system would be subject to DARWINian selection. However, it seems difficult to find a plausible model for such a fundamental reorganization process to occur at that stage. It can be imagined that a cell-membrane-like envelope is produced by the system. Such an envolope would need to be quite sophisticated for being favorable and no way has been indicated of how it could originate with an acceptable probability.

In the present approach copying errors are suppressed at a very early stage by formation of aggregates of interlocking molecules. Also in this case the copying-error catastrophy is omitted by functional integration of different replicating molecules. However, no new principle, besides DARWIN's principle, is used to achieve that integration. The driving force for cooperation and formation of more and more complex systems is a consequence of a continuous adaptation of the systems to environmental conditions (the chances of survival are increased by cooperation).

The difficult problem of compartmentation of a complex network of chemical reactions in homogenous solution does not appear in the present approach. In all stages of evolution the systems are in compartments allowing their DARWINian evolution (first compartments are pores in a rock formation, later simple selfmade envelopes and much later sophisticated cell membranes). There is no need in this case for a sudden appearance of a sophisticated selfmade membrane; such membrane results from a long evolutionary process.

We do not see why selforganization should have an advantage to start in a homogeneous or quasi homogeneous solution. Monomers can diffuse easily through the pores in a rock formation, while the strands carrying information are kept in the favorable region — and this is an advantage. Furthermore, the porous structure stimulates selection and evolution of longer strands from short strands since such strands can populate larger porous regions. The pores are crucial in the development of aggregates. Increasing independence of the porous region by developing a selfmade envelope is a strong driving force towards the evolution of a cell membrane and a translation machinery.

Why Universality of Genetic Code?

The universality of the genetic code is often considered as a fundamental enigma. It should be noted however that the systems were so simple at the time when the competition for the best coding machinery was in progress, that an appropriate system could adapt to the ecological niche of any other system and eliminate that system. Therefore the descendants of one form could win out in the course of numerous conquests.

Suppose that in one location on the earth systems with code C_1 evolved and in another location systems with code C_2 that later invaded the region of C_1-systems. Such a situation implies that the stage of evolution of the C_2-systems was higher than that of the C_1-systems, because the invasion of the C_1-region required a higher degree of emancipation of the C_2-form from its environment. Two situations may have arisen. (1) The newly arrived C_2-systems were able to exploit the raw materials in the new region and displaced the C_1-systems. (2) The C_1-systems were better adapted to the region considered than the C_2-systems, better utilizing the available raw materials. As a consequence a host-guest relationship would establish itself, in which the C_1-systems serve as suppliers to the C_2-systems. Such a relationship would not be stable, however, because of the higher degree of organization of the C_2-systems. Forms of the C_2-systems would evolve that would utilize the raw materials directly and thus more efficiently than was possible with C_1-systems acting as intermediaries. Henceforth the presence of C_1-systems no longer would be required and they would disappear. It is also conceivable that machinery needed by the host for the processing of raw material was slowly replaced by more efficient machinery by the guest. The host then would slowly degenerate and disappear. A stable coexistence in a host-guest relation can develop only at an evolutionary level in which both systems are sufficiently complex and structurally different so that they continue to depend on each other — the guests do not benefit by evolving towards independence from the hosts and the hosts are unable to evolve towards elimination of their guests.

Concluding Remarks

The cardinal problem concerning the origin of life is the question of how one can imagine the development of simplest systems that could adapt to their environment — systems consisting of a few macromolecules capable of mutual cooperation. A necessary requirement for this process is the existence of a spatially and temporally structured environment, needed to keep the potential building blocks of such systems from diffusing away from the region of importance and to drive the replication, assemblage, and disassemblage of the aggregates. A purpose of this account is the stimulation of experiments and of theoretical efforts towards the improvement and the refinement of the model described. The present model should be a possibility demonstrating principle ways to overcome the fundamental barrier. Basic to the approach is its paradigm: The search for detailed models, not for general principles of selforganization. This paradigm is based on the notion that the crucial event is a very particular cooperative phenomenon, the formation of an adaptable device possessing a selfreplication and translation machinery and this phenomenon cannot be grasped by general considerations.

References:

[1] DARWIN, C.: Letter to Hooker, dated Febr. 1, 1871 (Ref. from M. CALVIN, Chemical Evolution, Clarendon Press Oxford (1969), p. 5).
[2] DARWIN, C.: The Origin of Species: A Mentor Book from New American Library, New York (1958), p.450.
[3] KUHN, H.: Angew. Chem. **18**, 838 (1972); Angew. Chem. Intern. Ed. **11**, 798 (1972).
[4] KUHN, H. and J. WASER: Angew. Chem. **93**, 495 (1981); Angew. Chem. Intern. Ed. **20**, 500 (1981).
[5] KUHN, H. and J. WASER, In: Biophysics — A textbook, eds. W. HOPPE, W. LOHMANN, H. MARKL, H. ZIEGLER, Springer Berlin, Heidelberg 1982; H. KUHN, In: Biophysik — Ein Lehrbuch, Springer Verlag Berlin, Heidelberg 1977, p. 187—197.
[6] KUHN, H., In: E. GEISSLER, J. H. SCHARF, W. SCHELER, M. KOCH: Diskretität und Stetigkeit von Lebensprozessen. Akademie-Verlag Berlin 1977.
[7] LOHRMANN, R., P. K. BRIDSON and L. E. ORGEL: Science **208**, 1464 (1980).
[8] WANG, A. H.-J., G. J. QUIGLEY, F. J. KOLPAK, J. L. CRAWFORD, J. H. von BOOM, G. van der MAREL, and S. RICH: Nature **282**, 680 (1979);
DREW, H., T. TAJANO, S. TAKANO, K. ITAKURA and R. E. DICKERSON: Nature **286**, 567 (1980); ARNOTT, S., R. CHADRASEKARAN, D. L. BIRDSALL, A. G. W. LESLIE, R. L. RATLIFF: Nature **283**, 743 (1980).
[9] MANDELKERN, M., N. DATTAGUPTA, and D. M. CROTHERS: Proc. Nat. Acad. Sci. USA **78** (1981), 4294.
[10] EIGEN, M. and R. WINKLER: Naturwiss. **68**, 217 (1981).
[11] HOPKINS, R. C.: Science **211**, 289 (1981).
[12] LABUDA, D. and D. PÖRSCHKE: Biochem. **19**, 3799 (1980); D. PÖRSCHKE and D. LABUDA: **21** (1982), 53.
[13] SHEPHERD, J. C. W.; J. Molec. Evolution **17** (1981), 94; Proc. Natl. Acad. Sci. USA **78**, 1596 (1981).
[14] EIGEN, M. and P. SCHUSTER: Naturw. **64**, 541 (1977); **65**; 7, 341 (1978); P. SCHUSTER: Prebiotic Evolution, In: H. GUTFREUND. ed. Biochemical Evolution, Cambridge University Press, Cambridge (U.K.) (1981).

Discussion

SCHUSTER: Let me state the points upon which the ideas of Prof. KUHN and of us diverge: 1. Prof. KUHN suggests that selforganization in constant and homogeneous environments is of no relevance for the problems discussed in context with the origin of life. We think that the properties of polynucleotides,

I have discussed in my lecture, are the basis of the evolution of macromolecules. Optimization and evolution do occur under constant conditions. What the environment does is the creation of variation. On this variants ultimately long term evolution operates.

2. The solutions and results we have derived are not restricted to stationary states. We are able to show that they hold for almost all reasonable time programs as well. The only difference: concentrations have to be replaced by relative concentrations. Selection takes place, cooperation occurs within the system, no matter, how the total size of the population changes. We claim, in contrast to Prof. KUHN, that the phenomena observed at stationarity will occur also in growing or shrinking populations.

3. There exists no physico-chemically sound model which can explain the origin of the genetic code and of the translation machinery. The various suggestions made have to be subject to experimental tests. These tests are very hard to conduct and more profound knowledge is required, in particular experimental information on polypeptide-polynucleotide interaction is necessary before detailed models are meaningful.

4. The concept of catalytic coupling through hypercycles has been proposed to overcome the error catastrophy. Complex formation without being intimately related to replication — in this case the system would grow according to some kinetics which is of the same type as hypercyclic coupling — does not solve the problem.

KUHN: I fully agree with Dr. SCHUSTER in his statement that experiments are important, but I do not see why detailed models should not be valuable at the present stage. In contrary, for proposing important experiments the steps in a model pathway cannot be detailed enough. A detailed model should not be seen as describing the actual path followed by nature, but the detailed consideration of an imaginable path is a method to recognize the logical framework, to see a possible way to overcome the basic barriers in a logic scheme. In more general consideration it is easy to overlook the crucial difficulties in the selforganization of matter.

I have indicated in my paper why I consider a structured enrironment as crucial: I do not see another possibility besides interlocking aggregates to solve the problem of avoiding accumulation of replication errors at this stage. Therefore a compartimentation (e.g. pores in a rock formation) is needed to bring component molecules together (e.g. a hairpin and its replicate hairpin).

I cannot see why the emergence of a translation machinery should be a fundamental enigma. I have indicated in my paper a detailed model for the formation of a simple translation device whose experimental realization would be a great challenge.

H. A. ROSENTHAL: I would like to remind you that we had exactly the same discussion 5 years ago in Kühlungsborn. I fully agree with the view of Dr. SCHUSTER. Evolution should have proceeded via a concord formation of oligo- or polynucleotides and oligo- or polypeptides. If you assume polynucleotides might have been formed by change, why not assume that polypeptides were formed at the same time. There is chemical binding of amino acids to oligonucleotides. Very primitive "machines" of interactions between both types of molecules might have evolved to build up structures with replication and, later, also of translation. The so-called hypercycle might have had a less complicated structure than that published by EIGEN 10 years ago.

MENDE: The two viewpoints of KUHN and SCHUSTER seem not to be strictly contradictory in my opinion. They only concern different stages of early prebiotic evolution. The question is how early did the translation aparatus evolve out of the unity of structure and function, how complex is the most simple translation apparatus and is it simpler than the pattern-matched aggregation mechanism, and how did such an initially very crude mechanism improve step by step to overcome error catastrophies. The often heard argument that hypercycles show the important property of hyperbolic growth kinetics is not valid because rate-coupled chainreactions also show hyperbolic growth and once-for-ever selection. The exponential tower which I showed in my lecture on Monday is an example of such a chain mechanism, and hypercycles emerge by joining the input and output of such a tower. So rate-coupled chains are plausible as prestages of hypercycles which could lead to agregation behaviour in the sense of KUHN.

My other point: in a complex environment with diffusion barriers the influence of waste products such as error copies, may become a significant factor. The results of life activity in proximity to a system with feedback to just that system (so called selfinduction) are very often important evolution factors which drive the system ahead.

ANTONOV: I'ld like to comment that von NEUMANN has estimated the quantity of information necessary for a self-replicating machine to function and multiply. It is very high and comparable to that encoded in *E. coli* chromosomes. Thus the experimental evidence discussed and various models suggested are unable to explain the origin of life as they deal with much simpler systems.

SCHUSTER: I would like to comment once again on the error problem: not the lethal mutants endanger the master copy. They will die out anyway. The slightly deleterious mutants are the real problem. The American biologist MULLER has put this dilemma into a nice metaphor, known as MULLER's ratchet: Slightly deleterious mutants dilute out the master copy until it has disappeared and then the whole process starts again with the best but one which in turn is replaced by the best but two etc. I understand that complex formation, as Prof. KUHN suggests, can exclude very inefficient variants like lethal mutations, but I cannot visualize how slightly less fitting molecules, and there is a great number of these mutants, could be discriminated. Complex formation, to my opinion, is no way to solve the error problem. It may and will have been important for other reasons.

KUHN: I do not see why Dr. SCHUSTER thinks that point-mutations are insufficient in filtering out the corresponding molecules. A single mutation in a hairpin would change a complementary base pair into a pair of equal bases and in many cases this would be sufficient for obtaining a strand that would not form a hairpin anymore and would then not be able to take part in the aggregate.
I do not see how the molecular machinery proposed by Dr. ROSENTHAL should be related to a hypercycle since a hypercycle is a spatially extended reaction network and is not a compact machinery.
The discussion by von NEUMANN is concerned with the emergence of a machinery in a single step, while the model discussed in my paper is the result of a series of many simple steps. A system which would result in a single step with exceedingly small probability can be obtained in a many-step process with almost certainty.

Sektion Physik, PB 04, Humboldt-Universität Berlin

Stochastic Aspects of the Evolution of Macromolecules

W. EBELING

1. Introduction

In the contributions by P. SCHUSTER and H. KUHN at this and at a previous Kühlungsborn-Colloquium two different models for the evolution of macromolecules on the early earth were presented. The aim of this contribution is to comment on a particular aspect of these models, namely the influence of stochastic effects. We are thinking in particular of random effects resulting from the discreteness of the number of molecules. Let us consider a reaction volume V filled with replicating units of different kinds. The number of units of kind i is necessarily discrete

$$N_i = 0, 1, 2, \dots, 10^{23}, \dots . \tag{1}$$

In general N_i is a macroscopic number of the order of the Avogadro number, but any error copy (mutation) starts from a state with only one or at best a few representatives. We shall show that in the initial stage following the appearance of a new mutant species, stochastic influences are of primary importance. Deterministic models should therefore always take stochastic considerations into account. The classical stochastic theory of populations (BARTLETT, 1960) or replicating molecules (EIGEN, 1971) is concerned mainly with linear growth processes. Here we shall show that the extension to nonlinear laws of growth opens some new and interesting evolutionary pathways (EBELING and FEISTEL, 1980; EBELING, SONNTAG and SCHIMANSKY-GEIER, 1981).

2. Linear growth rates

Let us assume that in the deterministic picture the number of molecules of a new species obeys the macroscopic equation

$$dN_n/dt = E_n N_n - K_0 N_n \tag{2}$$

For simplicity we assume that besides the new mutant a master species is present in large numbers $N_m \gg N_n$ and that the condition of a constant overall number holds:

$$N_m + N_n = N = \text{const} \tag{3}$$

Then the solution of eq (2) is

$$N_m(t)/N = \left\{1 + \frac{N_m(0)}{N_n(0)} \exp(-\delta n)\right\}^{-1} \tag{4}$$

where

$$\delta = (E_n - E_m)/E_m \tag{5}$$

is the selection advantage, which can be positive or negative, and

$$n = E_m^t \tag{6}$$

is the number of generations after the appearance of the mutant.

From eq. (4) it can be seen that the new species is either rejected,

$$N_n(t) \to 0 \quad \text{if} \quad \delta < 0 \quad \text{and} \quad n \to \infty ,$$

or amplified

$$N_n(t) \to N \quad \text{if} \quad \delta > 0 \quad \text{and} \quad n \to \infty .$$

The border between amplification or rejection is always sharp.

The picture changes completely in the stochastic description. We assume that the probability of an increase or decrease in the particle number by 1 is given by

$$W^+(N_n) = E_n N_n (1 - N_n/N) \tag{7}$$

$$W^-(N_n) = E_m N_n (1 - N_n/N) .$$

Now we can calculate the probability that a mutant that starts from $N_n = 1$ will survive at least n generations (ALLEN and EBELING, 1979; EBELING and FEISTEL, 1980)

$$P_{surv} = \delta : [1 + \delta - \exp(-n\delta)] . \tag{8}$$

This curve is shown in Fig. 1. One sees the probability that a mutant will survive one, five on an infinite number of generations dependent on its selective advantage. Evidently the probability of surviving one generation is only weakly dependent on the advantage. The sharp border that appears in the deterministic model (survival or extinction) is smoothed out in the more correct stochastic picture.

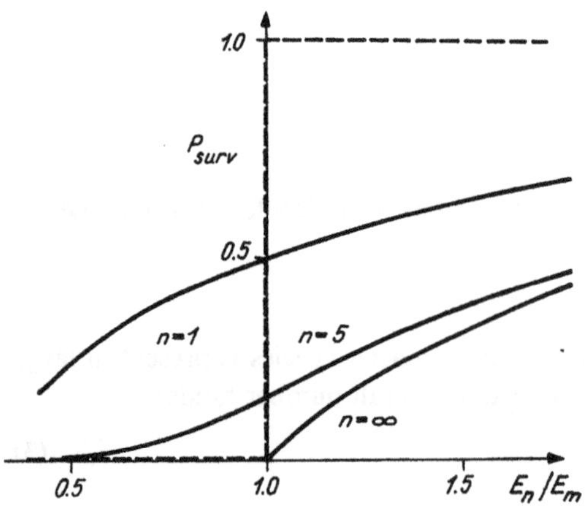

Fig. 1. The probability of survival of a mutant as a function of the relative selection advantage and the number of generations (E_n — reproduction rate of the new mutant species, E_m — reproduction rate of the master species). The number of individuals is assumed to be very large ($N = \infty$).

In the limit of infinite generations, $n \to \infty$, we find the probability that the mutant wins the competition:

$$\sigma = \lim_{n\to\infty} P_{\text{surv}} = \begin{cases} 0 & \text{if } E_n < E_m \\ 1 - E_m/E_n & \text{if } E_n \geqq E_m \end{cases}.$$

We have assumed so far that the number of individuals of the population is very large, $N \to \infty$. Let us consider now the case of infinite generations, $n \to \infty$, but finite number of individuals. Then the probability that the mutant will win the competition is given by (EBELING and FEISTEL, 1982):

$$\sigma = \sigma(\delta,N) = \delta:[(1 + \delta) - (1 + \delta)^{1-N}].$$

In the limit $N \to \infty$ we obtain the result given above. We see also that

$$\sigma(\delta,N) > \sigma(\delta,\infty).$$

In other words, the probability that the mutant wins the competition in a finite population is larger than in an infinite population. The probability that a neutral mutation ($\delta = 0$) wins the competition is

$$\sigma(0,N) = N^{-1}.$$

The probability that the master species present initially in $(N - 1)$ examples will win the competition is given by the complement

$$1 - \sigma(\delta,N).$$

We see again that the sharp distinction between better mutants ($\delta > 0$) and worse mutants ($\delta < 0$) disappears if fluctuations are taken into account.

3. Hyperbolic growth rates

The most important step in the evolution of biological macromolecules was the integration of polypeptides and polynucleotides to a functioning unit. EIGEN and SCHUSTER (1978) have shown that one mechanism capable of such an integration can be provided by hypercyclic couplings. The aim of this chapter is to investigate the influence of stochastic factors on the evolution of hypercycles. Let us first consider the competition between two hypercycles — the master copy initially present as N_m examples and a mutant hypercycle present in a rather small number $N_n \ll N_m$. Further we assume $N_n + N_m = N = \text{const}$. Then in certain approximation the behaviour of the mutant hypercycle is described by

$$dN_n/dt = B_n N_n^2 - K_0 N_n. \tag{9}$$

This equation possesses the stationary solutions

$$N_n^{(1)} = 0; \quad N_n^{(2)} = NB_n:(B_n + B_m); \quad N_n^{(3)} = N \tag{10}$$

where $N_n^{(1)}$ and $N_n^{(3)}$ correspond to stable states.

The competition between an established hypercycle (m) and its mutant (n) corresponds therefore to the dynamic behaviour of bistable systems. As a result of the hyperbolic growth the phase space $0 \leqq N_n \leqq N$ is split into two regions by the separatrix state $N_n^{(2)}$.

In general, mutations emerge at small intrinsic concentrations $N_n(0) < N_n^{(2)}$. Therefore even if they are a "better system" with respect to their kinetic properties they cannot win the competition with the established hypercycle since the trajectories cannot cross the separatrix $N_n^{(2)}$ which is an inpenetrable barrier in the deterministic theory. This behaviour changes drastically if fluctuations are taken into account. We shall show that now "better" mutants can win the competition with the established hypercycle if the reaction volume V is sufficiently small. This may be the case, for instance, in coacervates, pores or in absorbed layers on boundaries.

The stochastic process in the volume V with the total number of hypercycles $N = N_m + N_n$ is described again by transition probabilities, which are now given by

$$W^+(N_n) = B_n N_n N_m^2/NV \tag{11}$$

$$W^-(N_n) = B_m N_n^2 N_m/NV\,.$$

The states $N_n = 0$ and $N_n = N$ are absorbing states of the problem, i.e. the stationary solution is

$$P^0(N_n) = \sigma\delta(0,N_n) + (1 - \sigma)\delta(N,N_n)\,. \tag{12}$$

Here σ is the probability that the mutant will win the competition. If the mutant is present in only one example initially, we find (EBELING, SONNTAG and SCHIMANSKY-GEIER, 1981)

$$\sigma = \left(1 + \frac{B_m}{B_n}\right)^{1-N}. \tag{13}$$

Therefore, the probability that the mutant wins the competition is finite, increases with (B_n/B_m) and decreases exponentially with the total particle number in the system, i.e. with the volume. The probability of survival is shown in Fig. 2 as a function of the selection advantage $\delta = [(B\ /B_m) - 1]$ and the particle number N.

Clearly transitions to a state where the mutant survives and the master copy dies out can occur only in very small reaction volumes.

Simple estimations show that pores or coacervates with $N \lesssim 100$ and $\sigma \sim 10^{-3}$ were probably present under the conditions of the early earth.

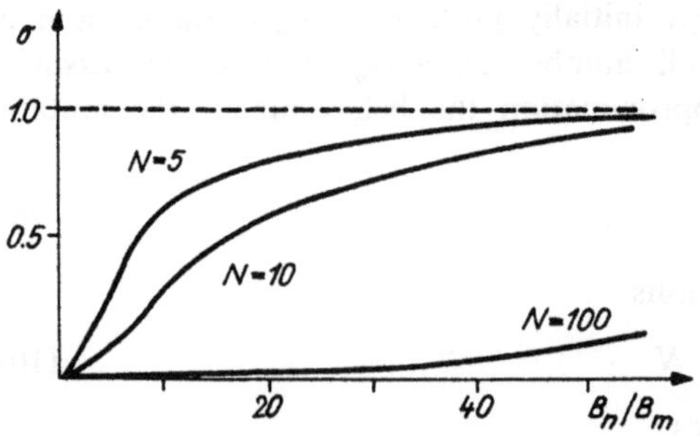

Fig. 2. The probability of survival of a mutant with a hyperbolic law of growth as a function of the relative selection advantage and the number of individuals in the volume (B_n — growth rate of the new mutant species, B_m — growth rate of the master species). The number of generations is assumed to be infinite ($n = \infty$).

We may conclude finally that the property of once-for-ever selection is lost in small systems. Hypercycles are able to evolve in small reaction volumes which form a kind of ecological niche for the new species. Following OPARIN we may assume that coacervates were the most important microreactors during the evolution of life in the early stages of earth. After winning the competition in a small coacervate volume the new mutant hypercycle may infect the whole solution by mechanical instabilities of the coacervate. Another way is the growth of coacervates filled with a "better" chemical composition (FEISTEL, ROMANOVSKY and VASILIEV, 1980; EBELING and FEISTEL, 1980).

4. Conclusions

In conclusion we underline the important role of stochastic effects in the evolution process. The small initial number of mutants leads, in the case of linear growth laws, to a much less definite selection behaviour than expected from the deterministic theory where the distinction between advantage and disadvantage is sharp. The neutrality with respect to small advantage is possibly related to the empirical fact that species always exist within certain limits of variation. In the case of hyperbolic growth, stochastic effects destroy the once-for-ever selection mode predicted by the deterministic theory. Better mutants can win the competition, crossing the separatrix, only if the reaction volume is sufficiently small, say $N \lesssim 100$. This may be the case, for instance, in coacervates, pores or in absorbed layers. After winning the competition in a small volume the new mutant may infect the whole system under certain conditions.

Finally we want to mention than hyperbolic growth rates are of fundamental importance not only for hypercyclic systems but also for a broad class of evolutionary processes.

References

[1] ALLEN, P. M., W. EBELING: On the stochastic description of a predator prey ecology. Université Libre de Bruxelles (1978). Biosystems (in press).

[2] BARTLETT, M. S.: Stochastic Population Models, Methuen, London (1960).

[3] EBELING, W., R. FEISTEL; studia biophysica **75** (1979), 131.

[4] EBELING, W., R. FEISTEL: Physik der Selbstorganisation und Evolution, Preprint 101, Akademie-Verlag Berlin 1982.

[5] EBELING, W., I. SONNTAG, L. SCHIMANSKY-GEIER: studia biophysica **84** (1981), 87 and Microfiche 1/37—54.

[6] EIGEN, M.: Naturwiss. **78** (1971), 465.

[7] EIGEN, M., P. SCHUSTER: Naturwiss. **65** (1978) 7, 341.

[8] FEISTEL, R., Yu. M. ROMANOVSKY, V. A. VASILIEV: Biofizika (Moskva) **25** (1980), 882.

[9] KUHN, H.: Selforganization in molecular aggregates and origin of life, this volume p. 171.

[10] KUHN, H.: Evolution der Makromoleküle. In: Diskretität und Stetigkeit von Lebensprozessen (E. GEISSLER, J. H. SCHARF, W. SCHELER, Hrsg.) Akademie-Verlag, Berlin 1977.

[11] SCHUSTER, P.: Grundlagen der Selbstorganisation biologischer Makromoleküle. In: Diskretität und Stetigkeit von Lebensprozessen (E. GEISSLER, J. H. SCHARF, W. SCHELER Hrsg.) Akademie-Verlag 1977 and contribution to this volume p. 151

[12] MENDE, W., M. PESCHEL: Messen—Steuern—Regeln **10** (1981), **12** (1981); **2** (1982).

[13] FEISTEL, R., W. EBELING: Biosystems **15** (1982), 291.

Department of Chemistry, Martin Luther University, DDR-4020 Halle/S., GDR

"Evolution of the Amphiphilic Pattern"

S. Hoffmann

Evolution of the molecular pattern of life has been described in terms of the development of species that exhibit within their structural and functional design both hydrophilic and and hydrophobic characteristics [6]. Dependent on the one-, two- and three-dimensional arraying of differently apted mesogens, the "omnipotent" primordial biopolyelectrolyte pool evolved into the specializations of informational, functional and compartmental components (Fig. 1). Yet the highly evolved biopolymeric structures of today have preserved beneath the cover of their specialization the ancient omnipotency of their early beginnings.

But the evolution of such an amphiphilic pattern — amphiphilic in the sense that it exhibits both hydrophilicity and hydrophobicity — appears to offer only one facet of a much more general picture, in which an amphiphilic pattern, originated within the tension between order and disorder, evolved in the creative dialectics of these two per se sterile

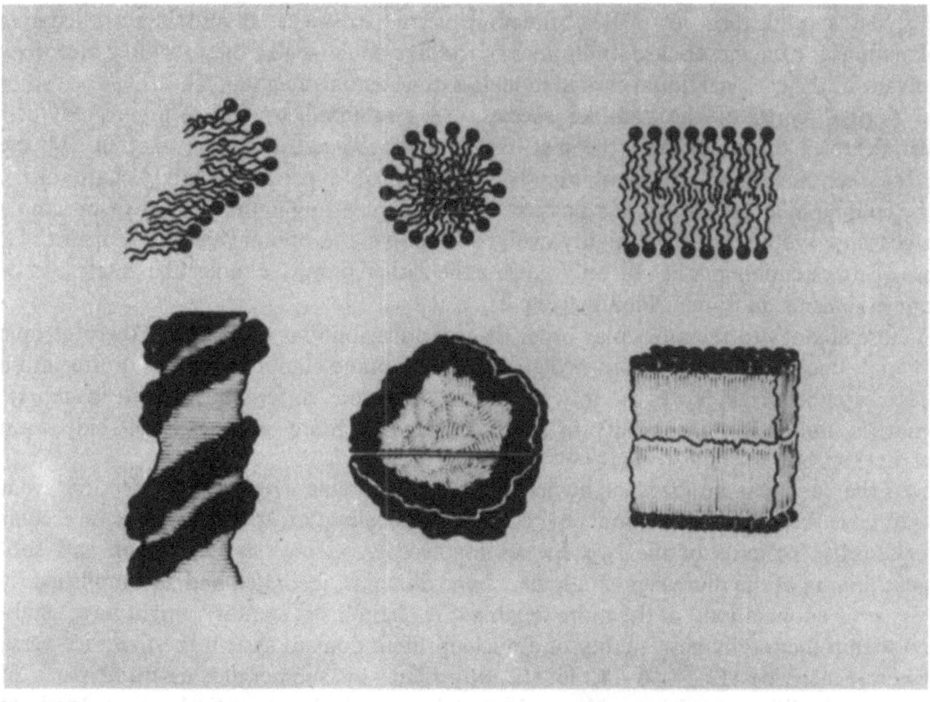

Fig. 1. One-, two- and three-dimensional variations of the principle hydrophilic-hydrophobic.

extremes. Just by imitating and reflecting the order-disorder pattern of their birth zones within the order-disorder gradients of their structural design, the very mesogenic constituents of this amphiphilic pattern provided not only important but even decisive prerequisites for the projection of individual molecular facilities into the structural and dynamic ensembles. In a continuing outburst of dynamics this amphiphilic pattern so far covers broading fields and enlarging areas of rapidly evolving "mesogenic" life.

In addition to the classical structural considerations on biopolymeric stereoelectronics and the attempts to generalize them to cover supramolecular organization, the hitherto uncommon phase view will open new and, in some respects, deeper insights into the complex processes of biopolymeric organizations. Certain aspects of their rather complicated "way of life" might even be modeled by the more transparent actionism of simple thermotropic synthetics [6—13].

These fascinating systems of quite extended "meso"-postitions between crystalline order and fluid disorder — a few decades ago the obscure object of queer outsiders, now advanced to become one of the most prominent themes of science — have commonly been characterized in terms of their thermotropic and lyotropic mesophase behavior [6, 16]. Both alternatives might, however, also, and perhaps better, be dealt with by regarding them as an amphiphilic pattern that acquires the abilities of self-organization and cooperativity by suitably aligning and organizing the specific order-disorder gradients of its mesogenic constituents (Fig. 2).

This order-disorder phase-view of mesogens not only provides a better understanding of regulatory effects in biopolymeric organizations, such as molecular hysteresis [6—13, 13, 18] and signal amplification [6—13] by directing mesophase areas. It also offers intriguing explanations for the mesophase similarities of rod-like and disc-like thermotropic mesogens — in view of the classical liquid crystal models a quite astonishing fact [3]. While the rather rigid "order"-cores of the rod-like species are solubilized by the terminals of their rather flexible "disorder"-tails, the disc-like "starfishes" gain a similar mobility by the help of their surrounding "swimming-belts" of flexible peripheral alkyl-chain arms.

The amphiphilic order-disorder pattern described here might thus range from simple thermotropic synthetics to the highly evolved biopolymeric organizations of current life. Some of its essentials might not only govern the rather primitive molecular interplays of of our organisms' molecular biology (Fig. 2).

In close analogy to the molecular order-disorder distributions of primitive thermotropic synthetics, nucleic acids as well as proteins and membrane components [6—9] offer more or less extended rigid cores in a wide variety of quite differently shaped backbone geometries, tailored to an infinity of specializations by more or less flexible side-chain patterns (Fig. 2).

Even the very first intimacy of nucleic acids and proteins within the nucleation event should have been brought about by complex annealing of the respective molecular order-disorder patterns of the two species involved. A vice-versa recognition and subsequent joining of the more rigid static backbone domains, mediated and accomplished by the dynamic cooperativity of the more flexible side-chain instrumentary, might have established within the totally new quality of a nucleoprotein domain system the first vice-versa polymerase activities [2—4, 6—8, 11, 12, 19]. Thus by submerging essential parts of their individualities, nucleic acids and proteins reached an entirely new level of promising development of their early informational and functional concepts.

By differently valuing the specific molecular order-disorder distributions of all those

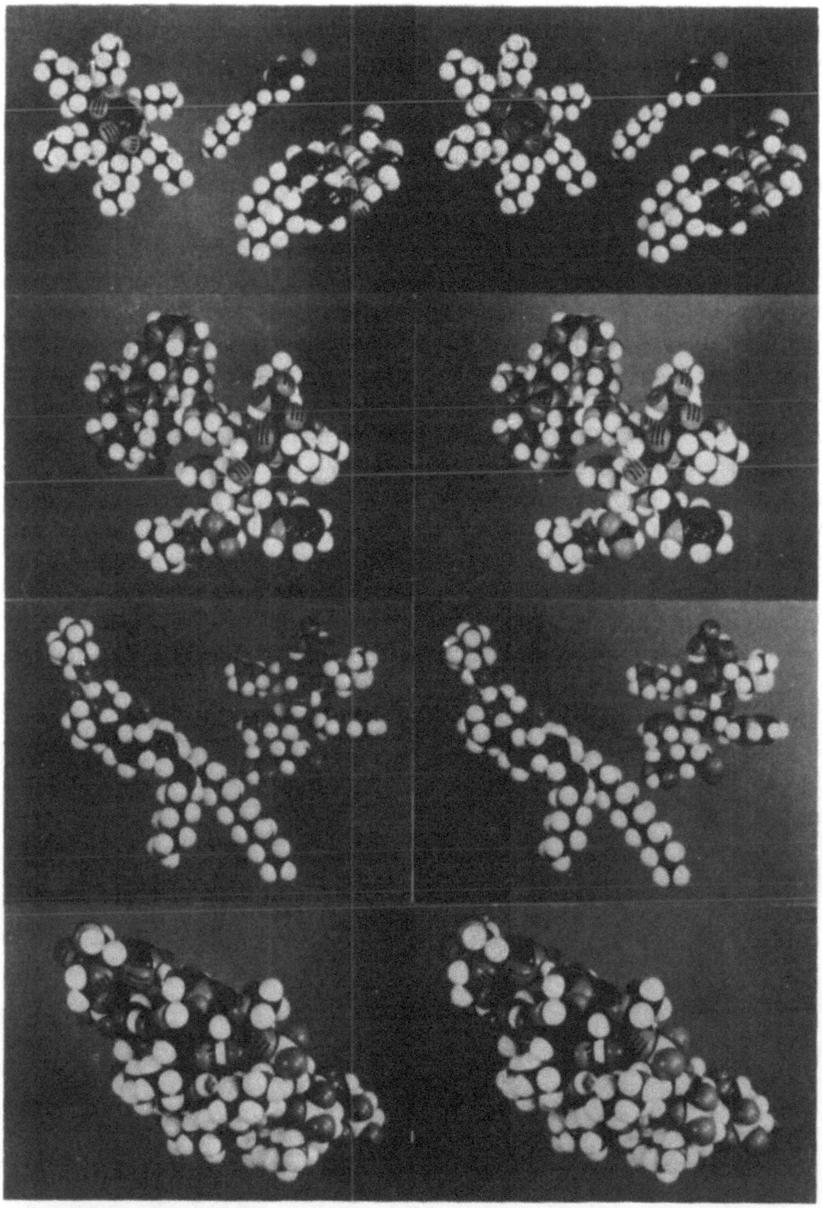

Fig. 2. Structural (bio)mesogen standards (top to bottom): disc- and rod-like (nematic, smectic) thermotropics, ssRNA and protein (α-helix and β-strand), protein (α-helix and β-strand) and lecithin, dsRNA(Olson-type) · $(L\text{-}Lys)_n$-interaction domain.

mesogenic individuals, entropy just in its tendency to minimize order and maximize disorder surprisingly should act as a driving force in building up cooperative and dynamic amphiphilic patterns with a pronounced tendency towards self-organizations and directionality (Fig. 2—6). The effectiveness and intelligence of the dynamic organizations so created and selected should have been from the very beginning a direct function of the evolving stereoelectronic sensitivity of the very mesogenic order-disorder design entropy should have acted upon.

As already implied by the structural and domain arrangement similarities of a thermotropic mesophase [1] and the helical mesophase array of a nucleic acid [20] in Fig. 3, the rather informational nucleic acids have long been looked upon as the static component of life, exhibiting, if at all, only a few possibilities for structural variability. This picture has changed more than even optimistic forecasts could have foreseen in the past [6, 8]. At present a broadening landscape enriched with a wide spectrum of nucleic acid geometries between the extrema of left- and righthandedness has opened to the amazed observer [21, 22] (Fig. 4). But even this for the future probably still enlarging structural and functional diversity will still be surpassed by the structural and functional richness of proteins [5, 7].

Though on a higher level of structural complexity and plasticity, which, of course, immediately offers higher levels of functional diversity, rigidity rather than flexibility in the case of nucleic acids and flexibility rather than rigidity in the case of proteins, as a first rough approximation, still reflect within their structural order-disorder design and derived mesophase facilities the specializations of the two biopolymeric kingdoms into the informational and functional components of the grand process. In order to secure sufficient flexibility and, moreover, fluidity in the case of membrane components [9], evolution had to avoid both extended backbone structures and spreading covalent fixations. But, though

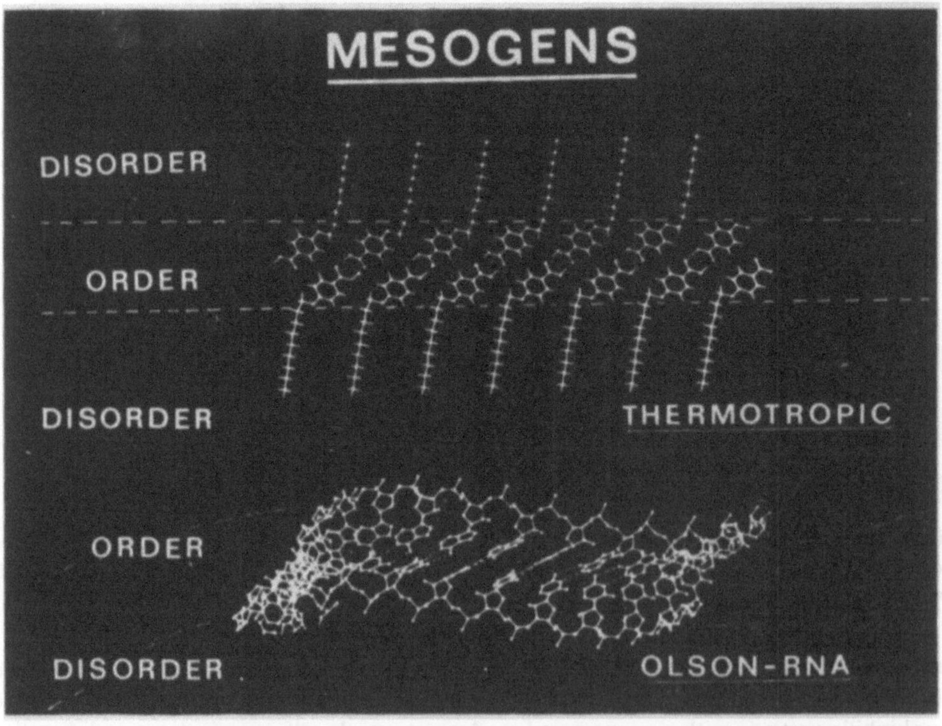

Fig. 3. Order-disorder distributions within the hypothetical mesophase arrangements of a simple smectic thermotropic and an RNA-duplex.

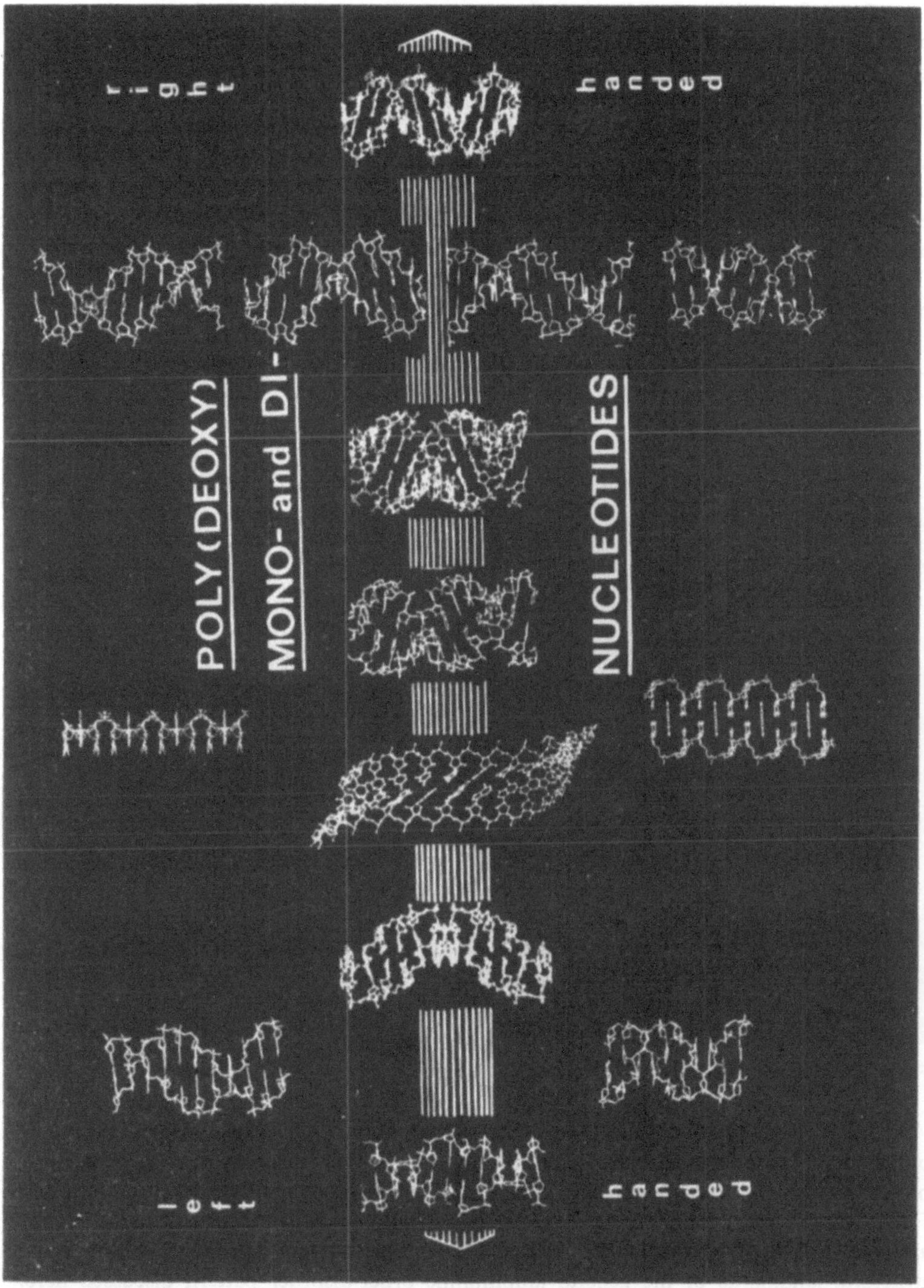

Fig. 4. Nucleic acid geometries between left- and right-handed variants.

superfluous for building up primitive bilayers, evolution secured for the rich populations of phospho- and glycolipids the very operation modes of entropy by "foreseeing" small glycerol-backbones. Thus, even in its dominating aspect of flexibility, the compartmental component of life fits well into the general picture.

In this view the amphiphilic pattern should exhibit functional activity by operating on specific order-disorder domains that might also be dealt with in terms of solute-solvent terms of solute-solvent relationships where solvent effects for rather rigid domains will be exerted not only by assistant partner molecules (as in the case of lyotropics) but also by the more flexible parts of the mesogenic molecules themselves (as in the case of thermotropics). With a hypothetical mesophase look of a simple thermotropic Fig. 5 primitively models the relatively ordered central arrays of rigid aromatics — the solute part of the system — and the relatively disordered, more flexible alkyl-chain domains — the solvent contribution of the system — within the complex arrangement of this "meso"-phase.

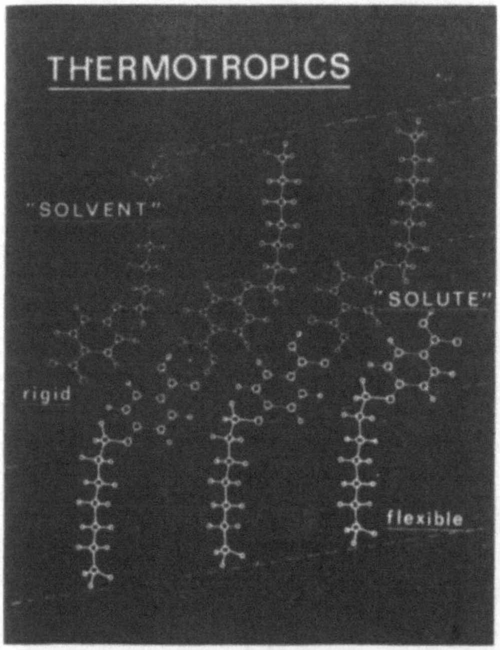

Fig. 5. Solute-solvent interpretation of a hypothetical smectic mesophase arrangement.

In this context the biologically important hydrophobic-bond organizations [23] are an outstanding, but nevertheless particular, example of a much more general order-disorder pattern of dynamic entropy-mediated amphiphilic mesogen organizations (Fig. 2 and 6). Fig. 2 tries to outline common order-disorder gradients, which have been valued, developed and screened for their capacity to build up suitable entropy gradients, within some structural landmarks of the general amphiphilic order-disorder pattern discussed here. Starting with simple and primitive experimentally screened disc-and rod-like thermotropics, ranging over the kingdoms of nucleic acids, proteins and membrane components, and ending up with a simplified example of nucleoproteinic organization [14, 15], it retraces

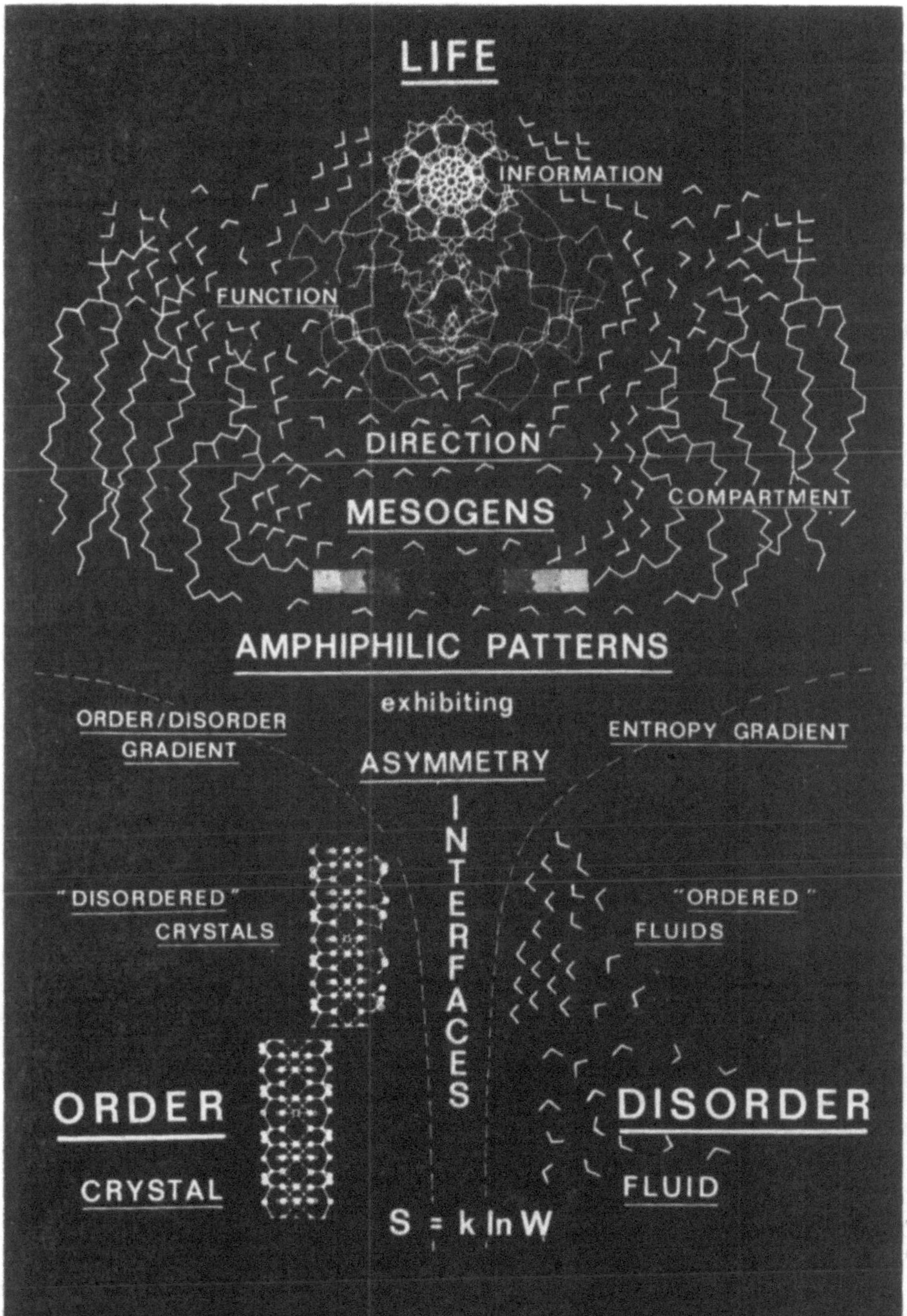

Fig. 6. DARWINian evolution of the amphiphilic pattern.

certain essentials of the structure-function relationships of those dynamic mesogenic organizations.

Born within the tensions between the two kingdoms of crystalline order and fluid disorder, molecularly imprinted by the very thesis and antithesis of their origin, first experienced in the creative dialectics of interchangeable values between "disordered" crystal surfaces and "ordered" water shells, and later on in the fluctuations of the conquered areas of their increasingly complex and permanently endangered world, mesogens managed the grand escape from the sterile extrema-futilities that dominated the boundaries of their extended interface-birth zones (Fig. 6). By intelligently reflecting the order-disorder thesis and antithesis of their origin within their individual molecular designs, the mesogenic constituents of the developing amphiphilic pattern succeeded in a fertile and creative synthesis, avoiding statics and hyperdynamics — the disadvantages of the extreme states that tried to govern their origin, and optimizing the promising advantages of the "meso"positions of ongoing ordered dynamics. It was the very molecular order-disorder design that rendered those mesogens sensitive to the amazing influences of entropy. Counteracting for these rather "intelligent" molecules in its usual eagerness and need for disorder, entropy compensated for their lost homelands of order and disorder with the gifts of self-organization, cooperativity and some sort of new-born directioned dynamic order within enlarging molecular ensembles and growing organizations.

And, like Mephisto in GOETHE's grand opus, who introduced himself as part of the power that always wants the evil while always doing good, entropy enabled these beloved playmates to propagate in a dramatic outburst of directed dynamics over the deserts of order and the abysms of disorder as an endangered, but incomparably beautiful, grand mesogenic pattern started in a grand out-set for far and unknown horizons.

References

[1] BRYAN, R. F., P. HARTLEY, R. W. MILLER, M.-S. SHEN: An X-ray study of the p-n-alkoxybenzoic acids, Mol. Cryst. Liq. Cryst. **62** (1980), 281—310.

[2] CARTER, C. W., J. KRAUT: Proposed model for interactions of polypeptides with RNA, Proc. Nat. Acad. Sci. USA **71** (1974), 283—287.

[3] CHANDRASEKHAR, S., N. V. MADHUSUDANA: Liquid crystals, Ann. Rev. Mater. Sci. **10** (1980), 133—155.

[4] CHURCH, G. M., J. L. SUSSMAN, S.-H. KIM: Secondary structural complementarity between DNA and proteins, Proc. Nat. Acad. Sci. USA **74** (1977), 1458—1462.

[5] DICKERSON, R. E., I. GEIS: Struktur und Funktion der Proteine, Verlag Chemie, Weinheim (1971), 1—120.

[6] HOFFMANN, S.: Molekulare Matrizen / Evolution, Akademie-Verlag, Berlin (1978), 1—115.

[7] HOFFMANN, S.: Molekulare Matrizen / Proteine, Akademie-Verlag, Berlin (1978), 1—294.

[8] HOFFMANN, S.: Molekulare Matrizen / Nucleinsäuren, Akademie-Verlag, Berlin (1978), 1—177.

[9] HOFFMANN, S.: Molekulare Matrizen / Membranen, Akademie-Verlag, Berlin (1978), 1—84.

[10] HOFFMANN, S., W. WITKOWSKI: Biomesogens and their models, In: Mesomorphic order in polymers and polymerization in liquid crystalline media (A. BLUMSTEIN, ed.) ACS-Symp. Ser. **74** (1978), 178—236.

[11] HOFFMANN, S.: Nucleinsäuremodelle, Z. Chem. **19** (1979), 241—249.

[12] HOFFMANN, S.: Steroid / DNA-Strukturkomplementaritäten, In: Wirkstofforschung 1980 (H. POSSIN, u. G. BARYSCH, eds.), KTB 2 (N 10), Wissenschaftspubl. Martin-Luther-Universität, Halle (1981), 35—94.

[13] HOFFMANN, S., W. WITKOWSKI: Biomesogen-Regulationen, In: Wirkstofforschung 1981 (H. POSSIN, ed.), KTB Wissenschaftspubl. Martin-Luther-Universität, Halle (1982) in press.

[14] HOFFMANN, S., W. WITKOWSKI: Das dsRNA-Erkennungsbild der Interferoninduktion, In: Wirkstoff-forschung 1981 (H. POSSIN, ed.), KTB Wissenschaftspubl. Martin-Luther-Universität, Halle (1982) in press.

[15] HOFFMANN, S., W. WITKOWSKI, S.-R. WASCHKE, K. WASCHKE, A. VECKENSTEDT: The dsRNA motif in interferons induction, Nucleic Acids Symp. Ser. **9** (1981), 239—241.

[16] KELKER, H., R. HATZ: Handbook of liquid crystals, Verlag CZemie, Weinheim—Deerfield—Beach—Basel (1980), 1—917.

[17] NEUMANN, E.: Molecular hysteresis and its cybernetic significance, Angew. Chem. int. Ed. **12** (1973), 356—369.

[18] NEUMANN, E., H.-J. NOLTE: Polyelectrolyte field effects in chemical and biological systems, In: Lectures on biological and chemical physics (M. BALABAN, ed.) Elsevier, Amsterdam (1980).

[19] NIR, S., R. GARDUNO, R. REIN, Y. COECKELENBERGH, R. D. MACELROY, J. T. EGAN: Display, manipulation and simulation of macromolecules, Int. J. Quant. Chem. Quant. Biol. Symp. **4** (1977), 135—141.

[20] OLSON, W. K.: Spatial configurations of ordered polynucleotide chains: a novel double helix, Proc. Acad. Sci. USA **74** (1977), 1775—1779.

[21] SARMA, R. H. (ed.): Nucleic acid geometry and dynamics, Pergamon Press, New York—Oxford—Toronto—Sydney—Frankfurt—Paris (1980), 1—424.

[22] SARMA, R. H. (ed.): Biomolecular stereodynamics, Adenine Press, Guilderland—New York (1981), (1981), **Vol. I**, 1—492, **Vol. II**, 1—518.

[23] TANFORD, Ch.: The hydrophobic effect and the organization of living matter, Science **200** (1978), 1012—1018.

Zentralinstitut für Molekularbiologie der Akademie der Wissenschaften der DDR, Berlin-Buch, DDR

Structural Analysis of Evolutionary Processes at the Level of the Single Gene

S. ROSENTHAL, D.-H. LIEBSCHER, T. RAPOPORT, and V. HAHN

As has been mentioned in the introductory lecture DNA recombinant technique and the rapid DNA sequencing procedure in combination with the structural and functional analysis of proteins will contribute very much to research in the field of evolution and stimulate new developments and discoveries.

The decisive progress lies in the comparative analysis of the single gene and their expression — in future possibly also of functional gene complexes — which allows an insight into the evolution of that gene. It also produces data on the behaviour of this gene during ontogenesis. The new methods permit the construction of living models which can be scrutinized under experimental conditions of the lab.

Insulin, a protein hormone, has, like haemoglobin and cytochrome C, long been a well documented subject of evolutionary research. There is no other hormone that reduces the sugar concentration in the blood. Insulin like material has also been found in insects [5] besides vertebrates.

Insulin was the first protein whose amino acid sequence was published. The existence of an insulin precursor protein had been demonstrated and its pathway to the active hormone (fig. 1) as well as the mechanisms of its intracellular transport and passage through the cellular membrane studied. Insulin belongs to the first objection of what is called genetic engineering. Microbially produced insulin of human type is already available [6]. Recently the first human insulin mutant in an adult onset diabetic was described [11]. The structure of the insulin gene was established by means of genetic

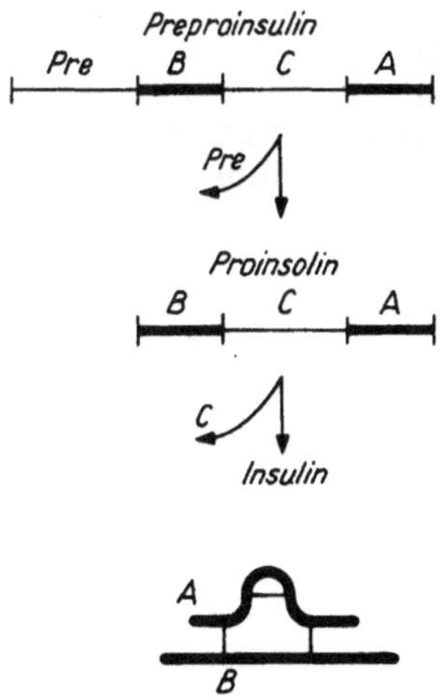

Fig. 1.

engineering [1]. However what happens after the insulin has been bound to the receptor molecule within the cellular membran is not well understood, nor is the effect of insulin on the metabolism of the cell. We do not understand, how insulin and its relatives bring about their influence on cell proliferation and growth stimulation. This brings me to what is interesting in the context of our DARWIN meeting, namely, can the variation of a single gene contribute or even lead to a qualitative jump in evolution?

We have been cloning the DNA copy of the messenger RNA of preproinsulin of the carp in *E. coli* and then analysing the nucleotide sequence and the amino acid sequence of the insulin [8, 9].

The majority of vertebrates possess one insulin species only. There are exceptions, like the rat, mouse, tunnyfish, toadfish and, a third fish species, bonito (fig. 2).

When switching to the level of DNA analysis it turned out in a first evaluation that man and chicken have one insulin gene within the haploid chromosome set, while the rat has two nonallelic genes established nearly 100 million years ago by gene duplication (fig. 3).

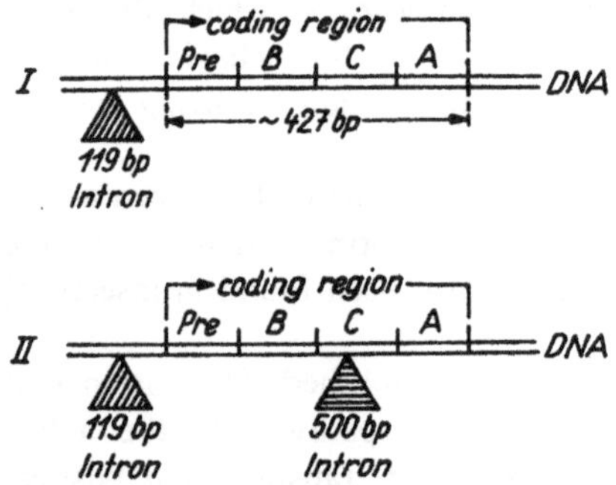

Fig. 2.

		B-2 B1 (A6)	B11		(A20) B19
carp		AsnAlaGlyAlaProGlnHisLeuCysGlySerHisLeuValAspAlaLeuTyrLeuValCys...			
human	—	— PheValAsn		Glu	
bovine	—	— PheValAsn		Glu	
rat I	—	— PheValLys	Pro	Glu	
rat II	—	— PheValLys		Glu	
other	—	MetArgThrSerArgArg	LysAsn .	AsnThr	PheSerAla
	—	ValAlaProThrGly	Asp		IleIle
		Tyr AlaPro	Hln		
		Lys			

	B20		B30
carp	...GlyProThrGlyPhePheTyrAsnProLys—		
human	GluArg	Thr	Thr
bovine	GluArg	Thr	Ala
rat I	GluArg	Thr	Ser
rat II	GluArg	Thr	MetSer
other	GlnAspAsp	TyrArgIleSerThrThrMet	
	ArgVal	AspAsnAspAsp	
	His	SerArgSerSer	
		Gln	Lys
		Pro	

(A6)
(A11)(B7) | (B19)
A1 | | A11 | A21

carp	GlyIleValGluGlnCysCysHisLysProCysSerIlePheGluLeuGlnAsnTyrCysAsn			
human		ThrSerIle	LeuTyrGln	Glu
bovine		AlaSerVal	LeuTyrGln	Glu
rat I/II	Asp	ThrSerIle	LeuTyrGln	Glu
other		GluGlyIle	ThrArgHisAsn	MetSer
		AsnThr	AspLysAsnAsp	
		ArgArg	Asn	

Fig. 3.

The rat preproinsulin gene II is the more ancient one. Like most of the eukaryotic animal genes, the human, rat II and chicken insulin genes possess one intron in the DNA-region of the 5′terminal nontranslated part of the mRNA (intron I) and another in the coding region for the C-peptide (intron II) [10]. Although it has been suggested that intron I has a functional role in the expression of the messenger, intron II seemed to have no function.

At the level of the protein structure of carp insulin our data show [9] that all amino acid residues essential for the biological activity, e.g. binding site to the insulin receptor and for the dimerization of the insulin are conserved in carp insulin (fig. 4). The amino acid sequence determined displays some unusual features of carp insulin. The B-chain is longer at the N-terminus by two residues compared with mammalian insulins, and the presequence is shortened by two amino acids. These differences reflect changes of the splitting site into the polypeptide chain by the signalase, a protease cutting off the presequences of secretory proteins. This is a very old and stable enzymatic mechanism.

Our carp insulin and all other known insulins of bony fishes have about 16 and 21 amino acid substitutions compared with mammalian and hagfish insulins, respectively. Thus, the evolutionary distance between the insulins of the most primitive fishes and the telost fishes is at the best as great as that between the the latter and mammalian insulins. The insulins of salt water teleost fishes are remarkably similar to each other (about 4 amino acid substitutions) but differ greatly from the insulins of fresh water teleost fishes (about 13 substitutions). The distance between the insulins of salt and fresh water fishes is not smaller than between those of fishes and birds.

These data concerning all these mutations and variations with — in the DNA copy of the carp insulin mRNA might be summarized in the following way (fig. 5)

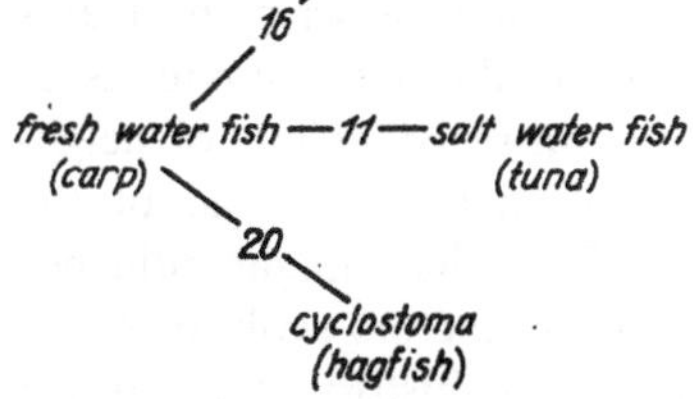

Fig. 4.

- The carp insulin cDNA shows a remarkable GC richness in the same manner as other eukaryotic genes. This is an expression of the nonrandom usage of codons.
- There exist mutations at the third position of triplets which do not affect the amino acid composition and are detectable at the nucleic acid level only.
- Other mutations lead to amino acid exchanges which do not influence the function of the hormone.
- A third group of mutations will lead to the exchange of amino acids which are no longer compatible with the function of the hormone.

	GC-content of DNA (%) Total DNA	Genomic part of Insulin gene	Coding part of B-chain
Human	38	64.8	55.5
Rat	40	55	53.3
Carp			54.7

Fig. 5.

	B4					B13				B20						
carp	GlnHisLeuCysGlySerHisLeuValAspAlaLeuTyrLeuValCysGly															
carp	CAGCACCTGTGTGGATCTCATCTAGTCGATGCCCTCTACCTGGTGTGTGGT															
chicken		C		C	C	CT	G	G	G	T						A
rat II		T		T		CT	G	G	A	T						G
rat I		T		TC		C	G	G	G	T	G					G
human	A		C	C	A	C	G	G	A	T			A		C	G
rat I		T		TC	T	C					G					G
chicken	8	C		CT	C	T		13			C		14			A
rat I	G	T	T	TC	T				G		G		G		T	
human	A	G	C	CT	A		11		A		C		A		C	
carp	G		T	A	T	T	A	C	T	C			G		T	T
human	A	8	C`	C	A	C	G	G	A	T		8	A	5	C	G
carp		G		AT		T	A	C	T	C	C					T
rat I	8	T		TC		C	G	G	G	T	G		14			G

Fig. 6.

We would like now to discuss the insulin-like growth factor. Insulin belongs to a family of related proteins like insulin growth factors (IGFI and II), relaxin and the serine proteases [2, 3, 4]. In vertebrates we find, in addition to the insulins, proinsulin-like compounds or polypeptides acting as tissue hormones stimulating the proliferative behaviour of neighbouring cells. This group of molecules the growth factors of nerve cells and of the epidermis includes. Man has at least 2 factors, IGFI and IGFII, which might influence the growth of the individual. Pigmies are defective in IGFI. These factors evolved on the basis of gene duplications quite early during, or shortly before, the history of vertebrate development. While the A-chain became 8 amino acids longer, the C-chain was shortened.

Making insulin out of proinsulin involves the activity and localization of a special protease which splits the prosequence of the entire molecule beside arginine, thereby forming the particular structure of the insulin.

By genetic manipulation monkey kidney cells received the genetic instruction to produce rat pre-proinsulin. These primate cells were like the oocytes of frogs, able to split the presequence but were unable to split proinsulin into insulin. Therefore proinsulin was secreted into the medium [7]. Similarly, fibroblasts are able, by an analogous reaction, to split hemaglutinins of influenza virus into the two active daughter molecules by cutting an arginine rich connecting peptide. In this case the specific protease is situated on the outer membrane surface and seems to be absent inside the cell. If speculations are allowed, we might conclude, therefore, that insulin evolution involved the development of a set of processes which include the appearance and expression of some intracellular specific proteases.

Conversely the development of insulin growth factors needs not only gene duplications and intragenic recombination but also the loss of the specific protease activity to split of the connective peptide.

References

[1] Bell, G. I., R. L. Pictet, W. J. Rutter, B. Cordell, E. Tischer, and H. M. Goodman: Sequence of the human insulin gene, Nature **284** (1980), 26—32.

[2] Blundell, T.: Conformation and molecular biology of polypeptide hormones, I. Insulin, insulin-like growth factor and relaxin, Trends in Bioch. Sciences **4** (1979), 51—54.

[3] Blundell, T. L., R. E. Humbel: Hormone families: pancreatic hormones and homologous growth factors, Nature **287** (1980), 781—787.

[4] Chan, S. J., S. C. M. Kwok, D. F. Steiner: The Biosynthesis of Insulin: Some Genetic and Evolutionary Aspects, Diabetes Care **4** (1981), 4—10.

[5] Duve, H., A. Thorpe, N. R. Lazarus: Isolation of material displaying insulin-like immunological and biological activity from the brain of the blowfly calliphora vomitoria, Biochem. J. 184 (1979), 221—227

[6] Bolivar, F., H. L. Heyneker, D. G. Yansura, R. Crea, T. Hirose, A. Kraszewski, K. Itakura, A. D. Riggs: Expression in *Escherichia coli* of chemically synthesized genes for human insulin, Proc. Natl. Acad. Sci. USA **76** (1979), 106—110.

[7] Gruss, P. and G. Khoury: Expression of Simian virus-40-rat preproinsulin recombinants in monkey kidney cells — use of preproinsulin RNA processing signals, Proc. Natl. Acad. Sci. USA **78** (1981), 133—137.

[8] Liebscher, D. H., C. Coutelle, T. Rapoport, V. Hahn, S. Rosenthal, S. Prehn, and R. Williamson: Cloning of carp preproinsulin cDNA in the bacterial plasmid pBR322, Gene **9** (1980), 233—246.

[9] Makower, A., R. Dettmer, T. A. Rapoport, S. Knospe, S. Prehn, P. Franke, G. Etzold, and S. Rosenthal: Carp insuline-amino acid sequence, biological activity and circular dichroism, Europ. J. Biochem. 122 (1982), 339—345.

[10] Perler, F., A. Efstratiadis, P. Lomedico, W. Gilbert, R. Kolodner, and J. Dodgson: The Evolution of genes; the chicken preproinsulin gene, Cell **20** (1980), 555—566.

[11] Tager, M., B. Given, D. Baldwin, M. Mako, J. Markese, A. Rubenstein, J. Olefsky, M. Kobayashi, O. Kolterman, and R. Poucher: A structural abnormal insulin causing human diabetes, Nature **281** (1979), 122—125.

Moscow State University, USSR

Genosystematics and Darwinism

A. S. ANTONOV

The modern synthetic theory of evolution (STE), darwinism of the 20th century, becomes the "central dogma" of the biology, although it has been and is severely criticized by many. Old theories of evolution are revived and new ones are suggested as alternatives to STE. By way of example a modern theory of "non-DARWINian" evolution may be mentioned.

The very fact of the wide-spread discussion indicates that at least some of STE theoretical contentions need further elaboration or be abandoned. What is the reason for the temporary crisis of Darwinism? As has been postulated previously by many authors, further development of STE is possible only basing on data obtained in the studies of Life at the four main levels of its organization: molecular-genetic, ontogenetic, populational and biocenotic. To thę disadvantage of STE, the "classical" biologists, its creators and ardent proponents are inclined to use in their theoretical analysis exclusively the results obtained by traditional methods, keeping apart in most of the cases the data obtained by molecular biology.

Meanwhile, it should be stressed that the studies of the molecular basis of evolution have now acquired a great importance. As a result of these efforts systematics has sprung a new branch; it has been suggested (ANTONOV, 1974) to call it genosystematics. The necessity and usefulness of such a term is prompted by the difference in the objects of investigations: phenotype in the case of "classical" systematics and genotype in genosystematics, the difference being of qualitative nature.

The problems of modern genosystematiccs may be divided into two groups.

1) those related to the evaluation of the mechanisms of DNA evolution.

2) those related to the evaluation of the interrelationships between genotype and phenotype evolution.

Data obtained by genosystematics may give a great impetus to the development of STE. It seems premature to declare that the elaboration of STE is completed and the main efforts should be undertaken in the development of its special parts (see, for example: ZAVADSKY, KOLCHINSKY, 1977). Not more justified seems to be the postulate that the main content and object of discussions concerning STE must be "the purely biological material" (TIMOFEEFF-RESSOVSKY et al., 1977). One should take into account that without a fresh spring of data discussions concerning STE for years move by the closed circle. Neither darwinists, nor their opponents can present a unequivocal proof for their theories.

It seems plausible that the reason for such a situation is in fact a "fatal ignorance" of most of the STE admirers in the recent advance of molecular biology (TIMOFEEFF-RESSOVSKY et al. 1977).

This **sancta simplicita** robs of such darwinists many arguments in their discussions with opponents, the situation being even more dramatic because genosystematics is ready to come at their rescue. Objectively, the data of genosystematics may be of no less importance for the development of STE than that of genetics half a century ago. I do not mean the results which are none other than a molecular description of phenotypically well-nown phenomena (i. e. chemical nature of genes. molecular basis of mutagenesis etc.). Data of fundamental importance are and will be harvested in the course of our studies of the principles of molecular organisation of genotypes and their functioning as well as of their evolution.

The first steps of molecular biology and genosystematics in this direction are of general interest. I should like to mention only a few of the most interesting discoveries in this field: the unraveling of the fine structure of genes, of the mechanisms of the formation of new genes in the course of evolution, various molecular aspects of genotype instability. The question is, how all these and many other facts correspond to the STE, based mainly on the results of the comprehensive study of phenotypes?

In a short presentation it is hardly possible to review all the new data, the more so that some of the results are discussed in other papers presented at this symposium; several review articles covering the field have been published recently. For this reason I shall mainly confine myself to the results obtained by my colleagues and myself in Moscow State University, which have been published in Russian, and thus could have avoided the attention of my foreign colleagues.

The first set of data and the relevant conclusions concerns such classical problems, as e.g. (I) the problem of the reality of higher taxa, (II) the equivalency of taxa of the same rank in the systems of various groups of higher eucaryotes, (III) sibling taxons, — all viewed as parts of a general problem of the mechanisms of divergent evolution.

Channelization of the Evolution of DNA

Evolutionary accumulation of genetic material was accompanied by its continuous modification under the action of diverse mutagenic factors. The nature of this process being stochastic, it would be logical to suggest that the variability of different parameters of primary structures of DNA should obey simple statistical laws due to the random nature of the mutational process. It turned out however that this is not the case. Analysis of variability of the nucleotide composition, of the distribution of pyrimidine clusters in DNA and of some other characteristics revealed quite different tendencies to exist. For example it was shown that equimolar DNA (50 mole per cent of GC pairs) are surprisingly rare. Comparison of some taxa of a class rank and higher has revealed that the DNA of the species of which they are composed most often contains 42 or 53 mole per cent of GC pairs. If one takes into account that DNAs of equimolar composition possess the maximal informational capacity, one should make the conclusion that Nature has not followed the principle of utmost parsimony when building genomes of the existing species. It is quite possible that the observed non-equimolar DNA composition, without being most economical from the point of view of recording the genetic information may provide the optimal disturbance stability of information transfer (ANTONOV, 1972; BELOZERSKY et al., 1972).

Similarly, the pattern of distribution of pyrimidine nucleotides in DNA chain was shown to be nonrandom; this was revealed by analysis of pyrimidine isopliths of different length in most diverse organisms (MAZIN, 1975). It has been shown that progressive evolution was accompanied by accumulation in DNA of long sequences composed of pyrimidine nucleotides only. In higher eukaryotes their proportion in DNA is markedly higher than in lower eukaryotes and prokaryotes.

The same conclusions has been drawn from nearest neighbour analysis of DNA structure. In other words, we are again dealing with the process which is quite similar to the one described above: as the evolution proceeds, the information capacity of DNA (per nucleotide) decreases. Increase in the degree of clustering of pyrimidines in total DNAs in the course of evolution may have been associated with symbatic changes in the degree of assymmetry in the complementary chains of their molecules. In microorganisms there is a distinct correlation between the degree of clustering and the degree of asymmetry of the complementary chains of their DNA molecules. The results of the analysis of cytochrome c cistrons are indicative of the similar changes in the DNA of higher forms: in the course of evolution there occurred accumulation of pyrimidine nucleotides in the sense H chain. The rate of this process in the groups of organisms of different degree of evolutionary complexity was most probably unequal and increased in the course of evolution. It is believed that the factor of selection, which was at play at the molecular level, is an increase in the disturbance stability of the protein synthesis (translation). It is known that cell-free systems of protein synthesis obtained from higher animal cells do possess higher disturbance stability than the same systems of prokaryotes.

The results described in this section may be interpreted as indicating that DNA evolution cannot be understood and described in terms of elementary stochastic processes of modification of its structure.

DNA × DNA Hybridization and Evolutionary Divergence

The method of molecular hybridization is one of that furnish the greatest amount of information about the evolution of nucleotides sequences in DNA. In the first approximation the accumulation of nucleotide substitutions, revealed in DNA × DNA hybridization reactions by the decrease in the per cent of homology of the DNAs compared, or, in other words, the process of fixation of mutational changes may be quantitatively described by Poisson distribution. One may suppose that, as the phylogenetic distance between the organisms compared increases, the degree of DNA homology will **gradually** decrease. On the contrary, data obtained clearly indicate that **there exist a discreteness in the DNA homologies** (MEDNIKOV et al., 1973). Further statistical analysis of these and some other results of vertebrate DNA × DNA hybridization experiments confirmed these observations MEDNIKOV, 1974). It appeared that among vertebrates 4 modal classes, 7 degrees of relatedness of DNA exist, the transgression of which is insignificant.

What does it mean for systematics and phylogenetics? It is tempting to conclude, basing on these results, that higher taxa of vertebrates are real entities, as well as that the commonly adopted hierarchy of taxa in zoological systematics is unjustifiably complex. It looks as if only the categories of classes, orders, families, genera and species may be used phylogenetically correct, natural systems. If so, some changes in traditional systems are inevitable.

Another conclusion drawn from these results is that not only sibling species, but also sibling taxa of a higher rank exist; for example, fishes should be subdivided not into two, but at least into four classes. Cartilagenious ganoids **Chondrostei** are to be considered, according to our data, as an independent class, since per cent of homologies in DNA of sturgeon, on the one hand, and shark, pink salmon and carp, on the other, is of the same order as similarity between sequences in DNA of chicken and lizard, man and salmon. Not less interesting are the results of the study of lungfish *Protopterus sp.* DNA (MEDNIKOV, ANTONOV, 1974).

In most of the present-day classifications **Dipnoi** are regarded as a subclass of the class of bony fishes — *Osteichthyes*. However, per cent of homologies in the DNA of *Protopterus sp.* and the DNA of sturgeon, perch or salmon is of the same order as that in the DNA of frog and pigeon.

We are sure that the process of division of the *Gnathostomata* into independent classes will not end here. Preliminary results obtained in the hybridization experiments with DNA of **Latimeria** as well as of bony ganoids (species of *Lepisosteus*) allow us to conclude that the number of classes of fishes is to be increased (KEDROVA et al., 1980). The discrepancy between the results of genosystematic studies and the generally adopted system may be readily explained: genotypic and phenotypic evolution rarely follows the same path. This conclusion has been drawn previously from evidence and finds additional support and quantitative estimates in the results of DNA hybridization. It goes without saying that an analysis of such situations may furnish a lot of information for the development of STE, the more so that recently very similar results were obtained in our experiments with plant DNA (VALIEJO-ROMAN et al., 1982).

Plant DNA evolves at a very high rate, much higher than that of vertebrates, and therefore we have analysed only intrafamily DNA hybridization data. Again, we see discreteness in the degree of homology in DNA of ferns, monocots and dycots and postulate that only species, genera and families are real evolutionary entities. For example, when the DNA of species from different genera of the same or different tribes are hybridized, the results are the same. One gets the impression that the category of tribe in plant systematics is artificial, as well as many other with prefixes "sub" and "super".

I cannot but mention that very similar results were obtained by B. M. MEDNIKOV (1976) in his study of quite a different system of information storage and transfer, human languages. It is worth saying that it was DARWIN (Origin of species, ch. XIV) who drew an analogy between linguistic and biological evolution as one of the first. Quantitative comparisons of various languages were made by B. M. MEDNIKOV using a modified glottochronological method of M. SWADESH. SWADESH suggested to compare a list of 100—200 words of the so-called basic vocabulary. The sooner the studied languages diverged, the fewer identical morphemes expressing similar entities are left in the lists. SWADESH assumed that over 1000 years 14 per cent of general words are lost from the diagnostic list; hence we may calculate that, e. g. ~ 1,3 thousand years have passed since the English and German languages diverged. This coincides with the VIth century C. E. — the epoch of King Arthur, when the Germanic tribes of Angles and Saxons invaded Britain.

In other words, SWADESH assumes that divergence of languages follows the same exponential law according to which DNA loses its homologous nucleotide sequences, the power being constant for all languages. As B. M. MEDNIKOV notes, this is the most arguable point in SWADESH's statements. The situation in the theory of evolution and in evolutionary genetics is absolutely similar. A number of scientists, and W. HENNING in the first place, believe that the evolution rate is so steady that ranks may be conferred to taxa just on the basis of their age. A critical review of this conception (cladistic) has been made by E. MAYR. Recently, the cladistic conception has been supported by admirers of the "non-DARWINian" evolution theory according to which evolution proceeds at the expense of accumulation of neutral mutations with almost steady rate.

So, the glottochronological method is an analogue of HENNING's cladistic theory and of the "neutralistic" conception. There is no need to explain what an appealing perspective it is to find out, basing on the steadness of the evolution rate, the time of divergence of both languages and living organisms. We are prone to think, however, that to assume the divergence rate to be steady in both cases would be an inadmissibly rough approximation. To date the moment of divergence basing only on an exponential equation without making use of other more or less accurate data is impossible. MEDNIKOV applied the lexico-statistic method of SWADESH with a different purpose in view: if the evolution of these two systems of information transfer is, by most rough estimations, based on similar mechanisms, could we using M. SWADESH's method, reveal in languages discrete distributions of relationship degrees analogous to what we had found in vertebrates and plants. Thereby he somewhat changed this method: by lack of correspondence of words SWADESH means only those cases when morphemes express different entities and he does not take into consideration "natural phonetic changes", i.e. he, in fact, compares not words as they are pronounced today, but restored word-forms. MEDNIKOV compared words as they are pronounced today, since the process of the loss of homologies in the language goes at the expense of both changes in the meaning of words, and changes of words themselves. The latter process is analogous to accumulation of mutations in nucleotide sequences, and here we have the same substitutions, deletions, insertions, in one case — of phonemes, in the other — of nucleotides.

Experimental lists of a 100 words each for a number of languages (Hindi, Ossetic, Latin, Italian, Spanish, Rumanian, English, German, Norwegian, Russian, Bulgarian, Czech, Polish, Serbo-Chorvatian, Lithuanian, Lettish, Finnish, Estonian, Mokshanian, Ersian, Mari, Udmurt, Komi, Hungarian, Nanaian, Evenk and Tamil) were analysed. The matrix of correspondences was built; the entire analysis was carried out by one person (B. M. MEDNIKOV) with one list of words. It was found that the relationship between the languages is also distributed in a discrete way. There may be distinguished 7 maxima, corresponding to groups of languages of various degree of evolutionary relatedness; the most interesting of them is group VII (5 per cent of correspondence) indicating that such distant families as Indo-European, Ural, Altaic and Dravidian are related to each other from the most ancient times. This brings to us vaguest aroma of the archiancient Nostratic ancestral language which was in use about 15 thousand years ago.

As a control object a large Turkish-Mongolian language group (Mongolian, Kalmyk, Touvinian, Khakass, Shorian, Yakut, Uigur, Uzbek, Kirgiz, Turkmen, Turkish, Tatar and Chuvash) was studied and identical results were obtained.

The analysis of the correspondences revealed conforms well to the historical and archeological evidence although discussion of these data would lead us too far astray. What is important for us in this case is that code structures of both genetic and linguistic information probably evolved according to similar laws. The research in this direction shows great promise.

What is the origin of the discreteness? Computer simulation experiments (MEDNIKOV, MENSHUTKEN, 1977) were performed based on the theory of STE. Polymodal, discrete distributions were obtained only in this case; if the programme was based on the "non-DARWINian" evolution concept, no discreteness was observed. Temporary, unstable discreteness was observed if the probability of "elimination" and "mutations" was taken as a periodical function of time of divergency (imitation of the concept according to which periods of rapid formation of new taxa were followed by much longer periods of adaptive radiation).

We may conclude that the theoretical aspects of STE concerning the mechanisms of divergent evolution find support in the results of DNA studies. Discreteness and hierarchy of relatedness is a result of evolution. Natural, phylogenetic systems must be built basing on the principle of hierarchy, as they have, contrary, to e.g. combinative systems, an evolutionary background.

Evaluation of the Role of Some Factors of Evolution

Since the early days of karosystematics, polyploidy has been believed to be among the leading factors of microevolution of *Angiospermae*. According to L. STEBBINS, up to 35% of species of angiosperms are polyploids. It is generally accepted that polyploidy increases the genetic variability of plant populations and thus facilitates adaptations; sometimes it results in saltatory species formation.

Changes in the biochemistry of plants following polyploidization have been studied for at least forty years with very little if any, attention paid to DNA, except its quantitative changes. We decided to study the influence of polyploidization on the rate of plant DNA evolution.

As follows from some earlier observations, DNA of angiosperms evolves at a very high rate. Compared to mammals (a group of approximately the same evolutionary age and genome complexity), angiosperms demonstrate a higher variability of GC-content of nuclear DNA and a lower degree of homology in nucleotide sequences of DNA of distantly related species.

Evolutionary processes based on polyploidy may be among the main factors that are responsible for this unique variability of plant DNA, as polyploids are quite abundant among plants and are very rare among vertebrates.

To verify this assumption, we compared the rate of accumulation on nucleotide substitutions in DNA of some closely related diploid and polyploid species belonging to genus *Seseli L.* (*Apiaceae*) (VALIEJO-ROMAN et al., 1979). We have found that repetitive DNA of plants from two different diploid populations of *S. condensatum* form the most thermostable hybrid duplexes; the thermostability is lower when DNA of closely related diploid species of the same genus are hybridized. At the same time, the thermostability of hybrid DNA molecules obtained in the experiments with the same fraction of DNA of diploid *S. condensatum* and diploid or hexaploid *S. mucronatum* proved to be different. The same results were obtained when the fraction of "unique" nucleotide sequences was hybridized. Hybrid *S. condensatum* (diploid) × *S. mucronatum* (hexaploid) DNA molecules display lower thermostability than the molecules obtained in interspecific reactions. The starting hypothesis was given an added support to by the results of experiments with labelled "unique" DNA of *S. mucronatum*: the thermostability of *S. mucronatum* (diploid) × *S. mucronatum* (hexaploid) hybrid DNA molecules proved to be as low as that of interspecific hybrids. Very similar results were obtained in hybridization experiments with repetitive DNA of *Aegilops* species (FLAVELL et al., 1979), although these authors did not treat them as we do. The data obtained are also in good agreement with the results of our study of the same genus and wheats (JANEVA, ANTONOV, 1977) and milfoils (KASHEVAROV, ANTONOV, 1982).

It should be taken into consideration that the nature of polyploid *Seseli* and *Achillea* species is unknown. At least some of them may be allopolyploids. One may suggest that the difference observed in ΔT between DNA of diploids and polyploids could be due to interspecific hybridization. Although such a possibility cannot be ruled out, the available data do not support this hypothesis. For example, DNA of diploid *Triticum boeoticum*, when compared with DNA of diploid aegilopses, reveals a high degree of homology, much higher than that observed when DNA of these diploids are compared with DNA of allopolyploid wheats. A detailed review of these results has just been published (ANTONOV, 1981).

By way of summing up, we may say that polyploidization apparently results in a higher rate of accumulation of nucleotide substitutions in plant DNA, probably due to a less pronounced pressure of natural selection on multigenetic loci of polyploids. Some other processes resulting in quantitative changes of DNA (i.e. selective amplification of some fractions of DNA or polynemization) may have had the same effect.

So we conclude that the rate of molecular evolution in plants could be far less stable than in other groups of eukaryotes where all these processes are less pronounced (e.g. mammals). All these observations should not be overlooked by plant protein taxonomists. If the rate of accumulation of nucleoide substitutions in plant DNA is so variable even in lower taxa one can hardly evaluate a protein phylema reflecting real phylogenetic relatedness of plant taxa. This may be one of the reasons for the well known discrepancies between the "classical" and "protein" phylemas of plants; on the other hand, the high rate of evolution of plant proteins is easily explained in the light of the ideas presented in this section.

Conclusion

Methods of genosystematics may be used to study both the problems of micro- and macroevolution. Some principles of STE find support in the results of genosystematics. This manifests the validity of the main concepts of Darwinism. The contemporary crisis of STE may only be overcome by joint efforts of biologists studying evolutionary processes at all levels of organization of the living matter.

References

[1] ANTONOV, A. S., (1972): In: "VII-th FEBS Meeting, Varna", abstr. N 435.
[2] ANTONOV, A. S., (1974): Uspekhi Sovremennoi Biologii, **77**, N 2, 31 (in Russian).
[3] ANTONOV, A. S. (1981): Priroda, **3**, 2 (in Russian).
[4] BELOZERSKY, A. N., A. S. ANTONOV, B. M. MEDNIKOV, (1972): In: "Stroenije DNK i polozhenija organizmov v sisteme" A. N. Belozersky and A. S. Antonov, eds. Moscow State Univ. Publ. House, M. (in Russian).
[5] FLAVELL, R. et al. (1979): Heredity, **42**, N 3, 309.
[6] JANEVA, Yu. N., A. S. ANTONOV: Genetica, **13**, N 4, 578 (in Russian).
[7] KASHEVAROV, G. P., A. S. ANTONOV: (1982) Botanicheski Journal, (in press).

[8] KEDROVA, O. S., N. S. VLADYTCHENSKAYA, A. S. ANTONOV, (1980): Molekularnaja biologia, **14**, N 5, 1001 (in Russian).
[9] MAZIN, A. L. (1975): Molec. Biol. 9, N 2, 252 (in Russian).
[10] MEDNIKOV, B. M., (1974): Journ. Obschei Biologii, **35**, N 5, 569; — 1976. Bull MOIP, biol. section, **81** (4), 134; — 1977. Divergencia genomov i nekotorie voprosy evolucionnoy teorii. Dr. Sc. thesis, Moscow State Univ., Moscow (all in Russian).
[11] MEDNIKOV, B. M., L. S. POPOV, A. S. ANTONOV, (1973): Journ. obschei Biologii **34**, N 4, 516 (in Russian).
[12] MEDNIKOV, B. M., A. S. ANTONOV, (1974): Doklady AN SSSR, **218**, N 6, 1475 (in Russian).
[13] MEDNIKOV, B. M., V. V. MENSCHUTKIN, (1977): Journ. Obschei Biologii, 37, N 2, 198 (in Russian).
[14] TIMOFEEFF-RESSOVSKY, B. V., N. N. VORONTSOV, A. V. JABLOKOV, (1977): Kratkij ocherk teprii evolucii, "Nauka", M. (in Russian).
[15] VALIEJO-ROMAN, K. M., M. G. PIMENOV, A. S. ANTONOV, (1979): Doklady AN SSSR, 245, N 4, 1021. (in Russian).
[16] VALIEJO-ROMAN, K. M., A. S. ANTONOV, (1982): Doklady AN SSSR, (in press).
[17] ZAVADSKY, K. M., E. I. KOLCHINSKY, (1977): Evoljucija evolucii, "Nauka", Leningrad. (in Russian).

Discussion

EBELING: I want to comment on your very interesting remarks on the similarities between the DNA-language and the human language. Evidently the basic principles for the evolution of human language were detected as early as about 130 years ago by the German scientist SCHLEICHER in Jena. We have learned this from the historical investigations done by Prof. SCHARF from Halle. Like in biological evolution, in linguistic evolution there is a guenerative principle (creation, mutation and exchange of words etc.) and an evaluative principle (valuation and selection). My question is, if DNA really forms language — at least in the formal sense — what is your opinion about the application of the methods of the formal mathematical theory of languages to the study of DNA-sequences?

RUSE: Charles DARWIN was not the first person to draw the analogy between organic evolution and linguistic evolution (at least, he was not the first person to draw the analogy publically in English). Herbert SPENCER drew the analogy in 1857 in his essay "Progress: its law and cause".

ANTONOV: Thank you for your notes, Prof. EBELING, and you, Prof. RUSE. I cannot but agree that it is better to say that Ch. DARWIN was "among the first" and not "the first" who compared the evolution of languages and living beings. If these two systems of information storage and transfer have evolved in extensively similar ways, then the methods of the formal mathematical theory of languages may be applied to the study of DNA evolution.

SCHUSTER: Prof. ANTONOV, I have enjoyed very much your results on the discreteness of the degree of homology found in the DNA of different species. I have two closely related questions: What is the significance of **four** levels of homology, why not three or five? What does this result tell us about the structure of the phylogenetic tree?

ANTONOV: The number of maxima may be different for different groups of organisms. The very fact of their existance as well as their number allow us to draw new types of phylogenetic tree based both on pheno- and genosystematic data.

ROSENTHAL, H. A.: Did you or anybody else hybridize DNAs from different *Bacillus* strains, for example from thermophilic and mesophilic strains, and what came out of these experiments?

ANTONOV: *Bacillus* is a very heterogeneous genus. Differences in DNA structures within it are so pronounced that in fact we have come to the conclusion that it is a conglomerate of genera. Thus the results of comparisons between the DNA of thermophilic and mesophilic strains are due mainly to this fact, not on their ecological characteristics.

BRODA: Do your linguistic considerations also apply to very distant and different languages like Chinese? One can read that some languages are not related at all to each other. This is probably nonsense.

ANTONOV: We did not study this group of languages, but as the trend revealed in the evolution of languages seems to be of general importance, it seems that the same regularities may be observed in the group of languages including Chinese if the principles of information storage and transfer are the same.

KLIX: One remark regarding the kinship among different languages. The results you receive depend on two different things, namely: 1. on the criteria you are using. You (or your colleague) compared the morphems. But language structures have far more dimensions: deep structure (basic grammatical properties), surface structures, degree of hierarchization etc. It is possible that other dimensions provide quite different types of kinships. 2. on the methods you use for comparison. Cluster analysis, factor analysis, hierarchical cluster analysis etc. are all possible methods. The results are not the same for each methods.

GHILAROV: You pointed out that, on the basis of DNA hybridization experiments, the class "fishes" can be divided into several separate classes. But this is also the opinion of such morphologists as L. S. BERG, who regarded "fishes" as several classes, and this has also been proved by classical methods of systematics.

ANTONOV: There is the only point in your statement which I cannot agree too. It is that the existance of several classes of fishes was "*proved*" by classical methods. I am a biochemist and nevertheless even I know at least five different systems of fishes, BERG's included. They all differ to some extent or another, and the system of BERG is not popular now among the ichthyologists: They prefer to subdivide fishes into two classes. The old system of MÜLLER seems to us to be the best from the point of view of genosystematics, but its subclasses should be raised to classes.

STEINMÜLLER: I would like to stress the differences between the evolution mechanisms in biology and linguistics which set rather narrow limits to the analogies of both fields. In biology, after speciation has taken place there is nearly no exchange of genes through the species barrier. In contrast, one language is always able to assimilate elements of another. There is exchange, no barrier.

ANTONOV: I cannot agree with your note because we have not quantitative estimates of lateral gene transfer processes in biological evolution, although the very possibility of such processes is evident.

F. VOGEL: The best hybridization is always achieved by satellite DNA. Is it possible that the clustering of hybridization percentages is caused by the presence of common satellite DNA species?

ANTONOV: No, we hybridized the "main band" nuclear DNA.

GRANTHAM: You did not show the distribution of nucleotide "clustering" (runs of one kind of base) for viruses. I reported on this in FEBS letters (December, 1980). We find that adenine is the base with the greatest tendency to occur in runs and that viral sequences are different from organism sequences.
We have also applied correspondence analysis to languages. Briefly, our results agree with yours in that languages can be grouped according to phonetic composition. In fact we found (in a study done some years ago) that the main feature separating languages is the content of consonants and vowels.

ANTONOV: We did not study viruses, but your results are very interesting.

Научный совет при Президиуме АН СССР по философским и социальным проблемам научким и мехники

Влияние Ч. Дарвина на современную концепцию биологического познания[1]

И. Т. Фролов (Москва)

Сила и значение того или иного учения в науке существенно определяется тем, как оно отвечает на вопросы, уже поставленные развитием теории и практики. Вместе с тем это учение приобретает тем большее исоторическое значение, чем больше оно ставит новых вопросов.

Ч. Дарвин не только решил многие поставленные до него вопросы, но, и, может быть, в большей степени вызвал к жизни новые, многие из которых мы решаем и по сию пору. Он оставил нам также некую загадку, тайну своего гения и творчества, некоторый образец, идеал ученого, и в этом состоит его влияние не только специально-научное, но и морально-этическое, нравственное. Он ярко выразил то, что сегодня входит в понятие „этос науки", а поэтому творчество Дарвина, я полагаю, должно рассматриваться и в этом плане.

Последнее имеет и более широкий философский смысл, требующий внимания и анализа. С этой точки зрения главное для нас — научный метод Ч. Дарвина. Мы знаем, что создатель „Происхождения видов" был обвинен в свое время Сэджвиком в отходе от известной традиции индуктивизма, хотя Дарвин всегда подчеркивал, что он пользовался „подлинно бэконовским методом".

Парадокс заключается, однако, в том, что в самом деле метод Дарвина не является индуктивным в бэконовском смысле. Дарвин создал **новый метод,** который и позволил ему совершить **революцию в науке о жизни,** подобную тем, которые впоследствии были сделаны Г. Менделем, введшим экспериментальный метод в генетике, и теми учеными, которые с помощью методов физики, химии и математики расшифровали код генетической информации и вызвали к жизни молекулярную биологию.

Суть метода Дарвина — в последовательном проведении **принципа объективности** и **историзма,** соединении индукции и дедукции, анализа и синтеза в биологическом исследовании, отказе от „телеолоческого мышления" и утверждении единства случайности и необходимости, статистичности действия естественного отбора. Мало оцененным является также сам **метод экстраполяции** понятия отбора, взятого из человеческой практики, и перенесения его в естественные условия, сопровождающееся устранением антропоморфизма, столь чуждого принципу объективности в науке.

Тем самым Дарвин оказал глубокое влияние на **методологию** современного биологического познания, на то, что условно можно назвать „биологическим мышлением", которое также претерпевает **эволюцию**, как и объекты, которые оно стремится постигнуть и выразить теоретически.

[1] В основу статьи положен доклад на 8 Кюлунгсборнском колловиуме по философским и этическим проблемам биологии „Дарвин сегодня" (8—12 ноября 1981 г.).

Биологическое познание сегодня движется к выявлению общих и фундаментальных свойств живого — его качественной специфики, системной целостности, инвариантности воспроизведения и известной направленности приспособления и развития, органической детерминированности процессов и проч., — исходя из которых можно строить единую теорию жизни, ближе подойти к ее основной цели — постижению сущности жизни. Последняя, существуя главным образом в качестве некоторого **регулятивного принципа**, имеющего, правда, объективные основания, направляет биологическое познание по пути анализа, редукции вплоть до молекулярного, субмолекулярного и субатомного уровней — и, вместе с тем, синтеза, интеграции, межуровнего системно-структурного исследования.

Сложность жизни порождает нелинейность познания, и оно само необычайно усложняется, а вместе с ним становится все более сложным и дифференцированным тот тип мышления, который ведет его к цели. Соответственно, можно сказать, что изменяется не только само по себе биологическое познание, но и характер мышления современного биолога.

Обращаясь к истории, мы увидим четкие рубежи, переломные моменты, развития „биологического мышления" и его изменяющиеся критерии. Мы увидим, в частности, как они определяли „биологическое мышление" до Дарвина и после него (описательно-классифицирующие и исторические подходы), что получило оно в результате внедрения эксперимента в биологию, а затем в связи с выходом на молекулярный уровень и т. д. Мы увидим картину, полную, на первый взгляд, взаимоисключающих тенденций и критериев „биологического мышления": организмоцентрических и популяционных, редукционистских и целостных, детерминистских и телеологических инвариантных и исторически изменяющихся. Но мы увидим такие, что именно они привели к тому, что называют **современным** „биологическим мышлением", в котором эти отдельные характеристики не только противопоставляются, но и не совмещаются механически, не выступают просто как „дополнительные", **а образуют диалектическое единство противоположностей.**

Вот почему современное „биологическое мышление" — это прежде всего диалектическое, фиксирующее сложную расчлененность и целостность биологического объекта, его устойчивость и изменение, развитие, противоречивое единство отдельных атрибутов, составляющих основу жизни и ее сущность. Оно соединяет в себе **эмпирический**, опытный характер и тенденции ко все большей **теоретизации**, а также аксиоматическому построению ряда своих подразделений. И одной из характерных форм ее в современной биологии является **математизация** знания.

Математизация биологического познания, усиливающаяся в особенности в связи с расширяющимся применением в исследовании живых систем теории вероятности и теории информации, приводит к тому, что присущая ему на протяжении веков исторического развития **наглядность** все более и более утрачивается. **Символический, условный** характер современного биологического знания, на первый взгляд, как бы отдаляет его от объективной реальности живой природы, взятой в ее цельности, в ее чувственно-наглядной форме. Однако оно, с другой стороны, позволяет глубже проникнуть в сущность жизни, и, следовательно, скорее и успешнее приближает познание к объекту, чем это было в классической биологии.

Уже сегодня биологическое познание и мышление характеризуется, следовательно, новыми чертами, соединяющими в себе элементы классической биологии с ее целостными подходами и новой, построенной на применении физико-химических и математических методов на молекулярном уровне.

Но что впереди? Куда идет биологическое познание? Пытаясь ответить на эти вопросы и как бы заглянуть в будущее, мы не можем не обратить внимание на многие свидетельства того, что речь в сущности идет о формировании нового типа науки о жизни.

Но для этого надо пройти длинный и трудный путь, в значительной мере связанный с преобразованиями в сфере методологии биологического познания. Реальная структура современного биологического знания такова, что в ней по большей части пока лишь **сосуществуют** элементы классической и новой науки о жизни, картезианские (редукционистские) и дарвиновские (целостные) подходы, молекулярная и организменная биология. Лишь тогда, когда в результате усиливающегося воздействия и, главное, — взаимопроникновения они сольются в единую науку о жизни, будут достигнуты непредвидимые сегодня результаты, и наступит подлинный „век биологии".

Это предполагает изменение положения и роли методологии, теории в структуре биологического познания, а также новые формы ее связи с практикой. Последнее наглядно проявляется уже сегодня в ряде фундаментальных процессов, которые во все большей мере привлекают к себе внимание ученых. Полагается при этом, что достижения, в частности, физико-химической биологии, приведшие к возникновению и развитию молекулярно-биологических исследований, генной инженерии, открывают перед человечеством начало новой эры — биотехнологии. Развитие биотехнологии будет иметь поистине всемирно-историческое значение для судеб нашей цивилизации, так как эта новая технология в отличие от традиционной — механической и химической — позволяет, помимо всего прочего, свести до минимума вредные выбросы в окружающую среду, наиболее экономно использовать природные энергетические ресурсы, принципиально по-новому решать продовольственную проблему (новые пути селекции и сельскохозяйственного производства и пр.), создать новую медицину, включая лечение наследственных заболеваний с помощью „хирургии генов".

Короче говоря, биотехнология — это новый путь научнотехнического решения основных глобальных проблем человечества сегодня и в будущем. В особенности же большие перспективы открываются перед биологическим познанием в его связи с практикой в тех областях, которые непосредственно обращены к человеку, его генетике, психофизиологии, структуре и деятельности его мозга. И здесь биологическая наука будущего приобретает еще один, может быть, самый существенный признак, создающий по существу сам новый тип науки о жизни: речь идет о ее **соединении с человеком** и тем комплексом дисциплин (естественнонаучного, обществоведческого и гуманитарного профиля), который занят его изучением, о такой структуре, которая включала бы в сферу биологического познания не только чисто **исследовательские**, но и ценностные подходы, регулирующие исследование, исходя из **социальных целей и гуманистических идеалов**.

Марксизм утверждает принципиальное единство научного познания и ценностей (включая социальные, мировоззренческие и этико-гуманистические). Основной принцип научного познания — его объективной истинности — не противоречит ценностным отношениям как внутри самого научного познания (я имею в виду не только логическую структуру знания и аксиологический базис его методологических норм, но и философско-мировоззренческую направленность, понимание общих целей и соответствующих им средств научного познания), так и в его отношении к обществу, соотносящему науку в целом со своими целями, выступающими как утилитарно-практические и нормативные или идеальные критерии достигаемой в итоге оценки.

Ценность науки связана не только с ее социальным статусом, но определяется и внутренней структурой познания. Это придает более широкое значение действию социально-этических и гуманистических принципов (регулятивов) современного научного познания, определяет во многом то, что сегодня называют **этикой науки,** должно неизбежно учитываться при построении этических кодексов ученого и их соотнесении с общими этическими ценностями человечества.

Последнее позволяет, как я думаю, снять провозглашаемую некоторыми учеными дилемму науки и морали, их якобы „комплиментарность" (как следствие разрыва объекта и субъекта в европейской культуре) и пр.; постулировать новый тип науки, в котором ее отношение к человеку, ценностям и гуманизму будут находиться в единстве и взаимодействии. Глобализация научных проблем и неизбежность их комплексного, системного решения, **новый этос науки** и **новый гуманизм**, — вот, по моему мнению, компоненты такого типа науки, который сегодня видится лишь как **идеал** или **цель,** имеющие определенное **эвристическое** и **регулятивное** значение.

В отношении к этим вопросам зачастую проявляются крайние подходы: от пессемизма и алармизма до неумеренного оптимизма. Но, пожалуй, опаснее всего — сциентистский конформизм, который под видом реализма пытается навязать обществу равнодушие, отрешенность от творческих поисков, уповая на авторитаризм и пассивность. Мы не идеализируем современное состояние науки, но, вместе с тем, видим в ней силу, которая при соответствующей социально-этической и гуманистической ориентации может эффективно служить целям человека и человечества.

Эта диалектическая идея принадлежит марксизму, но она стихийно пробивает себе дорогу и в тех философско-методологических интерпретациях науки, которые не связаны с марксизмом, а иногда и враждебно относятся к нему. Логика познания, сама жизнь заставляет ученых приходить к этой идее.

Последнее видно на примере ежегодно организуемых (уже в 10-й раз) международных конференциях, посвященных единству наук. И это отчетливо выражено, например, в опубликованной в 1980 году в журнале „Сайенс" статье президента американской ассоциации Джеймса Бэннера (Младшего) „Сходство естественных и гуманитарных наук"[2]. В статье говориться о докладе комиссии по гуманитарным наукам, финансируемой Фондом Рокфеллера, который посвящен состоянию дел и роли гуманитарных наук в жизни американского общества[3]. Доклад комиссии утверждает, что гуманитарное знание несет такую же ответственность перед естествознанием, какую естествознание несет перед гуманитарными науками. Важно сегодня не только провозглашать, но и реализовать практически идею единства и взаимодействия наук и гуманистической культуры в целом. Этой цели служит и созданный в 1980 г. Научный совет при Президиуме АН СССР по философским и социальным проблемам науки и техники.

Главной задачей совета является координация деятельности научно-исследовательских учреждений в разработке философских и социальных проблем науки и техники, дальнейшее развитие и осуществление ленинской идеи союза философии и ест-

2 "Science", vol. 211, No 4477, 1980.
3 The Humanities in American Life. Univ. of California Press, Berkley, 1980.

ествознания, обеспечение комплексного, системного подхода к этим исследованиям, а также их более тесная увязка с целями общественного развития СССР и других стран социалистического содружества, с научным освещением вопросов глобального характера, философское осмысление естественных, общественных, технических наук в их единстве и взаимосвязи.

Основные направления координационной деятельности Совета исключительно многообразны. К ним относятся:

— исследование мировоззренческих и методологических функций марксистско-ленинской философии в развитии науки и техники на современном этапе; развитие и практическое осуществление идеи союза философии и специальных наук; критический анализ буржуазных концепций современного научно-технического развития;

исследование роли науки как социального института в условиях НТР, места и роли науки в современном обществе, в системе культуры, методологических проблем социальной ответственности ученых, нравственных факторов научной деятельности, социально-психологических проблем научного и научно-технического творчества;

— исследование взаимосвязи человека, науки и техники; влияния НТР на развитие человека, анализ философско-психологических аспектов роли человека в процессе труда и человеческих потребностей, изучение проблемы гармонического развития личности, гуманистических аспектов НТР; анализ философских и социальных аспектов взаимоотношений человека и среди его обитания;

— разработка с позиций материалистической диалектики системного подхода к изучению глобальных проблем человечества, их социальных аспектов, роли науки и техники в их решении, принципов глобального моделирования;

— исследование будущего человечества, соотношений тенденций развертывания НТР и перспектив развития противоположеных общественных систем — социализма и капитализма; разработка методологических предпосылок и принципов моделирования, поискового и нормативного прогнозирования научно-технического прогресса, критика буржуазных футурологических концепций.

Одним из важнейших шагов в реализации этой комплексной научной программы явилась подготовка и проведение 22—24 апреля 1981 года в Москве Третьего всесоюзного совещания по философским вопросам современного естествознания, на котором в серии докладов ведущих ученых нашей страны рассматривались и проблемы, прямо относящиеся к нашей теме: о взаимосвязи между наукой, техникой и производством (доклады академиков Б. Е. Патона, А. М. Прохорова, Н. Г. Басова и др.), о взаимодействии общественных, естественных и технических наук, дифференциации и интеграции научного знания (доклады академиков Б. М. Кедрова, М. А. Маркова, Н. М. Эмануэля, академика АН УССР К. М. Сытника, члена-корреспондента АН СССР М. С. Асимова и др.), о сущности и социальных последствиях научно-технической революции, ее связи с человеком, о социально-этических и гуманистических вопросах современной науки и техники (доклады академиков А. А. Баева, Д. К. Беляева, Н. П. Дубинина, В. А. Энгельгардта, академика АМН СССР Н. П. Бочкова и др.[4]).

Таким образом, практически подчеркивается, кроме всего прочего, **стимулирующая, каталитическая роль философии** в исследовании этой проблематики. Конечно,

[4] Автор также представил на Совещании свой доклад, который был посвящен социально-этическим и гуманистическим проблемам науки.

многие вопросы еще недостаточно ясны, поэтому нужны научные дискуссии вокруг них. Я думаю, что к важным и острым философским вопросам, которые здесь возникают, относятся и вопросы о критериях научной эффективности вообще и соответственно — оценка этой эффективности, с гуманистической точки зрения. Прогрессивно развивающийся процесс превращения науки в неспоредственную производительную силу общества имеет свои особенности, и не только, к сожалению, позитивно оцениваемые. В самом деле, подчеркивая критерий практической эффективности науки, мы можем впасть в опасную односторонность, поскольку наука, хотим мы этого или нет, тем самым все больше вырывается из культуры в целом, и соответственно — играет все меньшую роль как определяющий фактор гуманистической культуры. Это приводит к тому, что критерии культурной, гуманистической значимости науки отходят на второй план, и мы видим этот в так называемой дивергенции культур — научно-технической и гуманитарной — явлении, которое так резко обозначилось в современных условиях. Не случайно сегодня в виде протеста и в форме утверждения каких-то новых идеалов человеческой культуры обозначается как раз конвергенция культуры научно-технической и гуманитарной в широком смысле.

Мне кажется, что в этом вообще — спасение и выход из того довольно четко обозначенного тупика, в котором находится сейчас наша человеческая культура. Поэтому все больше мыслителей, философов современности ставят вопрос о необходимости единства наук и человеческой культуры, конвергенции науки и других сфер культуры. Только тогда наука может в наибольшей степени выявить и свою общекультурную, гуманистическую значимость, миссию и пр.

Таким образом, это хорошо, что мы подчеркиваем возрастающую эффективность науки, имея в виду производственную ее эффективность и связанное с этим превращение ее в непосредственную производительную силу. Это очень хорошо, когда развитие науки ставится во многих случаях в те разделы планов, где говорится о развитии производства, экономики и пр. Но не хорошо, если наука в результате этого в существенной степени изымается как культурный фактор общественного развития, когда она изымается из ее общей связи со всеми другими сферами человеческой культуры. Тем самым ослабляется и ее человеческое предназначение и гуманистическая значимость как определенной духовной ценности.

Поэтому мне и кажется, что важнейшими вопросами сейчас являются вопросы взаимодействия и единства наук и других сфер жизни, единства гуманистической культуры в целом с непременным развитием этой, на первый взгляд, менее эффективной, с практической точки зрения, сферы человеческой культуры, которую обозначают как гуманитарную культуру. Очевидно, возникает большая угроза для будущего развития человеческой культуры, если мы односторонне ориентируемся только на развитие „практически эффективной" науки. Гипертрофируя ее роль, мы можем подрезать жизненные нервы, которые движут развитием науки в целом, деятельностью ученых и их желанием или нежеланием вообще заниматься наукой, потому что в жизни нормального человека есть много разных возможностей заниматься различными вопросами. Наука может превращаться в какой-то бездушный Молох, калечащий ученых, и в этих условиях и сами они могут превращаться в „бездушных ученых". Поэтому гуманистические, ценностные аспекты развития науки сейчас очень важны для размышления ученых именно в силу тех причин, о которых я говорил. Они составляют неотъемлемый элемент современного мира науки.

Известно, что в дискуссиях по этим вопросам выдвигается зачастую идея о том, что наука сама по себе является гуманной, моральной силой, а потому она не нуждается в каких-либо этико-гуманистических установлениях, внешних по отношению к ней. Изначальный гуманизм науки совпадает в этой концепции с этичностью истины и процесса ее познания., независимом от социального контекста, в котором осуществляется этот процесс. Такое представление о соотношении науки и морали является попыткой возродить идеи и представления, отражающие классический образ науки, и они вряд ли приемлемы в современных условиях, когда наука стала не замкнутой цеховой организацией, как это было еще в XIX — начале XX вв., а **социальным институтом**, тысячами нитей, связанными с жизнью человечества. Кроме того, колоссально возросла и усложнилась **сфера технологии**, технического применения науки, произошло резкое разделение „чистых" и прикладных исследований. „Классический" взгляд стремится отделить эти последние от науки вообще, чтобы сохранить хотя бы видимость своей незамутненности, отвести от науки обвинения в ее причастности к таким технологическим применениям, которые не могут быть не оценены как нарушения принципов гуманизма и морали по крайней мере частично.

„Классический" взгляд на соотношение науки и морали, игнорирующий **взаимодействие** этических ценностей познания с общими — социальными по природе своей — этическими ценностями человечества, в существенной степени детерминирующими также направление и характер познавательных усилий ученого, а не только его поведение в качестве члена научного сообщества, не может служить надежным компасом в современных условиях. Скорее наоборот, он может дезориентировать ученых, снять с них моральную ответственность в острых и сложных ситуациях антигуманного технологического использования науки, поскольку они могут успокаивать свою совесть тем, что непричастны к такому использованию и что это вообще не наука.

В этой связи хотелось бы заметить, что если обратиться к истории проблемы, касающейся соотношения науки, морали и гуманизма, то я не сказал бы, что в ней наблюдается бурное развитие ее исследования по восходящей линии, и я не сказал бы, что современная наука в значительно более глубокой и разносторонней форме ставит эти проблемы, чем это было раньше. Мне кажется, что даже в истории современной науки происходят какие-то волнообразные движения и даже возвращение к уже достигнутому в обсуждении этих вопросов, причем в разные периоды времени работают различные стимулы и „пусковые механизмы" при обсуждении проблемы науки и морали, гуманизма. Даже вспоминая короткую историю обсуждения социально-этических и гуманистических проблем, связанных с генной инженерией, и то мы можем сказать, что она имела свой пик в середине семидесятых годов, а сейчас наблюдается явный спад в обсуждении этих проблем.

Я думаю, что это требует внимательного анализа со стороны историков науки. В этой связи я хочу высказать такую, может быть, спорную мысль: очевидно, обращение к этим проблемам не связано прямо с возрастанием чувства социальной ответственности среди современных ученых. И, очевидно, то, что обозначается таким взрывом интереса к социально-этическим и гуманистическим проблемам науки и падение его детерминируется какими-то другими причинами, может быть даже более сильными, чем чувство, которое находится внутри нас и характеризует состояние социально-этической ответственности ученых в тот или иной исторический отрезок времени.

И здесь я вижу обратную сторону того, что наука сейчас стала многокомпонентным, многофакторным явлением. Она поистине стала социальным институтом современного общества, и поэтому в своем развитии и обращении к этическим и гуманистическим проблемам она также зависит от особенностей тех или иных периодов истории общества, и этим прежде всего и объясняется, я думаю, взлеты и падения интереса к этим проблемам. Если взять далекую историю и сравнить ее с современностью, то я бы это обозначил не как возрастание чувства ответственности ученых, а как возрастание все большей зависимости науки от общества, его идеологии, политики и в этом смысле все большее следование науки и ученых за теми поворотами, взлетами и падениями, которые переживает общество и которые диктует политика и идеологическое развитие. Поэтому во многих случаях мы имеем дело не с возрастанием чувства ответственности современного ученого, а скорее с **возрастанием чувства вины**, чего не было на предшествующих этапах развития науки, даже в такое недалекое время классической науки XIX века.

Так ли это — вопрос, конечно, дискуссионный. Мне хотелось бы только снять некоторый ореол большей моральной ответственности современных ученых по сравнению с ранее жившими и сказать, что, может быть, не происходит такого прогресса и такого развития в моральном плане, может быть, следует более позитивно высказываться в этом плане в пользу ученых, которые жили в прошлом. Мне кажется, что даже в среднем моральный, гуманистический уровень ученых, например, XIX века, был выше, чем у современных ученых. И это опять-таки связано с теми особенностями, которые имеет современная науки. Последняя стала массовым производством, тогда, как наука прошлого рекрутировалась в основном из людей выдающихся способностей, а не средних, как это имеет место сейчас, в массовом научном производстве.

И это имело определенное отражение в развитии науки, включая ее социально-этическую и гуманистическую проблематику. Во всяком случае, обращаясь к литературным, историческим, автобиографическим и пр. источникам, я для себя, поскольку я занимаюсь этими вопросами, нахожу зачастую в прошлом больше вопросов для размышления, чем тогда, когда я изучаю материалы, относящиеся к современной науке. В этой связи, я полагаю, очень многое зависит от того, как вообще понимается наука, как она реально предстает на разных этапах своего развития. Сейчас много говорят о разных типах и образах науки, но хотелось бы подчеркнуть, что кардинальный для нас гуманистический вопрос о том, **для чего** наука — был поставлен очень давно. Может быть сегодня он в некоторых случаях ставится и решается наименее адекватно тому, как он должен ставиться и решаться. Это не означает, что происходит деградация мысли по сравнению с предшествующей постановкой вопроса в истории науки. Но я повторяю, что решается этот вопрос менее адекватно, чем он должен решаться на современном уровне развития науки. И если мы обратимся к науке XIX века, мы можем сказать, что хотя социально-этические и гуманистические проблемы науки обнаруживались тогда реально во многом в зачаточном виде, но они выявлялись в философии в более сильной форме, чем это делается сейчас.

Что касается современности, то я еще раз повторяю, что, может быть, мы сейчас видим большее влияние политики и идеологии, чем этики и морали, и, может быть, мы многое относим к этическим явлениям, тогда как на самом деле они имеют чисто политическую и идеологическую природу, и это подавляет самостоятельные этические решения ученых, а оно принимается за такие решения, причем обозначается порой как драматическая позиция ученых в этических вопросах науки, что, может быть, не совсем правильно.

Вывод отсюда один: необходимо развивать **самостоятельные** поиски и решения современных социально-этических проблем науки, учитывая в то же время опыт прошлого и критически соотнося его с современностью. Это позволит избегать повторения того, что уже отброшено в истории мысли, и идти вперед в постановке и решении новых вопросов. Речь идет при этом не только о восходящих еще к Сократу просветительских представлениях о том, что знание и добродетель неразрывны, но и их антиподе — руссоистстких концепциях об их несовместности[5], получивших свое предельное выражение в современной „критике науки", в концепциях „контркультуры", в представлениях о дилемме науки и морали и т. п.

В этом последнем случае обнаруженное еще Кантом отсутствие прямой, однозначной связи между знанием и моралью трактуется, (например Г. Стентом) как их некоторая „комплиментарность", присущая якобы именно европейскому мышлению, в котором наблюдается разрыв между объектом и субъектом. Эту „комплиментарность" можно устранить лишь обратившись к восточной философии, где нет такого разделения субъекта и объекта, как в европейской[6].

К тому, что мне уже приходилось говорить об этом, возражая, в частности Г. Стенту, я хочу добавить, что, по-видимому, и мы сами порой слишком резко отделяем себя от традиций науки, называемых „просветительскими" и слишком резко подчеркиваем, что-де наука сама по себе не является „ни добром, ни злом", что все исключительно определяется социальными условиями ее функционирования и т. п. Мне кажется, что такой „этический нейтрализм" недопустим. Социально-этические и гуманистические принципы (регулятивы) современной науки, я полагаю, нельзя трактовать как нечто внешнее для нее, действующее лишь на уровне научного сообщества или общества в целом. Здесь, видимо, как и в исследованиях истории науки недопустимы крайности „интернализма" и „экстернализма".

Этика науки, следовательно, не может быть построена, не только исходя из самого научного познания (то есть норм, на которые оно опирается в своем движении к объективной истине), но и как такая регламентация поведения ученого, которая учитывает его исключительно в качестве члена научного сообщества и общества в целом. С другой стороны, было бы неправильно противопоставлять эти подходы, исходя из некой дилеммы науки и морали. Этика науки должна огранично соединять их, и основой здесь выступает марксистское учение о единстве научного познания и ценностных подходов, о науке как социальном институте, сущностное предназначение которого определяется гуманистическими идеалами, оказывающимися, следовательно, доминирующими в отношении „чистого" исследования.

Конечно, эти отношения (а тем самым и „первоосновная", принципиальная сущность науки как силы моральной и гуманной по природе своей) могут адекватно проявляться в соответствующих им моральных условиях и, наоборот, извращаться в других, неблагоприятных для их проявления. Это значит, что практически, реально определяющими во многих случаях оказываются как раз социальные условия функционирования науки. Однако мы не можем ставить пропасть между сущностью и явлением, как не можем игнорировать всякие различия между ними. Мы не можем

[5] Ж. Ж. Руссо в своем трактате „Способствовало ли возраждение наук и искусств очищению нравов?" выдвинул, как известно, эту идею. (См. Жан-Жак Руссо. Избранное. М., 1976 г.).

[6] Эти идеи Г. Стент развил в своем докладе на ХУI Всемирном философском конгрессе (1978 г.). См. об этом в моей книге „Жизнь и познание". М., 1981 г.

лишать перспективы развития гуманных этических ориентаций науки даже в неблагоприятных социальных условиях, как не можем считать, что в благоприятных — вообще никаких этических и гуманистических проблем науки, ответственности ученых якобы не возникает.

Речь, следовательно, может идти лишь о том, что сами по себе, в отрыве от соответствующих социальных условий, этические и гуманистические принципы науки не являются самостоятельными, они не могут выполнять роль единственного и главного регулятива научного познания. Поэтому иллюзорными являются представления о возможности саморегулирования науки, исходя исключительно из внутренних для нее и сообщества ученых моральных обязательств и гуманистических оценок, исключающих всякие государственные и общественные регламентации научных исследований. Ведь, если при капитализме, где цели и ценности науки и общества находятся порой в антогонистическом противоречии, а потому всякое государственное или общественное регулирование исследований воспринимается учеными как „внешний контроль", чуждый им и подавляющий свободу этих исследований, то при социализме, развивающемся по пути к самому гуманному и свободному обществу — коммунизму, в принципе недопустимо возникновение таких противоречий, и если они порой проявляются, то только лишь в результате нарушения этих принципов. Это наглядно иллюстрируется историей с „феноменом лысенкоизма", о чем так много говорят сегодня на Западе. И это видно на примере исследований в области генной инженерии, переживших множество трагических моральных коллизий в западной науке, которые невозможны при социализме, где гуманистические цели науки и общества находятся в единстве, что открывает колоссальные возможности для свободного научного поиска и, вместе с тем, направляется по пути, благоприятному для человека и общества.

Таким образом, исходя из концепции приципиального единства познания и ценностей, включая в последние не только этические, но и социальные, мы и говорим о социально-этических принципах (регулятивнах) научного познания, о социологии и этике науки, в которой приоритет гуманистических подходов определяется не изолированно от реальных социальных условий того или иного общества, а в тесной связи с ним. Это, мне кажется, может служить основой и нашего отношения к возможности и необходимости созданция единых этических кодексов, призванных регулировать научное познание в областях, затрагивющих жизненные интересы современных и будущих поколений людей. Хотя сегодня ведутся интенсивные исследования в этой области, и уже предложено немало интересных проектов, их роль я оцениваю в основном как стимулирующую, каталитическую. В настоящее время главная проблема заключается, на мой взгляд, в том, чтобы осуществлять более эффективный контроль за исполнением уже принятых социально-этических и правовых установлений, кодексов и соглашений.

Следует отметить, что говоря об этических кодексах ученых, я включаю сюда не только те из них, сила которых ограничивается лишь моральными установлениями, подобно „клятве Гиппократа". В настоящее время существует также целая система правового регулирования научных исследований, например, экспериментов на человеке и т. п. Ученые многих стран все настойчивее ставят вопросы, касающиеся выработки глобальных критериев, имеющих не только моральную силу, в отношении ряда научных исследований, которые представляют опасность для человека и жизни на Земле. Конечно, при этом не должен наноситься ущерб развитию науки, свободе научного поиска, однако она и здесь, по-видимому, должна пониматься как осознанная необходимость. Будущее покажет, что здесь можно сделать, но ясно одно: наука не должна выходить из-под контроля общества, она не можст развиваться как сила, несущая угрозу человеку и человечеству. Всякие формы софистических ухищрений, направленных на то, чтобы приуменьшить опасность бесконтрольного развития науки, глубоко аморальны, антигуманны и бесперспективны, так как они могут привести к гибели человечества, а значит-крушению самой науки.

Поэтому в современных условиях все чаще обращаются к проблемам социологии и этики научного познания (в особенности жизни и человека), причем это обращение осуществляется с разных философских и мировоззренческих позиций. И здесь иметь в виду, по-видимому, и ту активность, которую проявляют в этих вопросах наши идеологические противники, пытающиеся доказать, что Марксизм якобы вообще „равнодушен" к ним в силу его „сцентистского антигуманизма", игнорирования проблем человека, сферы наврственности и пр. Существует и более общая тенденция „дополнить" в этих вопросах науку религиозно-этическими принципами, дающими якобы ей единственно надежную ориентацию.

По-видимому, мы сегодня, обращаясь к социально-этическим и гуманистическим проблемам науки, можем довольно четко определить не только поляризацию сциентистских и антисциентистских подходов, но и внутри последних выделить научные, ложенаучные и антинаучные концепции, реалистические установки и многообразные варианты пессимизма, алармизма, конформизма и утопического оптимизма. Конечно, и с позиций реализма, мы обязаны видеть не только трудности и опасности, создающие угрозу будущему человечества, но и ясно обосначенные позитивные тенденции, вселяющие оптимизм и надежду. Эти объективные тенденции, выявленные и развиваемые марксизмом, стихийно охватывают современную науку в общемировом, глобальном масштабе, но сознательно и целенаправленно они реализуются в том типе науки, который существует в социалистических странах. Социально-этические и гуманистические, нравственные основы социалистического типа науки оказывают возрастающее влияние на мировую науку, в которой все больше проявляются сознание отвественности за будущее человека и человеческой цивилизации, критический подход ко всему, что может привести к ее катастрофе. Ведь, как говорил А. П. Чехов, „сколько погибло цивилизаций . . . потому что в свое время не было хороших критиков".

Надо надеяться, что минует это трудное и опасное время и, несмотря на все несовершенство наше и цивилизации, созданной с помощью науки, человечество устремится вперед — к истине и добру, которые для нас сливаются с коммунистической перспективой. Конечно, одной надежды мало, нужны и соответствующие действия, вдохновляемые новыми идеаями, новым пониманием науки и гуманизма, который дает марксизм, его, его филисофия — любовь к мудрости и человеку как носителю ее, выполняющая свою интегративную, критическую и ценностно-регулятивную, аксиологическую функцию в современной науке. И в этом, кстати говоря, научная и практическая ценность филисофии, (хотя она в принципе является самоценной и как самодостаточный элемент культуры не нуждается в утилитарном определении своего значения).

Резюмируя сказанное выше об отношении человек-наука-гуманизм и имея в виду перспективу исторического развития, можно сказать, перефразируя известное изречение Протагора, что именно человек является „мерой всех наук", и только такая гуманистическая ориентация создает прочную основу его будущему и человеческой культуры в целом. Последнее имеет и более широкое значение, так как гуманизм не связан только с наукой. И с этой зрения в равной мере опасны как абсолютизация значения науки в жизни человека и человечества, в развитии гуманистической культуры будущего, так и попытки принизить ее, а то и попросту „развенчать", представить в качестве антигуманной силы. Подлинный смысл наука приобретает лишь в связи с другими формами человеческой деятельности, образующую или материальную и духовную культуру человечества.

Огромную роль в этом плане играют трудовая деятельность человека, искусство, сфера нравственности. В каждой из них по-своему проявляются творческие, сущностные силы человека и человечества, и каждая из них является своеобразной гранью многоцветной, всесторонне и грамонически развитой личности. Стре́мясь достигнуть этой целостности и гармонии, мы обращаемся к наиболее развитым творческим формам жизнедятельности личности, к вершинам человеческого духа, нравственного мира творчески деятельной личности — „фаустовскому человеку", каким пытался изобразить его великий Гёте. Но для этого не нужна, конечно, „фаустовская сделка" с наукой, „демон" которой отнюдь не требует отдать ему человеческую душу, совесть и свободу человека. Поэтому нравственные, гуманистические основы науки и всей человеческой культуры приобретают все болшее значение сегодня и в будущем, так как, по-видимому, без этого возможны только мрак бескультурия и небытия, духовный и физический крах человечества.

Итак, следовательно, современная ситуация в науке о жизни, у истоков которой стоит Ч. Дарвин, породила много новых проблем, ряд из которых не имел такой остроты во времена Дарвина и вообще в XIX веке и даже в первой половине XX века. Решение этих проблем требует объединения усилий ученых — представителей разных наук, включая философию.

Потому-то так важно сегодня не только провозглашать, но и **реализовать практически** идею единства и взаимодействия наук и гуманистической культуры в целом, о чем уже говорилось выше.

Ленинская идея союза философов и ествествоиспытателей получает сегодня, таким образом, новое развитие, соответствующее современному уровню научного познания.

Когда-то Шиллер, обращаясь к естествоиспытателям провозгласил:

„Вражда да будет между вами!
Ваш преждевременен союз!
В своих исканьях разделитесь —
Сама к вам истина придет!"

Эрнст Геккель, приводя эти слова Шиллера в своих „Мировых загадках", сказал, что теперь, наконец-то, естествоиспытатели и философы встретились и могут объединиться.

Последующие события показали, что это далеко не так. Выражаясь словами К. Маркса, у естествознания и философии налицо была воля к объединению, способность же отсутствовала[7].

Заслугой В. И. Ленина является то, что он выдвинул идею о союзе философов и естествоиспытателей тогда, когда были уже не только воля, но и способность

[7] См. К. Маркс с Ф. Энгельс. Из ранних произведений. М. 1965, стр. 595.

к такому союзу, созрели его объективные предпосылки. Правда, субъективные факторы могут оказаться порой сильнее объективных. И это относится не только к деятельности философов, но и естествоиспытателей. Истекший период дал хорошую и поучительную проверку нашей способности к такому единству и совместной работе. Не все ее выдержали. Но общее дело союза философов и естествоиспытателей растет и развивается, находя свое новое выражение в тенденциях **единства и взаимодействия наук**.

Последнее выступает уже для нас, не только и не просто как комплексная **проблема**, но как главное стратегическое **направление**, **программа** исследований, в частности методологических проблем жизни и ее познания, в которых органически соединяются как гносеологические, так и социально-этические, гуманистические подходы.

Поэтому Дарвин сегодня — это для нас не только созданная им теория эволюции и новый научный метод; это и вся сумма новых проблем жизни и ее познания, вызванная силой его гения и тех, кто развивал и развивает его дело.

University of Guelph, Ontario, Canada

The Ideology of Darwinism

M. Ruse

According to a popular view of science, a major characteristic distinguishing between science and other products of the human mind, like literature or philosophy or religion, is that science is in some sense "objective". Something like literature, it is held, gives evidence of the wishes, desires, beliefs, hopes, aspirations, values, of the people who produce it. For instance, "Oliver Twist" makes very plain the views of Charles Dickens, about the state of the poor and the conditions of the work-houses at the threshold of the Victorian era. Similarly, Plato's "Republic" is very revealing about the author's views on such subjects as the moral value of literature, and the proper place of women.

But, science supposedly is different. It is not about the scientist at all; but, about the "real" world. It tells one how things are; not how one would like them to be, or how they ought to be. After all, either planets go in eclipses or they do not. It is no use wishing they go in squares, if they don't. And this is a truth for everyone, be they young or old, male or female, Canadian or German, capitalist or socialist.

Even in the social sciences, an area where people have deep-held convictions, the aim and realizable goal can, and ought to be, objectivity. Listen to Ernest Nagel, ohne of the chief spokesmen for this viewpoint:

> There is a relatively clear distinction between factual and value judgments, and . . . however difficult it may sometimes be to decide whether a given statement has a purely factual content, it is in principle possible to do so . . . [I]t is possible to distinguish between, on the one hand, contributions to theoretical understanding (whose factual validity presumably does not depend on the social ideal to which a social scientist may subscribe), and on the other hand contributions to the dissemination or realization of some social ideal (which may not be accepted by all social scientists). (Nagel 1961, pp. 488-9).

Admittedly there have always been some who have spoken against these claims. Marxists, for instance, and, increasingly, feminists. The latter, particularly, have taken aim at much of the psychological and sociological work of this century, especially that directed at explaining male/female differences. They argue that such work is permeated through and through with a grossly prejudiced attitude against females. (See Ruse 1981 for details).

1 In this paper, I shall quote very extensively from the writings of the scientists I discuss. I do this deliberately, so that the reader can see for him/herself the truth of may claims about the very "flavour" of scientific writing.

2 In this paper, I use the term "Darwinism" in a very broad sense.

However, most "real" scientists, that is to say, most physical or biological scientists, have sat smugly on the side, while these attacks were being launched. After all, the attacks were not directed at them! Indeed, physicists, chemists, biologists, and others, have rather welcomed such attacks. They merely serve to confirm already-held beliefs that the social sciences are not genuine sciences, after all. Certainly, the social sciences are not things worthy to feed at the research-grant trough, along with the big fellows. Such things as the social sciences, namely things which are value-impregnated, belong in the humanities, along with philosophy and literature and so forth.

What I want to suggest is that such complacency on the part of the older-established sciences may not be warranted. At least, what I want to suggest is that if we turn to one of the finest scientific creations of them all, the DARWINian theory of evolution through natural selection and its various extensions and modifications, we shall find that it is just as much a reflector and propagator of values, as anything to be named in the "softer sciences", or in the humanities for that matter.

I do not know whether what I have to say will prove to be the thin end of a wedge; a wedge that demonstrates that values and ideology go through every part of every science. I do know that what I have to say holds of evolutionary biology, in the past and present — and, probably, always will in the future. Moreover, half-defiantly let me say that I do not see the presence of values in biology to be a flaw — something to be regretted and eliminated, if possible. Indeed, making a virtue out of necessity, I rather welcome the normative nature of the science.

To carry through my plan of showing the values in DARWINian Science, I start by turning briefly to the past. Then, I face the present. And, finally, I offer a brief analysis of my factual findings.

Victorian Values

Think for a moment of some of the values and ideals of DARWIN's contemporaries, the Victorians. (For details, see RUSE 1979a; HOUGTON 1957). At least three sets of ideas or beliefs spring readily to mind. First, there was the whole socio-economic Weltanschauung, to use a rather clumsy phrase. Victorian Britain was capitalist, with some members benefitting mightily, but with most very much at the mercy of their "superiors". It was a world of machines: how else had Britain climbed to such world domination, as she raced ahead thanks to her industrial revolution? And, it was a time which put great demands on all, "successful" as well as the rest. To go ahead in Victorian England, one needed luck, guts, and an incredible capacity for hard work.

And, the point we should remember here is that, by and large, those who kept up with the swim, approved of Victorian society. They liked capitalism, thinking indeed that it was God-ordained. Remember the hymn that we all (i.e. all Englishmen!) used to sing as children:

The rich man in his castle, the poor man at his gate. God made them high and lowly, He ordered their estate.

Don't blame the capitalist for the present state of affairs. Thank God!

Next we have the attitudes toward inferiors, especially women and foreigners. There was general agreement that women simply are not the same as men — they just cannot do the things that men can. For some, this was the end of the matter:

Man for the field and woman for the hearth:
Man for the sword, and for the needle she:
Man with the head, and woman with the heart:
Man to command, and woman to obey:
All else confusion. (TENNYSON 1847, 5, lines 437-41)

Others gave women a greater role. Man is the creature of the brain. Woman is the creature of the heart. "*For woman is not undevelopt man,/But diverse.*" Hence, "*In true marriage lies/Nor equal, nor unequal. Each fulfills/Defect in each*". (Ibid, 7, 259—60, 284—6).

But, even "liberals" took a sterner line when it came to foreigners: "Wogs begin at Calais". The non-British are the "frogs", the "spicks", the "yids", the "niggers". These vile terms speak for themselves. General opinion was that there is a scale, more or less falling away, as one gets farther from the English coast. One progresses down through the French and Germans, past negroes, until at the bottom one reached the pathetic wretches at the bottom of South America, the Tierra del Fuegians. Actually, there was some scope for choice. Some put the Tierra del Fuegians at the bottom, with the Irish hust above. Others put the Irish at the bottom, with the Tierra del Fuegians just above.

Listen to the popular writer Hugh MILLER, who combines this position with the remarkable claim that Jesus of Nazareth was a kind of honorary Briton.

> Now, all history and all tradition, so far as they throw light on the question at all, agree in showing that the centre in which the human species originated must have been somewhere in the temperate regions of the east, not far distant from the Caucasian group of mountains. All the old seats of civilization, — that of Nineveh, Babylon, Palestine, Egypt, and Greece, — are spread out around this centre. And it is certainly a circumstance worthy of notice, and surely not without bearing on the physical condition of primaeval humanity, that in this centre we find a variety of the species which naturalists of the highest standing regard as fundamentally typical of the highest races of the globe. . . .
> It walks, however, the boards of our Parliament House here in a very respectable type of Caucasian man; and all agree that nowhere else in modern Europe is it to be found more true to its original contour than among the high-bred aristocracy of England, especially among the female members of the class. . . .
> Let me next remark, that the further we remove from the original centre of the race, the more degraded and sunk do we find the several varieties of humanity . . . till at the extremity of the (American) continent we find, naked and shivering among their snows, the hidous, small-eyed, small-limbed, flat-headed Fuegians, perhaps the most wretched of human creatures. (MILLER 1856, pp 250-4)

But goodness, are we not forgetting somebody?

> People who are remarkable for open projecting mouths, with prominent teeth and exposed gums; and their advancing cheek-bones and depressed noses bear barbarism on their very front. In Sligo and Northern Mayo the consequences of the two centuries of degradation and hardship exhibit themselves in the whole physical condition of the people, affecting not only the features, but the frame. Five feet two inches on an average, — pot-bellied, bow-legged, abortively featured, their clothing a wisp of rags, — these spectres of a people that were once wellgrown, able-bodied, and comely, stalk abroad into the daylight of civilization, the annual apparition of Irish ugliness and Irish want. (Ibid, p. 257)

Finally, let me draw your attention to the fact that the Victorians were, above all else, God intoxicated. They thought religion. They believed religion. They behaved religion. They argued endlessly about such vital matters as the reality of eternal damnation, and they worried incessantly on such crucial issues as the acceptability of Sunday trains (or "Sabbath trains", as low-church evangelicals inevitably referred to them). That which in some sense was the direct creation of God, or reflected God's glory, was good. That which did not, was not.

Victorian biology

Now, it is very easy to show that these values and ideals of the Victorians (let me refer to this as the "Victorian ideology") got reflected into the science of the time. The whole capitalist ethos is, of course, part and parcel of the Malthusian claim about the inevitability of human struggles for existence, given the propriety for population numbers to increase geometrically and for food supplies to increase only arithmetically. (MALTHUS 1826).

And, certainly, Malthusianism was used by other scientists, biological scientists, in such a way as to show their approval of the phenomenon. For instance, SPENCER (1857) wrote of an "essential beneficence" at work in the world, and he saw the MALTHUSian struggle forcing organisms up the evolutionary scale, until finally we get the evolution of *Homo Sapiens*.

Again, attitudes towards women and blacks come through in the writings of many. Listen to Adam SEDGWICK and David BREWSTER on the subject: "The mould in which Providence has cast the female mind, does not present to us those rough phases of masculine strength which can sound depths, and grasp syllogisms, and cross-examine nature" (BREWSTER 1844, p. 503). Rather, woman has a "soft and gentle temperament", "quick appreciation of character", and "instinctive knowledge of what is right and good" (SEDGWICK 1845, p. 4).

And, we need look no further than SPENCER for views on the savages. As one might expect, given that "the civilized European departs more widely from the vertebrate archetype than does the savage" (SPENCER 1857, 1, 57), SPENCER had little trouble showing that the epitome of evolution is *Homo britannicus*, and that "it is more especially in virtue of having carried [the] subdivision of function to a greater extent and completeness, that the English language is superior to all others". (SPENCER 1857, 1, 57).

Nor are the Irish forgotten. They rank somewhere along with the herring in their tendency to multiply.

> And here it must be remarked, that the effect of pressure of population, in increasing the ability to maintain life, and decreasing the ability to multiply, is not a uniform effect, but an average one. In this case, as in many others, Nature secures each step in advance by a succession of trials, which are perpetually repeated, and cannot fail to be repeated, until success is achieved. All mankind in turn subject themselves more or less to the discipline described; they either may or may not advance under it; but, in the nature of things, only those who **do** advance under it eventually survive. For, necessarily, families and races whom this increasing difficulty of getting a living which excess of fertility entails, does not stimulate to improvements in production — that is, to greater mental activity — are on the high road to extinction; and most ultimately be supplanted by those whom the pressure does so stimulate. This truth we have recently seen exemplified in Ireland. (SPENCER 1852, p. 267)

Finally, as in everything else, we find that the writings of the scientists reek of religious views. Some of the discussions are very blatant, indeed. For instance, much of the work of the geologist William BUCKLAND (1823) was directed towards showing that there was a monstrous deluge, covering the whole earth, not so very long ago. But, even those who had gone beyond Biblical literalism, constantly brought religion into their scientific writings, showing that they were proud and pleased that science underlines religious belief.

The Swiss-American paleontologist, Louis AGGASIZ, for instance, laboured to show that there is a three-fold parallelism in the organic world — in living organisms, in embryological development in the fossils — with a progression, from primitive to complex. Expectedly, he made much of the fact that such a progression, culminating in man, shows God's creative powers: He prepared the way for man, by stocking the earth with a succession of creatures, each in turn best suited for the state of the world, as it cooled from its original incandescent form. (See OSPOVAT 1976).

Analogously, let me draw attention to the fact that scientist after scientist explicitly linked his analyses of the organic world, to God's purposive designing nature. The eye is like a telescope, therefore, just as the telescope has a telescope-maker, the eye has an eye-maker (or eye-Maker). This, of course, is the teleological argument, much beloved by natural theologians. But it was used by scientists, in scientific writings, also. A favourite example of mine, much-quoted in the nineteenth-century, occurs in the middle of a sober paper published in the Transactions of the Royal Society, where the anatomist Richard Owen (1834) is discussing the adaptations of the kangaroo for feeding its young. At one point, in sheer admiration, Owen declaims that such features are "irrefragable evidence of creative forethought".

DARWIN's Values

The points I have just made need no labouring. Time and again, we find that those scientists interested in biology and geology (particularly paleontology), let their Victorian ideology shine forth — proudly. Unashamedly, they let the reader know their values, ideals, prejudices, and fears. But, what I want to point out also — and this is really important — is that this self-same ideology can be found in the writings of Charles DARWIN himself!

The father of Darwinism was a Victorian, and he was just as ready as his fellows, to let the world know, through his science, just what he thought. And, his thoughts were much the same as anyone else! Do not think that, because DARWIN was a brilliant pioneer in evolutionary biology, that he was any less a product of his time and place than his fellows.

Start with the socio-economic question. It was hardly to be expected that the grandson of Josiah WEDGEWOOD, one of Britain's leading industrialists, would reject the social system within which he lived and worked. Especially since, thanks to the family pottery works, it was not necessary for DARWIN to seek paid employment. And indeed, DARWIN's approval of the system does come through. First to mind is DARWIN's wholesale use and endorsement of the MALTHUSian social theory, which DARWIN generalizes:

> A struggle for existance inevitably follows from the high rate at which all organic beings tend to increase. Every being, which during its natural lifetime produces several eggs or seeds, must suffer destruction during some period of its life, and during some seasons or occasional year, otherwise, on the principle of geometrical increase, its numbers would quickly become so inordinately great that no country coulÈ support the product. Hence, as more individuals are produced than can possibly survive, there must in every case be a struggle for existence, either one individual with another of the same species, or with the individuals of distinct species, or with the physical conditions of life. It is the doctrine of MALTHUS applied with manifold force to the whole animal and vegetable kingdoms; for in this case there can be no artificial increase of food, and no prudential restraint from marriage. Although some species may be now increasing, more or less rapidly, in numbers, all cannot do so, for the world would not hold them. (DARWIN 1859, pp. 64-4)

At this point, one suspects that the defender of the value-neutrality of science will object. It is certainly true that DARWIN uses Malthusianism as a model or foundation, but this is not to say that DARWIN approves of Malthusianism. Indeed, in an important sense, he clearly does not. MALTHUS preaches an extremely conservative doctrine, maintaining that change is impossible DARWIN turns MALTHUS on his head, arguing that change is inevitable!

But, whilst there is certainly some truth in all of this, when DARWIN comes to consider our own species, *Homo Sapiens*, it is very clear that he is far from disapproving of an unrestrained MALTHUSian struggle. Consider, for instance, DARWIN's discussion of the "benefits" of modern medicine,

> With savages, the weak in body or mind are soon eliminated; and those that survive commonly exhibit a vigorous state of health. We civilised men, on the other hand, do our utmost to check the process of elimination; we build asylums for the inbecile, the maimed, and the sick; we institute poor-laws; and our medical men exert their utmost skill to save the life of every one to the last moment. There is reason to believe that vaccination has preserved thousands, who from a weak constitution would formerly have succumbed to small-pox. Thus the weak members of civilised societies propogate their kind. No one who has attended to the breeding of domestic animals will doubt that this must be highly injurious to the race of man. It is surprising how soon a want of care, or care wrongly directed, leads to the degeneration of a domestic race; but excepting in the case of man himself, hardly any one is so ignorant to allow his worst animals to breed. (DARWIN 1871, 1, 168)

And, when it comes to the virtues of capitalism, all pretence at value-free inquire vanishes completely.

> In all civilised countries man accumulates property and bequeaths it to his children. So that the children in the same country do not by any means start fair in the race for success. But this is far from an unmixed evil; for without the accumulation of capital the arts could not progress; and it is chiefly through their power that the civilised races have extendend, and are now everywhere extending, their range, so as to take the place of the lower races. Nor does the moderate accumulation of wealth interfere with the process of selection. When a poor man becomes rich, his children enter trades or professions in which there is struggle enough, so that the able in body and mind succed best. The presence of a body of well-instructed men, who have not to labour for their daily bread, is important to a degree which cannot be over-estimated; as all high intellectual work is carried on by them, and on such work material progress of all kinds mainly depends, not to mention other and higher advantages. (Ibid, 1, p. 169)

What about nature's inferiors namely the non-British and women? Again, DARWIN shows himself to be at one with his fellow country-men. On the one hand, we learn that the British will beat the lesser Turks in the struggle for existence, and that:

> The care-less, squalid, unaspiring Irishman multiplies like rabbits: the frugal, foreseeing, selfrespecting, ambiotious Scot, stern in his morality, spiritual in his faith, sagacious and disciplined in his intelligence, passes his best years in struggle and in celibacy, marries late, and leaves few behind him. (DARWIN 1871, 1, 174, quoting W. R. GREG)

On the other hand, we learn that the DARWINian mechanism of sexual selection leads to some fairly Sedgwickian differences between male and female. "Man is more courageous, pugnacious, and energetic than woman, and has a more inventive genius. (In compensation), woman has greater tenderness and less selfishness." (DARWIN 1871).

Finally, there is the question of religion. Here, one might think one runs right into trouble, if one tries to show that DARWIN's work reflected orthodox religious belief.

And, it is indeed true that if one looks in DARWIN's writings for Christian support, one will be disappointed. However, DARWIN was never an atheist, and it was not until after the "Origin" that his faith crumbled away to agnosticism. Until that point, hc was — like so many Victorians — a *deist*, believing in a God who works through unbroken, natural regularities.

And, the "Origin" itself proclaims this belief, proudly. Consider the closing passage:

> It is interesting to contemplate an entangled bank, clothed with many plants of many kinds, with birds singing on the bushes, with various insects flitting about, and with worms crawling through the damp earth, and to reflect that these elaborately constructed forms, so different from each other, and dependent on each other in so complex a manner, have all been produced by laws acting around us. These laws, taken in the largest sense, being Growth with Reproduction; Inheritance which is almost implied by reproduction; Variability from the indirect and direct action of the external conditions of life, and from use and disuse; a Ration of Increase so high as to lead to a Struggle for Life, and as a consequence to Natural Selection, entailing Divergence of Character and the Extinction of less-improved forms. Thus, from the war of nature, from famine and death, the most exalted object which we are capable of conceiving, namely, the production of the higher animals, directly follows. There is grandeur in this view of life, with its several powers, having been originally breathed into a few forms or into one; and that, whilst this planet has gone cycling on according to the fixed law of gravity, from so simple a beginning endless forms most beautiful and most wonderful have been, and are beeing, evolved. (DARWIN 1859, p. 489-90)

What more needs to be said? The work of Charles DARWIN is as value-laden as a novel by Charles DICKENS, or a sermon by John Henry NEWMAN!

Evolutionism today: Sociobiology

But, what of the present? Do we see contemporary ideologies reflected in the work of evolutionists? I think we undoubtedly do, no less than Victorian ideology was to be found in DARWIN's work. Let me make this point by referring briefly to three areas which have engaged the attention of evolutionists in the past decade — and which, indeed, still engage the attention of evolutionists.

First, and most obviously, there is the sociobiology debate. (For details, see RUSE 1979b). Since the time of the "Origin", evolutionists have realised that a comprehensive analysis of organic origins and of the present state of organic nature must cover, not just the actual morphological forms of organisms, but also how they (animals in particular) behave and relate to other organisms and to the inorganic forces of nature.

Unfortunately, however, despite sterling efforts by DARWIN himself, the study of behaviour lagged somewhat behind other areas of evolutionary interest — for fairly natural reasons, given that it is far easier to measure and record morphological features, than behavioural features. The red eye of a fruit fly is easily identifiable, but the strange sexual behaviour of the same fruit fly could as readily be an artifact of its cage as a reflection of its evolutionary heritage.

However, since about 1960, the study of animal behaviour from an evolutionary perspective — both in fact-gathering and in theory-creating — has exploded. Now, we have a full-blown science of "sociobiology", pushing ahead vigorously. It is, of course, still part of the overall evolutionary family, by virtue of its dependence on the central DARWINian principles of evolution through natural selection working on small random mutations.

In itself, sociobiology is hardly that controversial — I think all evolutionists welcome its coming, and its application to some of the more thorny questions of animal behaviour. For instance, a problem which has attracted the attention of evolutionists since DARWIN, is that of the sterile castes which one finds among certain insects, the Hymenoptera (ants, bees, and wasps). At one stroke, through the notion of "kin selection", sociobiologists have been able to show the precise reasons for such sterility, and why it occurs in almost no other species of animal (WILSON 1975).

However, sociobiology has attracted attention in another respect: attention which is a lot less favourable. A great many sociobiologists, having competed their discussions of the animal world, turn at once to our species, arguing that we too, particularly in our social behaviour, are no less a product of our genes as fashioned by natural selection, than are other animals. Consequently, in much that we do and think, we show our evolutionary heritage. It is, for instance, argued that in our attitudes towards strangers — fearing them and treating them hostily — we show how our ancestors, who lived in small bands threatened by other small bands, were selected to treat non-insiders very warily indeed (ALEXANDER 1979).

It is argued also that our sexual attitudes reveal our biology: men can fertilize at will, and hence have biological urges making them want to do this, whereas females are left (literally) carrying baby, and thus have biological urges making them most cautions about sex. As one eminent sociobiologist, E. O. WILSON writes:

> The anatomical difference between the two kinds of sex cell is often extreme. In particular, the human egg is eighty-five thousand times larger than the human sperm. The consequence of this gametic dimorphism ramify throughout the biology and psychology of human sex. The most important immediate result is that the female places a greater investment in each of her sex cells. A woman can expect to produce only about four hundred eggs in her lifetime. Of these a maximum of about twenty can be converted into healthy infants. The costs of bringing an infant to term and caring for it afterward are relatively enormous. In contrast, a man releases 100 million sperm with each ejaculation. Once he has achieved fertilization his purely physical commitment has ended. His genes will benefit equally with those of the female, but his investment will be far less than hers unless she can induce him to contribute to the care of the off-spring. If a man were given total freedom to act, he could theoretically inseminate thousands of women in his lifetime.
>
> The resulting conflict of interest between the sexes is a property of not only human beings but also the majority of animal species. Males are characteristically aggressive, especially toward one another and most intensely during the breeding season. In most species, assertiveness is the most profitable male strategy. During the full period of time it takes to bring a fetus to term, from the fertilization of the egg to the birth of the infant, one male can fertilize many females but a female can be fertilized by only one male. Thus if males are able to court one female after another, some will be big winners and others will be absolute losers, while virtually all healthy females will succeed in being fertilized. It pays males to be aggressive, hasty, flickle, and undiscriminating. In theory it is more profitable for females to be coy, to hold back until they can identify males with the best genes. In species that rear young, it is also important for the females to select males who are more likely to stay with them after insemination.
>
> Human beings obey this biological principle faithfully (WILSON 1978, pp. 124-5).

And, in many, many other respects we betray our simian origins. Even homosexuality is put down to evolution: the homosexual is likened to the sterile worker ant!

Now, it is not my aim here to evaluate the truth-status of human sociobiology. I have, indeed, already had much to say about this matter elsewhere. What I want to do is to draw your attention to the ideological element within the sociobiology debate. I do not mean simply that the opposing scientists had different ideologies — although this is true — but also that these ideologies were reflected in the very scientific positions that they took.

Start with the critics. Several of the leading critics of sociobiology were avowed Marxists, who, by their own admission, have tried to let their theorizing be influenced by their Marxism.

> As working scientists in the field of evolutionary genetics and ecology, we have been attempting with some success to guide our own research by a conscious application of Marxist philosophy. (LEWONTIN and LEVINS 1976, p. 34)

And in this case, there was such influence.

As good Marxists, many of the critics were committed, not simply to the essential equality of every human, but to a much stronger thesis about identity, and about the causes of human social and behavioural differences. They believe that what makes one man a doctor and another a lavatory attendant are not innate differences, but social forces: whether one is born into one sort of family, or another; whether one is born into one sort of society, or another. Like the Jesuits and the behaviourists, they believed that we are all born essentially the same, and that differences then come about, because of impinging environmental causes.

Hence, the critics opposed human sociobiology, because (rightly) they saw it as denying such identity and such environmentalism. And a major way in which they opposed human sociobiology was by arguing that all the empirical and theoretical evidence points to their own position! One representative quote must suffice:

> We have pointed out that adaptive evolutionary arguments can be rendered totally untestable by postulating genes wherever the theory requires them and by telling 'just-so' stories about adaptation. There does exist, however, one possibility of tests of such hypotheses, where they make specific quantitative predictions about rates of change of characters in time and about the degree of differentiation between populations of a species. Population genetics, the analytical base on which evolutionary explanations are built, makes specific predictions about such rates, based upon estimates of mutation rate, population sizes, selection intensities, and migration patterns. In addition, there are hard data on the degree of genetic differentiation between human populations for biochemical traits . . .

Both the allowable rates of genetic change in time, from theory, and the observed genetic differentiation between populations, are too small to agree with the very rapid changes that have occurred in human cultures historically and the very large cultural differences observed among contemporaneous populations.

To take a particular case, we can consider the rise and fall of Islam. In the sixth century the Arab world was made up of poor, pastoral people. In 633, Mohammed appeared and within 200 years Islam was the supreme power in the West, both politically and culturally. Music, poetry, mathematics, science, historiography, all flowed in profusion from Baghdad and Cordoba, the two great capitals of power and culture in the 10th century. But by 1492 the last Moors had been ejected from Spain and the Arab people were again a poor pastoral culture. This cycle, occupying less than 30 generations, was too rapid by orders of magnitude for any large change by natural selection. It would appear that really significant cultural changes occur by some nongenetic means and that an understanding of human history and fortunes must be sought where it has always been, not in genes but in culture.

The same problem arises from the immense cultural differences between contemporaneous groups, since we know from studies of enzyme specifying genes that at least 85% of that kind of human variation lies within any local population or nation, with a maximum of about 8% between nations and 7% between major races (ALLEN et al 1977, pp. 27-8).

Additionally, the critics went right in and used (for want of a better term) philosophical arguments, designed to show that (from a Marxist perspective) many aspects of human sociobiology are unacceptable. For instance, repeatly, the leading geneticist R. C. LEWONTIN accused the sociobiologists of the cardinal sin of "reification", that is to say, of making up a concept or category (in the mind), and of then assuming that it must have real existence.

> It cannot be assumed that any behaviour or institution to which a name can be given necessarily has an existance as a real thing subject to the laws of nature. Is entrepreneurship a real category for which there are genes and upon which natural selection operates as an entity, or is it an arbitrary construct, historically determined and useful as a way of describing human socioeconomic activity? The same question applies to religion, kinship, altruism, and so on. Any historical view of social thought, as well as any sophisticated comparative ethnography shows immediately that the categories of description of social institutions are historically contingent. What did "religion" mean to the Greeks (they had no word for it and as a separate social category it did not exist for them) or "revenge" to the Tasaday? Is "violence" real, or is it a social construct without any one-to-one correspondence to an actual physical act (what is "verbal violence" or a "violent exception")? Nothing has a more obfuscating effect on the understanding of society (itself a reification!) than the total confusion between categories of thought and real objects. (LEWONTIN 1976, p. 24).

Clearly, for LEWONTIN, science which does not fit within his own Marxist paradigm is inadequate.

But, what about the sociobiologists? Were they value or ideology free? The critics claimed not. They argued that sociobiology was permeated through and through with a vile, racist, sexist, capitalist view of humans, and of their society.

> For more than a century, the idea that human social behaviour is determined by evolutionary imperatives and constained by innate or inherited predispositions has been advanced as an ostensible justification for particular social policies. Determinist theories have been seized upon and widely entertained not so much for their alleged correspondence to reality, but for more obvious political value, their value as a kind of social excuse for what exists. (ALLEN et al 1977, p. 3)

Continuing their attack, critics cited Konrad LORENZ who, supposedly, began with geese and ended with HITLER's filthy racial doctrines of extermination.

Somewhat dogmatically, again primarily because I have written extensively on these matters elsewhere, let me say that I do not find the critics' charges well-raken. There is, for instance, no sociobiological implication that the people's of one race are better, or worse, or indeed that much different, from the people's of another race. And, although it is indeed true that the sociobiologists see differences between male and female, there are no implications that males are better than females. Indeed, in the human qualities that really matter — sensitivity, love, intelligence, wit, loyalty — females, if anything, seem to do better than males.

Does this then mean that I see no ideology in human sociobiology? No indeed! I think the science of the sociobiologists to be no less value-laden, than that of their Marxist opponents. Time and again, sociobiologists go beyond their evidence, in order to draw conclusions they find ethical or otherwise acceptable and desirable. For instance, although the sociobiologists certainly do not think that we are all identical, a major claim by them is that we are all united by much the same types of genetic causal mechanisms.

Consider, for instance, the following passage from WILSON's major work, "Sociobiology: The New Synthesis".

> The building block of nearly all human societies is the nuclear family . . . The populace of an American industrial city, no less than a band of huntergatheres in the Australian dessert, is organized around this unit. In both cases the family moves between regional communities, maintaining complex ties with primary kin by means of visits (or telephone calls and letters) and the exchange of gifts. During the day the women and children remain in the residential area while the men forage for game or its symbolic equivalent in the form of barter and money. The males cooperate in bands to hunt or deal with neighboring groups. If not actually blood relations, they tend at least to act as "bands of brothers." Sexual bonds are carefully contracted in observance with tribal customs and are intended to be permanent. Polygamy, either covert or explicity sanctioned by custom, is practiced predominantly by the males (WILSON 1975, pp.553-4).

To be quite frank, WILSON has little or no evidence for this claim. Certainly, he has nothing approaching direct evidence, where, for instance, one might try to identify the actual genes involved. What then is his rationale for saying what he does. No more or no less than a hope that, at some later point, it may be proven true! The belief that, under the skin, humans are united by a common biological nature, forged in the fire of natural selection. A link between the New York businessman and the Kalahari bushman. And, if this is not to be beyond strict, factual, "objective" science, into the world of values and ideology, I do not know what it is.

Evolutionism Today: Genetic Variation

I move on. Next, I want to consider brilliant work that LEWONTIN has been doing in his own right, and not merely as a critic of some other scientists' efforts. I refer to the fascinating work performed by LEWONTIN in genetics, on the interface between the physical and the biological sciences, over the amount of genetic variation one finds in natural populations.

Consider the genes of an organism, divided as they are into two sets, on opposing chromosomes. On any one chromosome, and on its corresponding mate (at any one "locus"), one has two genes or "alleles", and thus through any particular interbreeding population one has sets of such alleles, which can and do occupy the same loci. Since the 1930's and the development of population genetics — since the development of the "core" of modern DARWINian evolutionary theory — evolutionists have differed over these sets of alleles. Some, particularly H. J. MULLER and his followers, argue that there is not much variation over loci. At most, one finds one or two usually deleterious, mutant forms.

> The classical theory assumes that at nearly every locus every individual is homozygous for a "wildtype" gene. In addition, each individual is heterozygous for rare deleterious alleles at a handful of loci, on the order of a hundred out of tens of thousands of genes. Some very small proportion of the population will be so unfortunate as to be homozygous for a rare deleterious gene and thus be severely handicapped, and this proportion will be much increased in the offspring of matings between close relatives. (LEWONTIN 1974, p. 24)

Opposing this "classical hypothesis", we have the "balance" school of DOBZHANSKY, which argues that the sets of alleles are large, and that consequently there is much variation at loci. Any population, therefore, one has a large reserve of different genes.

The balance hypothesis in its most extreme form postulates, in direct opposition to the classical hypothesis, that individuals from sexually reproducing, cross-breeding populations are heterozygous at nearly every one of their loci, and only rarely will a locus be homozygous except in the offspring of closely related mates. This extreme view is expressed . . . as follows: "Subject to the limitations imposed by chance elimination of alleles, by mating of close relatives, and by the finite number of alleles at a locus, we feel that the proportion of heterozygosis among gene loci of representative individuals of a population tends toward 100 percent". (LEWONTIN 1974, p. 25). (See fig. 1)

	Two randomly chosen individuals	
Classical hypothesis	+ + + m + ... + + + + + + + ... m + (+ = wild-type allele; m = deleterious mutant)	+ + + + + ... m + + + + + + ... + +
Balance hypothesis	$A_3B_2C_2DE_5 \ldots Z_2$ $A_1B_7C_2DE_2 \ldots Z_3$ (A_1, A_2, = different alleles)	$A_2B_4C_1DE_2 \ldots Z_1$ $A_3B_5C_2DE_3 \ldots Z_1$

Fig. 1

At the purely biological level, it is virtually impossible to resolve the classical/balance dispute. Far too many distorting factors intervene. In a master-stroke, therefore, LEWONTIN turned the problem into one of molecular genetics, developing the technique of "gel electrophoresis". This enabled him to measure protein differences in organisms, and since such differences are functions of the DNA carried in the nuclei of cells, by making the connection between DNA molecules and MENDELian genes, LEWONTIN had a direct and powerful insight into the genetic differences between organisms. And such differences there are too! There is incredible variation between the alleles at loci, within the members of an interbreeding population. In short, the balance hypothesis seems to win hands down.

But, not so fast! The balance hypothesis supposes that the variation is not just chance, but a function of selective pressures (like that brought about by so-called superior heterozygote fitness). In other words, the variation is, in some sense or senses, linked through natural selection to adaptive advantage. What supporters of the classical hypothesis now argue is that the undoubted molecular variation which exists in populations may well have absolutely no effect on the organism at the level of selection! In other words, protein variations are adaptively neutral, and their existence and percentages "drift" randomly in populations. Selection itself continues to involve just a very few features in a population, at any one time.

Expectedly, balance supporters fight back. For instance, they suggest that protein and other features do indeed have the capacity for selective effects, and that this is shown by the fact that artificial selection can bring about great changes in virtually any direction in virtually any organism. (See AYALA and VALENTINE 1979; RUSE 1982) But, my interest here is not in the replies and counter-replies per se, but in LEWONTIN's own reaction to the whole debate, and most particularly to the way in which he treated the debate in his very important discussion, "The Genetic Basis of Evolutionary Change". He uses the whole controversy as an excuse — or rather, as a way — to put across his Marxist message.

Thus, we have LEWONTIN's belief in the contradictory nature of phenomena and of our understanding, and how resolution or escape can be achieved only by a dialectical process. "A . . . use of a Marxist approach is in the analysis of how a science has got itself into apparently irresolvable contradictions". (LEWONTIN and LEVINE, p. 63) Explicitly, LEWONTIN links this with the problem of variation in populations. "In evolutionary genetics at the present time there are serious contradictions between observations on the genetic variation within species and all standard explanatory theories". (Ibid) Hence, the only satisfactory understanding is holistic and integrative. "Marxism stresses the wholeness of things, both between organism and surroundings and within organism". "Marxism stresses the integration of phenomena of different levels of organization". And, most fully, LEWONTIN writes as follows:

> One way in which a Marxist viewpoint can inform scientific work is by encouraging an alternative paradigm to the analytic Cartesian method. Such an alternative stresses system properties as the primary objects of study as opposed to the conventional view of the main effect of separate elements, to which are added, as a secondary refinement, the interactions between them. (Ibid, p. 62-3)

In line with this position, LEWONTIN first presents the debate in terms of ever escalating, but unresolvable conflict: balance hypothesis; classical hypothesis; molecular balance hypothesis; molecular classical hypothesis. Then, at the end of his discussion of the classical/balance dispute, in effect LEWONTIN says "a plague on both your houses". The only way in which resolution of the dispute can be achieved is by taking a new, holistic approach.

> A complete description of a gas in phase space is utterly hopeless because the dimensionality of the phase space is too great, but the macroscopic law $PV = nRT$ collapses the dimensionality of the phenomenon into three eminently measurable dimensions — pressure, temperature, and volume. (LEWONTIN 1974, p. 318)

Analogously, by working with the whole genetic unit, it might prove possible to capture insights hidden in work with individual genes. No wonder LEWONTIN has felt able to say: "As working scientists in the field of evolutionary genetics and ecology, we have been attempting with some success to guide our own research by a conscious application of Marxist philosophy".

Evolutionism today: Paleontology

As throughout this paper, my aim is not to judge either the science or the ideology. Rather, it is to point to the existence of the latter in the former. And, as we come to the third fact of contemporary evolutionary thought to be discussed, we shall see this relationship yet again. What I want to direct your attention to is the controversy presently raging in paleontology about the nature and process of evolutionary change.

On the one side, we have the traditionalists, the so-called "phyletic gradualists". They argue that evolution is smooth and continuous, and that the chief causal process is the DARWINian mechanism of natural selection. On the other hand, we have the rebels, the neo-saltationists, otherwise known as the "punctuated equilibrists". They argue that change is sudden, followed by long periods of relative stability: "stasis". Whenever a new species or higher group is produced, there is an essentially random element in the process. A new group is not necessarily better adapted to its conditions, than the older group from which it came. The fossil record, therefore, is somewhat jerky, with gaps representing real switches rather than inadequacies in the record as understood by us.

Whether, overall, one can see some direction in the record is a matter of some debate amongst the punctuated equilibrists. There are those who claim that, through the process of "species selection", one gets a form of higher level selection, which does give some overall adaptive content to the record (STANLEY 1979). Others claim that the record is virtually totally random, and indeed they try to show that a computer programmed for randomness produces a reasonable facsimile to the record left by nature (GOULD 1980).

In a very real way, what is going on in paleontology is reducible to matters of brute empirical fact. Are there gaps in the fossil record? Does one see non-random trends in the record? Does stasis prevail universally? And so forth. But additionally, as always in evolutionary studies, one discovers an ideological element. Quite openly, one of the leading punctuated equilibrists, Stephen Jay GOULD, admits to his Marxism, and lauds the way in which his science is informed by his beliefs, and how conversely his beliefs are bolstered by his science. (For details, see RUSE 1982)

Specifically, GOULD identifies three points where his paleontology and his Marxism interact, and where he feels drawn towards punctuated equilibrism because he is a Marxist — and where, no doubt, his Marxism guided his paleontological theorizing. First, GOULD likes the saltationary, jerky nature of the record, because this accords well with what he believes about the nature of historical change. One should expect **revolutions**, rather than gradualism. And, GOULD is quick to point out the gradualism of Darwinism is a (supposed) direct function of Charles DARWIN's liberal Victorian beliefs, about the desirability of gradual, smooth, progressive change. In short, punctuated equilibrism is to the fossils, what orthodox Marxism is to human history. Change is sudden, abrupt, drastic, rather than slow, smooth, and continuous.

Second, GOULD sees puntuated equilibrism as emphasizing the whole, integrated organism. Either something belongs to another. Essentially, there are no intermediaries (or, if there are, they are very unstable). And, again, GOULD feels that he has a connexion between his science and his ideology: a reductionist position will not do. We must consider the organism as an indivisible, whole entity. Moreover, we must think of it as an historical entity: the features that an organism has are not simply functions of present selective forces, but also reflections of the organism's past. The way that an organism used to be, sets definite limits on the ways that it could change and the way it is now.

Third and finally, GOULD's science is connected to his views on man. If the hypothesis of punctuated equilibria is correct, then either an organism is a member of *Homo sapiens*, or it is not. There are no semi-men or women. But, this being so, then all humans are — in some very crucial sense — biologically identical. Essentially, there are no variations of note between humans, for there are no variations of note between any members of the same species. And, this view of human nature, as GOULD himself proudly notes, is precisely what is environmentalist position on human differences demands, expects, and wants.

In short, what I argue is that through and through GOULD produces and endorses a view of paleontology which is molded by, and conversely supports and proclaims, a view of the world he holds dear. We are offered the fossil record as seen through the lens of Marxism. And, with this final point made about the ideology of modern evolutionary biology, my examination of the actual views of scientists is complete.

Conclusions

What remains to be said? Let me try to sum up, or rather, without being bound by reasonable scholarly conventions and feeling obliged to present only those conclusions which are justified by argument which has gone before, what should be said about the relationship — proper or improper — between values and ideology and science? I want to say a number of things.

First, anyone who thinks there are no values, no ideology, in biology — past or present — is pathetically naive. Biology drips with as many wishes/wants/desires/urges, as many exhortations towards right actions, as a sermon by LUTHER or WESLEY. In a sense, I think that when a philosopher like Sir Karl POPPER (1972) pontificates about "objective" knowledge, implying that the scientist and only the scientist can discern the unvarnished, untained, unemotinal, un-value-laden truth, his self-serving naivity verges on the immoral. All I can do is to repeat the exhortation of Oliver CROMWELL: "I beseach you, in the bowels of Christ, think it possible you may be mistaken".

Second, let me try in a tentative manner to unpack the way or ways in which values enter into science. To begin, there is the simple act of choosing a problem as worthy of study. Surely the fact that virtually every biologist at some point or another turns to *Homo sapiens* tells us something? The implication is that there is something special about our own species — an implication no less obvious than SPENCER's explicitly putting us at the top of his evolutionary process.

Again, when people like WILSON or LEWONTIN start to look at questions to do with race or sex or caste, or (in WILSON's case) sexual orientation, the implication is undoubtedly that these are matters about which we ought to care. There are questions of pressing social importance.

Next, I think we get values in the whole manner in which the ideas and concepts of a theory are presented. What I have in mind here is a wide range of items which collectively come under what one might refer to as the "clothing", or perhaps even the "skin" of a theory. (I like biological metaphors!) I mean for instance the very Language which is used to talk about a theory. WILSON for instance was criticized as a sexist for referring to *Homo sapiens* as "man". I mean also the pictures which are used to illustrate a point. Again WILSON was criticized because his pictures supposedly portrayed males more prominently than females. I think in this case the objection fails, but the general point is well-taken.

Undoubtedly another item which must be included in the clothing of a theory are the examples, metaphors, and analogies which are used to support and clarify the theory. Yet another point is the way in which the theory is presented or laid out: the structure, if you like. LEWONTIN's way of discussing the various hypotheses about genetic variation shows and supports his commitment of the Marxist dialectic.

I think now you will have the general drift of what I have in mind. There may perhaps be more items operative here, but I am sure I have touched on some of the main ones. You will remember just above that I equivocated on whether to refer to these items as the "clothing" or the "skin" of a theory. This was not just a nice point of literary style. In a sense, both terms talk about something on the surface — not part of the real, central body — and this is certainly what I see as distinctive about the items I have listed above, vis-a-vis the theories in which they appear. WILSON's pictures, for instance, are more "peripheral" than say his mechanism of kin selection. Without kin selection, sociobiology as we know it today would not exist. One can and does publish books on sociobiology without pictures.

However, there is a crucial difference between clothing and skin. Clothing one can take off, and although one might feel awkward about appearing in none at all, one can change clothes and still be the same person. Skin, however, may be on the surface, but is not something changed — at least, were one to change one's skin one would be a different person, even though one would not be quite so different as if one were to change one's sex.

Hopefully, the point I am trying to make in my rather drawn-out manner, is starting to become clear. Prima facie it would seem that all the items I have listed could be changed and the theory itself would not be changed. One could for instance have different pictures. (This is the "clothing thesis".) But I am not quite so sure that matters are this simple (i.e. in respects I incline to the "skin thesis"). Influenced by Max BLACK's (1962) view of metaphor, I wonder if change of language, example, and so forth, would not run deeper. I will not argue the point here but simply leave my listeners with the information that had WILSON's final picture, that illustrating modern *Homo sapiens*, been a nice rosy picture of one of the dictators of this century, then I for one would not have written a book sympathetic to sociobiology.

Continuing my list of ways in which values enter into science, I argue that the very statements of science itself reflect and endorse values. Now, matters are not absolutely straightforward here. I am still sufficiently old-fashioned to think that one can draw a distinction between value claims ("you ought to eat spinach") and factual claims ("spinach contains iron"), and that generally science is composed of the latter — although incidentally straight value claims do sometimes occur in scientific writings (for instance, DARWIN's endorsement of capitalism in his Descent of Man), and I am not altogether comfortable at an a priori ruling that they are not "really" part of science.

But, these value claims apart, what I would argue is that many of the claims of biology are supportive of particular value positions, and that without the scientific factual claims the positions would be unsupported. In this way, I see even the factual claims as having value connotations. The sort of situation I have in mind is one which occurs in debate about the morality of vegetarianism. Both sides might agree that one ought not eat persons — their disagreement is over the personhood status of (say) cows. Here factual claims enter into the debate, as when for instance an unregenerate meat-eater like myself argues that biology shows that cows do not have the level of self awareness of humans. (The debate might also involve factual claims about human bodily needs for certain proteins.)

You might think that although all I am saying is true, it is a little bit trivial. No one denies that scientific claims can be used to support moral positions. But it is hardly to say that scientific claims are value-impregnated in any interesting sense. However, my position is a little stronger than this. Science constantly outstretches its reach, in the sense that given a number of basic facts, one can spin any number of explanatory hypotheses to account for them. The MULLER/DOBZHANSKY debate illustrates this point. As I see matters, the same facts were available to both of them. They just wanted to use the facts to support different hypotheses — were the undoubted cases of variation common, as DOBZHANSKY thought, or uncommon, as MULLER thought? My claim is that science becomes value impregnated in that a scientist wanting to argue a particular moral position and therefore needing a particular factual claim, takes his/her limited range of facts and uses them to support the precise claim, even though the facts could support a different claim, one which could indeed go against his/her moral position.

Incidentally, although I did not bring this matter up in the main body of my paper, I think DOBZHANSKY's science was value-impregnated because, wanting to argue to the existential singularity and worth of each one of us, he took the facts to support a claim that every individual in a population is different. I think MULLER's science was value-impregnated because, wanting to argue to the superiority of some over others, he took the facts to support a claim that there is a normal, average form within a population, and essentially there can only be falling away from this.

Please note that I am not now throwing everything to the wind, arguing now that science is totally "subjective"; that a biologist can believe and argue precisely what he/she pleases. One still has all the checks on hypotheses that one had before. A claim supporting a value-position cannot violate the empirical evidence, and still can and must be judged by other criteria of good science, like simplicity, consilience, and so forth. My point is simply that there is a place where a biologist's values can and do enter into the hypotheses of his/her science.

The final place where I see values entering into biology is in what I like to follow the KANTians in calling "regulative principles" (RUSE 1980). By these I mean the standards and criteria to which theories must conform, if they are to be judged "good" science, or indeed if they are to be judged "science" at all. I would say that many of the nineteenth-century evolutionists were showing their approval of the machine in particular and the Industrial Revolution in general in their endorsement and support of evolutionism, and in their insistence that a proper approach to biological origins must be mechanistic and conform to unbroken law. If you do not believe me, listen to that enthusiastic evolutionist, Reverend Baden POWELL, on the subject: "Precisely in proportion as a fabric manufactured by machinery affords a higher proof of intellect than one produced by hand; so a world evolved by a long train of orderly disposed physical causes in a higher proof of Supreme intelligence than one in whose structure we can trace no indications of such progressive action". (Baden POWELL 1855, p. 272).

Again, I see values in the way that Spencer and others wanted to see progress reflected in their scientific theories. This belief in a way functioned as one of the parameters of their theorizing. Or in this century, I think LEWONTIN's Marxism functions as a regulative principle, inasmuch that for him the proper approach to science is one which stresses the integrative, holistic nature of the world. This belief is one which imposes itself, as it were, on what for him is constitutive of good science. "Marxism stresses the wholeness of things, both between organism and surroundings and within organism" (LEWONTIN and LEVINS 1976, p. 62).

I confess that I have not altogether got the notion of a regulative principle completely sorted out in my own mind. Although I think they are different, there could perhaps be overlap between an actual claim in science and the conditions for making such a claim — at least, there might be difficulty in sorting the two out. Were SPENCER's views on progress actual scientific claims that he wanted to make, or were they conditions which he laid on his scientific claims? I am sure that such principles are rather fuzzy: after all, they are generally unarticulated, and even (or especially) articulation brings conflict — what is "true" Marxism?

Also, I expect that we get different factors and elements entering into or influencing such principles for different people. Take DARWIN. I think he was influenced by the machine-metaphor. We know he endorsed capitalism. and after all he was able to lead the life he did because Grandfather WEDGEWOOD was a leading light in the Industrial Revolution. But also his approach to organic origins and his conviction that they must be explained through law came from an urge to be the NEWTON of biology.

> Astronomers might formerly have said that God ordered each planet to move in its particular destiny. In same manner God orders each animal created with certain form in certain country, but how much more simple and sublime power let attractions act according to certain law, such are inevitable consequences — let animal he created, then by the fixed laws of generation, such will be their successors. Let the powers of transportal be such, and so will be the forms of one country to another. — Let geological changes go at such a rate, so will be the number and distribution of the species!! (DARWIN 1960, B, pp.101-2)

I do not find at all improbable the suggestion that in trying to be a Newtonian DARWIN thought he was extolling things British rather than French. Do not forget that DARWIN (at the time of the discovery of his theory) was close to WHEWELL. Indeed, I do not find improbable that DARWIN thought he was extolling Cambridge!

In sum therefore, I find biology — past and present — to be value-impregnated through and through. In conclusion, let me ask whether all of these values in biology are a good thing? Or should we regret their presence and try to minimize them as much as possible? I have to confess that the values do not worry me. Indeed, I rather rejoice in them!

In a sense, this is rather making a virtue out of necessity. So much biology, past and present, contains so much by way of ideology and values that for a philosopher to fulminate against them would be a classic case of telling the tide to turn around (not that the futility of a project ever deters philosophers). I certainly think that we philosophers have an obligation to point out values and ideological assumptions. But as far as regretting them, I would say that the past and common sense both suggest that value and ideological commitments lead, not to bad science or non-science, but to the very best science.

DARWIN, WILSON, LEWONTIN, GOULD — these are men absolutely at the top of their field, and they were driven to their heights in major part because of their underlying value-commitments. Precisely because he is a Marxist, LEWONTIN is motivated to try new approaches to incredibly difficult problems in heredity. And the same can be said of the others. This is not to say that one cannot or ought not attack the values at certain times. The scientist has no special immunity. What it is to say is that we should thank God that science, like every other significant result of human effort, is part of and relevant to the real world. Ideology and values in biology: past, present, **and** future.

References

ALEXANDER, R. D. (1979): Darwinism and Human Affairs. (Seattle: Washington University Press).

ALLEN, E. et al (1977): Sociobiology: a new biological determinism. In Sociobiology Study Group of Boston (ed), Biology as a Social Weapon. (Minneapolis: Burgess).

AYALA, F.-J., and J. W. VALENTINE (1979): Evolving: The Theory and Process of Organic Evolution. (California: Benjamin, Cummings).

BLACK, M. (1962): Models and Metaphors. (Ithaca, N. Y.: Cornell University Press).

BREWSTER, D. (1844): "Vestiges, etc.," North British Review, 3, 470—515.

BUCKALND, W. (1823): Reliquiae Diluvianae. (London: Murray).

DARWIN, C. (1859): Origin of Species. (London: Murray).

— (1871): Descent of Man. (London: Murray).

— (1960): DARWIN's notebooks on transmutation of species, ed G. de BEER. Bull. Brit. Mus. (Nat. Hist.) Ser., 2, 7.

GOULD, S. J. (1980): Is a new and general theory of evolution emerging? Paleobiology, 6, 119—30.

HOUGHTON, W. E. (1957): The Victorian Frame of Mind. (New Haven: Yale).

LEWONTIN, R. C. (1974): The Genetic Basis of Evolutionary Change. (New York: Columbia University Press).

— (1977): Sociobiology — a caricature of Darwinism. In: F. Suppe and P. Asquith eds. P. S. A. 1976. (Lansing, Mich: P. S. A.), 2, 22—31.

LEWONTIN, R., and R. LEVINS (1976): The problem of Lysenkoism. In: M. and S. ROSE eds. The Radicalisation of Science. (London: Macmillan), 32—64.

MALTHUS, T. R. (1826): An Essay on the Principle of Population, 6th ed. (London).

MILLER, H. (1856): The Testimony of the Rocks, or Geology in its Bearings on the Two Theologies. Natural and Revealed. (Edinburgh: Constable).

NAGEL, E. (1961): The Structure of Science. (London: Routledge and Kegan Paul).
OSPOVAT, D. (1976): The influence of Karl Ernst Von BAER's embryology, 1828—1859: A reapraisal in light of Richard OWEN's and William B. CARPENTER's 'Palaeontological application of von BAER's law'. J. Hist. Biol., 9, 1—28.
OWEN, R. (1834): On the generation of the marsupial animals, with a description of the impregnated uterus of the Kangaroo. Phil. Trans., 333—64.
POPPER, K. R. (1972): Objective Knowledge. (Oxford: Oxford University Press).
RUSE, M. (1979): The DARWINian Revolution: Science Red in Tooth and Claw. (Chicago: University of Chicago Press).
— (1979): Sociobiology: Sense or Nonsense (Dordrecht: Reidel).
— (1980): Philosophical aspects of the Darwinian Revolution. In: F. WILSON (ed) Pragmatism and Purpose. (Toronto: University of Toronto Press).
— (1981): Is Science Sexist And Other Problems in the Biomedical Sciences. (Dordrecht: Reidel).
— (1982): Darwinism Defended: A Guide to the Evolution Controversies. (Reading, Mass.: Addison-Wesley).
SEDGWICK, A. (1845): Vestiges, etc., Edinburgh Review, 82, 1—85.
SPENCER, R. (1952): A theory of population, deduced from the general law of animal fertility. Westminster Review, n. s. 1, 468—501.
SPENCER, H. (1857): Progress: its law and cause. Westminster Review.
STANLEY, S. M. (1979): Macroevolution: Pattern and Process. (San Francisco: W. H. Freeman).
TENNYSON, A. (1847): The princess. London.
WILSON, E. O. (1975): Sociobiology: The New Synthesis. (Cambridge, Mass.: Harvard University Press).
— (1978). On Human Nature. (Cambridge, Mass.: Harvard University Press).

Discussion[1])

GEISSLER: First of all I would like to say thank you very much for your very stimulating lecture. Your lecture fits so well into the framework of this conference that 2 of the 3 major points you made were also major points in the introductory remarks by Herbert HÖRZ and myself, which you unfortunately could not hear. Nevertheless I would like to comment on each of your points.
First of all I don't believe that the economic background and the family tradition of Charles DARWIN is important for any discussion on the validity of the DARWINian theories. And I think that you didn't intend such a conclusion either. Notwithstanding, I am convinced that this information, for instance in respect of DARWINS view on the range of Irish people in a scale of development is a good example for the responsibility of scientists and for the possible misuse of scientific results and/or papers.
Secondly, of course, I agree with you that at another Kühlungsborn Colloqium on "DARWIN today" in a hundred years time, if this earth still exists then, it is to be hoped that creationists will be not mentioned. But I believe they are not merely a curiosity but also a real danger: E. O. WILSON in one of his latest books quotes the results of a survey showing that more than 90% of US citizens believed in God or a similar Designer. Thus, the soil is ready to take creationist seed.
In our introductory remarks we quoted a Georgia judge who declared Darwinism to be the source of all evils on this earth. Although it was not explicitly stated, I got the impression that this judge as well as other creationists finally believe that Darwinism is not only an invention of Satan but also the cause of Marxism. And therefore I could imagine that these people might someday call for participation in a crusade against DARWINism and communism. This is why I am convinced that we have to oppose creationism actively.

1 In the lecture he presented during the meeting Prof. RUSE discussed also problems of so called scientific creationism, while other topics mentioned in the paper submitted for publication were only briefly mentioned or even omitted during the lecture. Therefore there are some discrepancies between RUSES paper and the content of the following discussion.

Finally, I would like to refer to your comment on the ideas developed by GOULD and LEWONTIN. You mentioned that both scientists are MARXists. That may be. But this does not necessarily mean that the ideas they express are Marxist ones. As we said in our introduction, the question of whether evolution is gradual or saltatory seems to be undialectic and, hence, non-Marxist! We are rather of the opinion that evolution takes place both smoothly and in jumps.
Furthermore, the conclusion drawn by GOULD and mentioned by you that human beings are alike in genetical terms is not Marxist but sociologistic and ignores the biological dissimilarity of man already mentioned by MARX and LENIN.

RUSE: Let me reply to your three points in order.
1. I think you are very old-fashioned in your view of science. I agree, of course, that there is an "objective" side to science. I am sure that the struggle for existence and natural selection exist, irrespective of the socio-political views of Charles DARWIN and Alfred Russel WALLACE. However, there is surely more to the story. The metaphors and examples used by DARWIN were important in the acceptance of his theory. And these did reflect his background, and the Victorian era generally. Let me put matters this way. What right have you got to say that natural selection is really part of DARWIN's theorizing, but that his views on the Irish were not?

2. I think you overstimate the threat from Creationism. Most people in the U. S. do not hold such extreme views as the Creationists. Indeed, many Christians and Jews think Creationism to be blasphemous, because today our reason guides us to evolution. Remember, DARWIN himself was a deist and was led to his evolutionism because he thought that a God who is an unmoved mover, working only through law, is the highest kind of God.

3. Whether you approve of the GOULD-LEWONTIN form of Marxism is one thing. The point is that they think they are Marxists, and this belief influences their work.

GEISSLER: Regarding your second comment I would like to quote Isaac ASIMOV: "It is religion that recruits their (e. g. the creationists, E. G.) squadrons. Tens of millions of Americans who neither know or understand the actual arguments for — or even against — evolution, march in the army of the night with their Bibles held high. And they are a strong and frightening force, impervious to, and immunized against, the feeble lance of mere reason" (New York Times Magazine, 14. 6. 1981). I also diagree with your opinion regarding the influence of Marxism on GOULD's speculations regarding evolutionary processes. Of course ideology influences scientists' plans and conclusions. But the very experiments, the collection of the actual data are not grossly influenced by the Weltanschauung of the individual scientist. They depend on the facts. And we, too, have to decide which facts have been forwarded and how they fit into the frame of evolutionary theory — regardless of whether these results have been obtained by Marxist scientists or not.

RUSE: I certainly do not deny the place of facts in science. Of course not! But I do claim that the values permeate and structure the science far more deeply and thoroughly than you are prepared to allow. In my opinion, the only science without values and so forth, is dead science. Remember, after all, that when COPERNICUS put forward his heliocentric theory, it was not really the facts that influenced him. He could not explain the heavens much better than PTOLEMY. What influenced him was his Pythagorean/Platonic urge to give the sun a major place in the universe.

LEY: The relations beween the social conditions, structure and superstructure of society are a complex amalgam of different branches of influence. If we wish to measure the results, we cannot declare that one particular twig of the bush composed of the elements is immoral. Thinking is contained in ideas or institutionalized concepts. The close connection between ideas and labour is well known to exist and be effective but it is not describable in terms of their real structures functioning in mechanistic causality and in stochastic sequencies, and your "effect volumes" stem from HEISENBERG's type of indeterminism. So you have a bush of ideologies in the development of DARWIN theory. In Germany there is Ernst HAECKEL spreading the concept of the "island" on the continent, and others did the same on in the different continents, and especially in North America. The differences between the multivalent contents various streams in group ideologies do not affect experimental and theoretical results, for instance in biology. The question of receiving Nobel prizes is not directly dependent upon specific divergent aspects of religious opinions held by other people who need some belief to cope with every day life. Surely Dr. RUSE is not a fundamentalist and he has not to describe the conditions under which fundamentalism can exist: He had not the time to talk about all antiscientific opinions now existing in this world and restricted himself to talking about Lamarckists and their protagonists in some countries. It is necessary to continue discussing the eternal LAMARCKanian breed existing in institutions and fighting against the basis of science and the application of modern ideas in medicine and industry.

If you are speaking about Marxist theoreticians in Cambridge, Dr. RUSE, different opinions are stated in the numerous publications by Marxists on the topic; I agree with Dr. GEISSLER that RUSE's man is not completely characteristic of Marxist thinking. But there are in all groups holding the different theories about intelligence some people which believe that all mankind has equal standards of thought. That is the product not of Marxism but of the bourgeois revolution in the 17th and 18th centuries and refers more to equality before the law than to the ability to have ideas or to act as innovators in theory, science and culture. And here are some people who know that Karl MARX like Charles DARWIN insisted that there are differences between the individuals of a single species, including *Homo sapiens*. And these differences are expressed not only when playing football or jumping at sporting events, but also in the laboratories and the R. and D departments of industry and in the individual ability to find innovations.

BRODA: Surely no scientist in this right senses could say that the genetic basis of all mankind now is identical.

RUSE: I agree. However, I do think that American Marxist biologists want to deny that the genetic differences between humans make any difference in their behaviours or in the roles that they play in life. In other words, humans are so genetically uniform that any differences are due solely to socio-economic factors.

HÖRZ: 1. Dr. RUSE adopts a superficial Marxist viewpoint when he states that the social background of a scientist is expressed necessarily in only one facet of his science. It is very difficult to study the relationship between science and philosophy, science and ideology.
2. The point made in the thesis that there are differences between people who oppose Marxism is wrong. (LENIN, research-programs show the natural and social differences in human individuals)
3. What are the idological reasons for the boom of creationism in the USA?

RUSE: 1. I presume that when Dr. HÖRZ calls my position "superficial MARXist" he is criticizing me. Is he objecting to my superficiality or to my Marxism? Perhaps my position is trite. All I can say is that many scientists and philosopher in the USA and Canada would disagree with my thesis that the ideology of scientists permeates their science.
2. I am quite sure that there are different brands of Marxism. My point is simply that one version of Marxism — that held by certain American biologists, is reflected in their science.
3. On the surface, the creationist movement stems from conservative Christianity. I suspect that beneath all of this lies the dislike of some Americans — right-wing, Southern, rural, less-well educated — for what they perceive to be the dominant elite (i. e. well-educated people in New York, Boston and like cities). Northerners support evolution; hence, the attack.

LEY: Religion and science are different subjects and are not amenable to joint consideration even if one considers only a single country or period in the development of science. When science and ideology diverge, the result may be both fundamentalism and modern genetics, both applications and the inability to find applications in science and agriculture.

JENSSEN: I found Dr. RUSE's concept of ideology imprecise. 1. Is it your opinion that ideological influences (whatever you mean by 'ideological') necessarily have consequences for the truthvalue of theories put forward by scientists; (ex.: it has been argued that HARVEY had a monarchist ideology and that this ideology played a (psychological) role in the formation of the theory of the circulation of the blood;

does this imply that the theory is bad or false (presupposing that monarchism is an inadequate social philosophy)?
2. You have mixed many different factors together (social background, religion, general philosophical opinions, methodological/heuristics/positions). But you cannot compare all these.

GOULD and LEWONTIN are (like MAYR) defending an anti-reductionistic methodology. You cannot just say that this is an unscientific (ideological) influence. At least: You cannot take that position without presupposing a traditional positivistic dichotomy between facts to be picked up and metaphysical speculation. Methodological rules, however, belong to neither the first nor the second category.

RUSE: I agree with just about everything you say, but I still think that I am right! Call it what you like — "ideology", "metaphysics", "philosophy" — there is a non-factual element in science which reflects scientists' hopes and desires. Morover, this non-factual element is very important in the acceptability and acceptance of a science.

KEYNES: 1. I think that the wealth of the DARWIN family is being somewhat exaggerated: I wish it were not! It is true that DARWIN never had to earn his living, but he was no more than comfortably off by 19th century standards. DARWIN's father did not inherit wealth, he earned a good income as a very successful physician, and managed his financial affairs skilfully so that he could hand money on to his children. DARWIN's grandfather, Josiah WEDGWOOD the potter, was certainly much richer, since he was a successful industrialist. But it should not be forgotten that as such he was a pioneer well ahead of his time in having built a model village for his workers, and taking a special interest in their welfare. He was also a distinguished applied scientist who designed instruments such as a pyrometer, and supplied the chemists of the day with their experimental equipment.
2. DARWIN's account of his scientific approach as truly Baconian is, like some other remarks in his "Autobiography", a bit misleading. As he wrote in 1883 in his "Third Notebook on Transmutation of Species": 'The line of argument often pursued throughout my theory is to establish a point as a probability by induction, and to apply it as hypotheses to other points, and see whether it will solve them'. There is a telling admission of his attitude in a letter to Henry FAWCETT, written in 1861, in which he wrote: 'About thirty years ago there was much talk that geologists ought only to observe and not to theorise; and I well remember some one saying that at this rate a man might as well go into a gravel-pit, and count the pebbles and describe the colours. How odd it is that anyone should not see that all observation must be for or against some view if it is to be of service!'

RUSE: 1. It still remaines true that DARWIN came frome a rich family by the standards of his time. The house at Downe that he and his wife bought and improved just after his marriage was not the house of a poor man.
2. I agree, and discuss the matter at great length in my bock, "The DARWINian Revolution" (Chicago: Chicago University Press 1979).

URBANEK: First of all I would like to assure Dr. KEYNES that by our standards Robert DARWIN, father of Charles, was a member of the working intelligentsia . . . and therefore there is no place for any uneasiness. Returning to the question of the ideological background of punctuated equilibria I would be inclined to think that the ideological component is there secondary and (that the) relation of the punctuated equilibria concept to Marxism is a very remote one. Only in their second paper ELDREDGE and GOULD confessed that they were flirting with Marxism — and this seems somewhat shocking to some colleagues in the West. This obscures, however, the fact that recent controversies on gradualism and punctuated equilibria are to a considerable extent a result of objective contradiction between neontological and palaeontological data. The idea of stasis, after all, is a classical palaeontological thesis which has been known for decades, to mention only the "typostase" concept developed by O. H. SCHINDEWOLF. What ELDREDGE and GOULD did is an extrapolation of this thesis from macroevolution to specialism. I am sure that GOULD must be familiar with SCHINDEWOLF at least as well as with MARX.
The significance of general scientific and extrascientific ideas for scientific discovery is difficult to evaluate and may easily be overemphasized. But this is certainly true — as LENIN once put it, one cannot live in a society and be free of that society.

GHILAROV: I have something to say on the character of evolution: it is neither possible nor correct to say either that the evolution is continuous or that it is interrupted by jumps. Philosophically considered, both are organically connected: the jump is only a break in graduality. Even groups of species in the taxonomy of recent organisms and in paleontology exhibit some differences. You mentioned that Darwinists incorrectly attach significance to some parts of evolution because the whole organism evolves. It is incorrect to counterpoint the whole and a part: in some periods of evolution some parts of the body achieve special significance (e.g. brain and hand in human evolution). I was amazed that creationism, especially "scientific" creationism, is allowed by the public opinion of scientists — geologists, biologists etc. I am really astonished at the obscurantism of such a "near marasmatic" situation.

RUSE: I must point out that orthodox scientists loathe scientific creationism and do all in their power to oppose it. I am afraid that one of the problems with living in a democracy is that sometimes crazy ideas do influence a number of people for a period of time. Fortunately, in a democracy we do have the means at hand to oppose such ideas, and this is what many evolutionists are doing. In this context let me note that my friend Stephen Jay GOULD is leading the fight against creationism.

NOVAK: I very much appreciate all of what Prof. RUSE has said, including what he told us about DARWIN's political views and position even if we still find, as Prof. KEYNES mentioned, that his works nonetheless contain very progressive ideas. But I think in sciences, we should consider each phenomenon — and this includes personal views — in relation to the environment, i.e. the space and time when it occurs. In this respect, what may be awkward and reactionary nowadays might well have been progressive in DARWIN's time. I agree, however, that in the present discussion in the USA it may be useful to emphasize the capitalist background of DARWIN.

As to GOULD's punctuated equilibria conception, I should like to emphasize, that the contradiction is not between gradualism and punctuated equilibria but between gradualism and emergentism. Dialectical materialism, as in other cases, seeks to form a synthesis of both and, in my opinion, GOULD's view comes quite close to this.

As far as the equality of people is concerned, we have to distinguish between two different things: 1. equality as the equal capabilities of all races and 2. the equality of individuals. In the former case, the present sciences have no doubt and today there is enough practical evidence for the equipotentiality of races today. When Australian aborigines were educated from birth in European families, they proved just as capable of finishing university training successfully as European children. There are differences, of course, between the capabilities of individuals. Contemporary science is producing more and more evidence, however, that most if not all of these differences are due not to genes but to both positive and negative influences during the first years of life. Various inhibitions acquired during this period are very difficult to eradicate in later life.

RUSE: I agree with much that Dr. NOVAK says. I find his final comment about the causes behind human differences to be very interesting and revealing, supporting my case about the ideological elements in science. Perhaps Dr. NOVAK is right. All I can say is that the empirical evidence at this point is not definite, and I suspect that Dr. NOVAK takes his strong stand for ideological reasons. In this I do not condemn him. He is simply being a scientist.

SCHUSTER: I have enjoyed Prof. RUSE's paper. It seems to me that his contribution cannot have been superficial since we are discussing it in such great detail and with strong emotions. I am sure I take a position which is unpopular in this audience when I speak in favour of the points he made. Scientific facts in general, and particularly in evolution theory, do not provide a complete basis for argumentation. There are gaps, and we have to fill these gaps when we talk about our results. The material for filling has to be taken from somewhere — from other fields of science we are familiar with, from our non-scientific education, from ideology etc. This was RUSE's point, if I got it correctly. Coming to the punctualism versus gradualism debate we should accept that the fossil record is by far not as good as to allow an easy decision. I had a chance to hear a famous British biologist, John MAYNARD-SMITH, lecturing on this subject recently. His opinion was that the records we have need substantial completion before we can draw any relevant conclusions. On the other hand, we do not know the mechanism of speciation either. There is plenty of room for personal interpretation, therefore, and the debate is conducted at length. I consider both Prof. GOULD and Prof. LEWONTIN to be brilliant scientists devoted strongly to their subject. If there were hard facts they would not deny their existence. When it comes to such vaguely defined properties like IQ as a measure of intelligence and the question of how much is inherited it is no wonder that we observe ideological bias. The last topic, creationism as a pseudoscience: it seems to me that this is a peculiar problem of the U.S. But what we observe in all Western countries I have visited is a strong antiscientific and often irrational feeling in the public opinion which makes it impossible, to give an example, to discuss the advantages and disadvantages of nuclear power stations rationally in public in my country.

RUSE: I agree strongly with everything that Dr. SCHUSTER has to say. I am very glad that he made the point that creationism may be a peculiarly American phenomenon, but it is clearly just part of a world-wide simplistic reaction against science.
Dr. GEISSLER has pointed out that 94% of Americans admitted on a Gallup Poll that they believe in some form of God. Do not, however, think that all or most support creationism. Many Christians and Jews think that creationism is anti-God and blasphemous: God gave us our reason and so we should use it. And reason points to evolutionism not creatinism.

GEISSLER: Of course I don't think that all believers in the States are ready to take part in a crusade against communism. But, on the other hand, religious people are much more ready to accept creationism than nonbelievers. Regarding the question of how evolution proceeds I must say that it is not correct to refer only to the outliers of both views. There are a lot of people who believe that evolution proceeds both smoothly and in jumps. And they base their views not only on the fossil record but on data obtained by molecular biologists and genetic engineers. In fact sequence analysis demonstrates clearly that exons of protein coding genes do evolve gradually, i.e. by base substitution, whereas introns and intervening sequences evolve saltatorially, i.e. by inversion, insertion and other gross rearrangements.

RUSE: Perhaps you are right. All I can say is that paleontologists today are divided, bitterly, over the nature of the fossil record, and that this division is, in part, a function of differing ideologies.

BRODA: Concerning one of my friend Peter SCHUSTER's remarks: The movement of the "greens" in Austria and elsewhere is directed against the misuse of hard technology, such as the destruction of nature in the Alps by hydropower development, and has nothing to do with creatinionism. On the contrary, these are two opposing ends of the spectrum.

RUSE: It is often the case that opposing ends find themselves back-to-back, and thus supporting virtually the same ends. Perhaps this is also the case here. (A nice example of opposing ends coming together is that GOULD's Marxist emphasis on the integrated functioning of the whole organism is virtually identical with the claim of the conservative Christian creationist CURIER about the correlation of parts demanding integrated functioning!)

Zentralinstitut für Philosophie, Akademie der Wissenschaften der DDR, Berlin

The Philosophical Approach to the Evolution of the Organisms in DARWIN's Theory and After DARWIN

R. LÖTHER

The development of human knowledge by natural sciences, social sciences and progressive philosophy is proceeding from static creationism to an evolutionist world view and materialistic and dialectical evolutionism. "Evolution" is a philosophical category denoting processes by which phenomena appear, change irreversibly, and disappear irretrievably. Thery are many more or less far-reaching conceptions of evolutionary processes on manifold subjects. Every conception of the evolution of anything anywhere must, at least implicitly, answer certain fundamental questions: whether evolution exists and, if so, how can man recognize it, how evolution takes place and what are its moving forces. The answers to these questions relate every conception of evolution to world outlook, epistemology and philosophical methodology. These relations show that the philosophical approach to evolution can be materialistic or idealistic, dialectical or metaphysical.

For as long as evolutionism and creationism coexist, consistent materialistic and dialectical conceptions of evolution on the one hand and static creationism on the other represent a polarity characterized not only by struggle but also by mediating conceptions. These arise not only as a result of emancipation from static creationism and confrontation with new knowledge, but also due to the reaction of creationism to evolutionism, to attempts to reconcile creationism with evolutionism, to the advantage of the former. At the same time creationism is forced to change and adopt forms more compatible with science with the aid of idealistic philosophy. Other than evolutionism and creationism there is no world outlook, only agnosticism. This may be a screen for a materialistic outlook in science that is opposed to creationism or it can favour creationism.

DARWIN's Philosophical Approach to the Evolution of the Organisms

Darwinism is not synonymous with evolutionism in respect of life. The evolutionist view of life began to emerge many years before the DARWINian scientific revolution of 1859. Even LINNÉ suspected that some species are "filiae temporis". Later there was hardly an eminent naturalist or philosopher who did not at some time play with evolutionist ideas with regard to life. And after 1859 even more concepts of evolution arose. A fundamental reason for the superiority of the DARWINian theory compared with its so-called forerunners and later non-DARWINian conceptions was DARWIN's philosophical approach to evolution. It is amusing to read how G. WICHLER [19. p. 195] credits DARWIN with "the absence of a motive to philosophical cognition" and testifies that "the results of his research work were not influenced by a religious or philosophical conception" [ibid., p. 196]. DARWIN's works show that his conception of evolution was materialistic, and his works reflect spontaneously the objective dialectics of nature to a high degree. His materialist point of view is confirmed by his comments on religion in his autobiography, letters and talks recorded by partners [cp. AVELING, DARWIN, 4]. These documents show, that DARWIN's self-characterisation as an agnosticist was a screen for his materialistic and atheistic outlook.

DARWIN's early writings, and especially his notebooks of 1837—39, the transmutation notebooks and the notebooks on man, mind and materialism, show that the formation of his theory of evolution coincided and interacted with the formation of his materialistic point of view. It was a development in thought which enabled Darwin to depart radically from the ideology of static creationism and natural theology — doctrines that still dominated in natural science in England — and reach the conclusion that matter is the stuff of all existence and mind is a product of brain. As S. J. GOULD [7, p. 69] resumes "the notebooks are clear on two additional points: First, DARWIN was interested in philosophy and aware of its implications. He knew that the primary feature distinguishing his theory from all other evolutionary doctrines was its uncompromising philosophical materialism. Other evolutionists spoke of vital forces, directed history, organic striving, and the essential irreducibility of mind — a panoply of concepts that traditional Christianity could accept in compromise, for they permitted a Christian God to work by evolution instead of creation. DARWIN spoke only of random variation and natural selection.

Second, DARWIN resolutely applied his materialistic theory of evolution to all phenomena of life, including what he termed 'the citadel itself' — man's mind. And if mind has no real existence beyond the brain, can God be anything more than an illusion?"

In connection with this question, GOULD refers to what DARWIN wrote in one of his transmutation notebooks: "Love of the deity effect of organization, oh you materialist! ... Why is thought, being a secretion of brain, more wonderful than gravity a property of matter? It is our arrogance, our admiration of ourselves" (GRUBER/BARRET [10], pp. 450—451). DARWIN's epigram also deserves emphasis: "PLATO ... says in Phaedo that our **'imaginary ideas'** arise from the preexistence of the soul, are not derivable from experience. — read monkeys for preexistence" (GRUBER/BARRETT 10, p. 290). Here, as GOULD (loc. cit.) remarks, "DARWIN cut through 2,000 years of philosophy and religion".

H. E. GRUBER points out that materialism at that time was more heretical and outrageous than belief in evolution. He describes the persecution of materialistic ideas during the late eighteenth and early nineteenth centuries and concludes: "In virtually every branch of knowledge, repressive methods were used: lectures were proscribed, publication was hampered, professorships were denied, fierce invective ridicule appeared in the press. Scholars and scientists learned lessons and responded to the pressures on them. The ones with unpopular ideas sometimes recanted, published anonymously, presented their ideas in weakened form, or delayed publication for many years" (GRUBER/BARRETT 10, p. 203). DARWIN not only knew some historical examples of this. He also had personal experience: at the University of Edinburgh he was a member of the Plinian Society where in 1827 his friend, W. A. BROWNE, read a paper presenting a materialistic view on life. After much debate, all reference to BROWNE's papers, including the record (from the previous meeting) of his intention to deliver it, were removed from the minutes. DARWIN learned the lesson, for he wrote in one of his notebooks: "To avoid stating how far I believe in Materialism, say only that emotions, instincts, degrees of talent, which are hereditary, are so because the brain of the child resembles parent stock" (GRUBER/BARRETT 10, p. 276).

"DARWIN was, indeed, a gentle revolutionary", GOULD [loc. cit., p. 70] states with reference to DARWIN's delay in publishing his theory until twenty years after 1838 and his assiduous avoidance of any public statement regarding the philosophical implications of his theory. "Yet the content of his work was so disruptive to traditional Western thought that we have yet to encompass it all. (G. G. SIMPSON marked the centenial of DARWIN's "Origin" in 1959 with an essay entitled "One Hundred Years Without Darwin Are Enough)" [GOULD, ibid].

DARWIN's concepts of chance and struggle (of chancefully inherited variation and natural selection in the struggle for survival) were not only contrary to the concepts of pre-established harmony and wise design in the doctrines of creationism and natural theology. They also reflect the spontaneously objective dialectics of nature and enabled DARWIN to conceive evolution as a process of the inherent motion of matter driven by internal contradictions. Variation and selection really are inseparable opposites in the process of evolution, and their interaction is the essential moving force of evolution. At the same time DARWIN's explanation of the moving force of evolution reflects the character of evolution as a statistical and selfregulating process in the dialectical unity of chance and necessity. This view has been the starting point for many misunderstandings concerning not only Darwinism but also of how the theory of evolution is related to mathematics and cybernetics and of biological thought in units above the organismic level such as the population and ecosystem. DARWIN, of course, was no conscious dialectical thinker, and his "Origin of Species" is "developed in the crude English style" (MARX). But DARWIN was a men who was conversant with nature, science and the experience of animal and plant breeders, and his materialistic world outlook gave him the unprejudiced objektivity to reflect on the dialectical aspects of objective reality.

Even today in some countries philosophical materialism, especially dialectical materialism. is subject to defamation and discrimination in public, and there is a lack of trustworthy information about materialistic philosophy. This is why some DARWINian evolutionists do not like to hear or read that the DARWINian theory of evolution is materialistic. "Materialism of any kind or atheism are not the concern of the theory of evolution. It is a scientific theory, which is verified . . .", wrote, for instance, G. H HEBERER [11, p. 78]. This statement can be used as an elementary example of a logical contradiction. It is exactly, because the theory of evolution is a verified scientific theory that it is concerned with materialism and atheism, since it interprets nature without fantastic creationistic and idealistic ingredients. Darwinism is not synonymous with evolutionism, but it is synonymous with the tendency to a (spontaneous or conscious) materialistic and dialectical approach to the evolution of organisms in biology. The main point of DARWINism in this connection, too, is the principle of natural selection.

On the philosophical approach to the evolution of organisms after 1859

The publication of DARWIN's "Origin of Species" in 1859 was a turning point in the history of biology and in the history of the idea of evolution. All conceptions of evolution and all anti-evolutionism in science and ideology during the subsequent years would have been inconceivable without the influence of Darwinism on intellectual life. In the further development of the evolutionary approach in biology there were three distenet, overlapping periods: the heroic period of the struggle against anti-evolutionism, the critical period of Darwinism and the period of the "synthetic" or "biological" theory of evolution.

Simultaneously, in the bourgeois world outlook the idea of evolution became familiar and shallow during the first period, then irrationalistic and epistemologically undermined, and finally it was attacked. There is a striking congruity between this ideological development and the transition from the capitalism of free competition to monopoly capitalism, the decay of capitalist society and the transition from capitalism to socialism.

The advance of the idea of evolution

The appearance of DARWIN's epoch-making work evoked a broad intellectual evolutionist movement in science and world outlook, at the centre of which was Darwinism and the struggle raging around it. The vehement controversy between the proponents of evolutionism, led by men like T. H. HUXLEY in England, E. HAECKEL in Germany and K. A. TIMIRYASEV in Russia, and the representatives of static creationism on the alternatives "Moses or DARWIN! — Creation or Evolution!" gave that time its heroic character — the more so because creationism was not only defended by weak arguments but also by repressive methods directed against the evolutionists. The result of this movement was that the idea of evolution and the idea of descent and the change of organisms had become familiar in the world outlock of bourgeois society by the end of the 19th century.

"There is perhaps nothing more significant of our age than the change which has taken place in our conception of the universe. The old notion of eternal unchangeableness, with occassional upheavals — the static view — has given way to the newer idea of progressive development in nature. Gradually, within the last half-century, this new interpretation, which regards all things from the evolutionary or dynamic point of view, has extended from one field of inquiry to another, creating everywhere fresh interests, giving novel aspects to old problems — colouring and transforming the whole purpose and meaning of modern life . . . Showing as it does the small beginnings of all things, their gradual transformation, step by step, towards higher stages of perfection, this new philosophy has opened out to us new possibilities, new visions and aspirations", wrote S. HERBERT [12, p. 1] enthusiastically.

This quotation indicates clearly how the idea of evolution became familiar in the bourgeois world outlook: as a concept of linear and gradual progress towards higher stages of perfection, in the form of plain evolutionism. In this view the evolution of organisms was associated with conceptions of the evolution of inorganic matter (cosmic evolution, geological evolution, atomic evolution, origin of life) on the one hand and of naturalistic conceptions of social life and "Social-Darwinism" on the other. This coherent picture of an evolving world was designed and interpreted from a basis of philosophical conceptions with a limited materialistic content, conceptions like Vulgar Materialism, H. SPENCER's "Synthetic Philosophy", and E. HAECKEL's "Monism", his „Naturanschauung von GOETHE, LAMARCK und DARWIN", the ingredients of which were indeed "borrowed" from these three autors and from romantic German „Naturphilosophie".

SPENCER's philosophy, for instance, was a very typical and influential representation of the affirmative philosophical approach to evolution, of the essentially non-dialectical evolutionism with a materialistic content of the bourgeois ideology of the second half of the 19th century. SPENCER's far-reaching influence is evident when we recall that we all know his phrase "survival of the fittest" and use the term "evolution" with a meaning the world began to receive in SPENCER's philosophy. After the collapse of preformationism the word "evolution" came to mean the opposite of its meaning in preformatistic embryology (cp. GOULD, 8, pp. 31—32). In the course of its change in meaning it was SPENCER who primarily used the term "evolution" as a synonym for "organic change" in his essay "The Development Hypothesis" (1852). In his "System of Synthetic Philosophy" (1862—1896) he broadened "evolution" to cover not only living beings but also all things and phenomena, and defined a general concept of evolution. SPENCER's general conception of evolution is expressed in his following formula: "Evolution is an integration of matter and concomitant dissipation of motion, during which the matter progresses from a relatively indefinite, incoherent homogeneity to a relatively definite, coherent heterogeneity; and during which the retained motion undergoes a parallel transformation" (cited by HERBERT, [12], p. 306). In conformity with the embryological origin of the word "evolution" and K. E. von BAER's description of embryonic development. SPENCER identified evolution with progressive change towards increasing complexity. Such evolution he conceived in a mechanistic-reductionistic way as the redistribution of matter and motion, and thereby blotted out the distinction between different spheres of the material world. Yet he denied an implied tendency in everything to become something higher, a preset internal program of evolutionary processes, and refered to their determination by interaction between inner and outer forces. At long last SPENCER set his partial and mechanistic materialism in an idealistic-agnosticistic frame, postulating one mysterious force behind all things and phenomena, an unknowable "absolute reality" transcending all human cognition. This was an effort to reconcile science and religion by providing a common belief.

Plain evolutionism included the evolution of organisms, and it contributed much to the rebuttal of static creationism. But what was the matter with DARWIN's theory of evolution and its main point, the principle of natural selection? As J. C. GREENS [9, p. 719] states, in DARWIN's view "struggle and chance, the antitheses of pre-established harmony and wise design, became the engines of organic change and the architects of such adaptations as could be discerned in nature. This was the last and harshest blow to the traditional view of nature . . . Most of DARWIN's contemporaries evaded the full force of the blow by transfering the element of wise design from the structure of nature themselves to the general systems of matter in motion which had produced those structures in the course of time." This manoeuvre gave rise to compromises with creationism in the deistic conception of God as law-giver of matter in evolution or by identifiying God with evolving nature and its laws in a pantheistic manner. The concepts of laws of nature as eternal laws of the dynamic type and plain evolutionism contained, or were open to, such compromises.

The tendency to evade the philosophical consequences of DARWIN's view, to harmonize the theory of evolution with mechanical determinism, plain evolutionism and theology led to minimization of the role of natural selection and to the assumption of some kind of directive agency or progressive tendency in the process of hereditary variation even among the DARWINians. And these endeavours were continued with much greater vigour by the proponents of anti-DARWINian conceptions of evolution in biology (cp. ZAVADSKY [20], SUTT [18]). From A. KÖLLIKER in 1864 until the different forms of so-called Neo-LAMARCKism, pre-DARWINian conceptions, in streamlined form and at variance with DARWINism, were reintroduced to these doctrines. The advance of evolutionism in this way after the DARWINian revolution was really much less an advance of DARWINism. Both the DARWINian and the anti-DARWINian camp contained reputable zoologists, botanists and palaeontologists.

The Critical Phase of Darwinism

At the turn from the 19th to the 20th century a revolutionary change took place in the concepts and methods of the life sciences (cp. ČIŽEK and JANKO [3]) in connection with the emergence of experimental and statistical methods in developmental physiology and Mendelian genetics. This transition from the biology of the 19th to the biology of the 20th century included a critical reexamination of the 19th century evolutionary approach in biology and the proposed explanations of evolution.

L. V. KELLOG [13, p. 5] resumed: "The fair truth is that the DARWINian selection theories, considered with regard to their claimed capacity to be an independently sufficient mechanical explanation of descent, stand to-day seriously discredited in the biological world. On the other hand, it is also fair truth to say that no replacing hypothesis or theory of species-forming has been offered by the opponents of selection which has met with any general or even considerable acceptance by naturalists. Mutations seem to be too few and far between; for orthogenesis we can discover no satisfactory mechanism; and the same is true for the LAMARCKian theories of modification by the accumulation, through inheritance, of acquired or ontogenetic characters. **Kurz und gut,** we are immensely unsettled."

Intemperate anti-DARWINians relegated Darwinism already to the past and exulted, "Darwinism now belongs to history like that other curiosity of our century, the HEGELian philosophy; both are variations on one theme: how a whole generation can be led by the nose", was the grotesque misjudgement — with to the HEGELian philosophy also — of H. DRIESCH [6, p. 355]. The former student of E. HAECKEL and A. WEISMANN also spoke of "the softening of the brain of the DARWINians." E. DENNERT [5, p. 4] took much pleasure in explaining that "we (anti-DARWINians) are now standing by the death-bed of Darwinism, and making ready to send the friends of the patient a little money to insure a decent burial of the remains."

Such disappointment and malicious joy, and there was much of them, reflected the position of Darwinism in biology at the turn from the 19th to the 20th century in regard of the changed philosophical approach to evolution in bourgeois ideology, which time took place at the same. They are manifestations of the "resignation of progress, history, cognition and truth" (BUHR and STEIGERWALD [2]), spreading in the bourgeois philosophy since A. SCHOPENHAUER and F. NIETZSCHE, for which even plain evolutionism, say nothing of Darwinism was too much. The scientific conclusion from the doubtful situation had a different character: "We are ignorant; terribly, immensely ignorant. And our work is to learn. To observe, to experiment, to tabulate, to induce, to deduce. Biology was never a clearer or more inviting field for fascination, joyful, hopeful work. To question life by new methods, from new angles, on closer terms, under more precise conditions of control; this is the requirement and the opportunity of the biologists of today" (KELLOG, op. cit., p. 387). And this was precisely the way to pass from the critical phase of Darwinism to its renewal in the "synthetic" or "biological" theory of evolution, to overcome anti-evolutionistic and anti-DARWINian nihilism, agnosticism and scepticism as well as autogenetic and ectogenetic conceptions of evolution.

Anti-Darwinism Today

Evolutionary biology, marching along in this direction, and the dominating antiscientific tendency of the philosophical approaches to evolution in bourgeois world outlook have diverged irrevocably since the end of the 19th century. Consequently the latter are inaccessible to all approaches outside of biology or to outsiders in biology. They continue to attack or evade the philosophical consequences of Darwinism in the old or a modified manner. Nowadays there still exist all kinds of anti-evolutionism and anti-Darwinism that have emerged since 1859 or even earlier. The anti-evolutionistic approach in the defence of static creationism has been given a further lease of life, for instance, by the Creation Research Society in the United States, an institution of ultraconservative protestantic fundamentalists. The Creation Research Society declares in its publications that the description of the creation of the world and of the organisms as given in the original text of the Bible is literally true in all details, and that man and all other organisms were really created by God in six days. This cannot be considered a mere curiosity, because obscurantists are having remarkable succes in repressing the teaching of evolution in national education in the United States. In Sacramento/Cal. they brought Darwinism to trial once again in 1981.

More subtle anti-evolutionism attempts to throw doubt on the objective reality of evolution by disqualifying the scientific theory of descent by episteomological means (cp. LÖTHER [15]. E. MAYR [17], for instance, postulated that the theory of descent is only an interpretation of data to suit an aprioric idea, an idea conceived prior to and independently of experience, and cannot be proved or refuted empirically a posteriori. Another example of the same tendency is the positivistic doctrine of the hypothetical past. This doctrine states that the theory of descent is only a subjective hypothesis which is useful for ordering biological data but is impossible to prove or refute: statements concerning the past can as a matter of principle be nothing more than intellectual constructions since in the distant past, nobody was there to observe events.

Noteworthy, last but not least, is the anti-DARWINian character of evolutionary doctrines worked out by bourgeois philosophers in the 20th century like the American pragmatists Ch. PEIRCE and W. JAMES, H. BERGSON, A. N. WHITEHEAD, P. Teilhard des CHARDIN and others, or fit into the framework of conceptions like Morphologic Idealism, Holism or Emergent Evolution. The border between this kind of evolutionistic philosophy and some anti-DARWINian outsider-conceptions of evolution in biology is fluid. This philosophical evolutionism is free of the materialistic content of the 19th century plain evolutionism (which also still exists), but is no deeper. Starting from the problem of the origin of new qualities, which plain evolutionism has been unable to solve, it uses idealistic, irrationalistic and mystic speculations to give a mystificated conception of evolution and cognition.

MARXian Philosophy and Darwinism

The idea of universal motion change (1813, Logic) was anticipated before its application to life and society. With regard to society it was proclaimed earlier (1847) than it was proved in its application to man (1859) (cf. W. I. LENIN [14], p. 131). HEGEL's "Science of Logic", the first two volumes of which came out in 1812 und 1813, the "Manifesto of the Communist Party" by K. MARX and F. ENGELS, written at the end of 1847 and published in February 1848, and DARWIN's "Origin of Species" here are emphasized as the decisive steps on the way which led to the dialectical conception of reality becoming universal and scientific in character. LENIN shared this high opinion of MARX and ENGELS of HEGELian dialecties and DARWIN's theory of evolution.

The founders of MARXism-LENINism conceived DARWIN's theory of evolution as an eminent natural-scientific basis for dialectical materialism and as the direct natural-historical prerequisite for the historical-materialistic conception of human society and its history. This is why MARX, ENGELS and LENIN confessed to DARWINism. ENGELS refuted in his "Anti-Dühring" E. DÜHRING's attacks not only against Marxism but also against Darwinism. MARX called himself a "sincere admirer" of DARWIN. ENGELS and LENIN regarded the scientific results of MARX and DARWIN as being of the same order and mutually complementary. The intimate connection between MARXian philosophy and DARWINism, founded by the classics of the scientific world outlook, continues and is an essential part of the alliance between dialectical materialism and modern science.

References

[1] AVELING, E., (1891): Die DARWIN'sche Theorie. Stuttgart.
[2] BUHR, M., and STEIGERWALD, R., (1981): Verzicht auf Fortschritt, Geschichte, Erkenntnis und Wahrheit. Berlin.
[3] ČIŽEK, F., and JANKO, J., (1981): On the Problems of Revolutionary Changes in Biology at the Turn of the 19th and 20th Centuries. Acta historiae rerum naturlium necnon technicarum, Special Issue 14. Prague.
[4] DARWIN, Ch., (1959): Autobiographie. Leipzig and Jena.
[5] DENNERT, E., (1903): Vom Sterbelager des Darwinismus. Stuttgart.
[6] DRIESCH, H., (1896): Die Maschinentheorie des Lebens. Biolog. Centralblatt, Vol. XVI.
[7] GOULD, S. J., (1974): DARWIN's Delay. Natural History, Vol. LXXXIII, No. 10.
[8] GOULD, S. J., (1977): Ontogeny and Phylogeny. Cambridge, Mass., and London, England.
[9] GREENE, J. C., (1959): DARWIN and Religion. Proc. Amer. Phil. Soc. Vol. 103, No. 5.
[10] GRUBER, H. E., and BARRETT, P. H., (1974): DARWIN on Man. New York.
[11] HEBERER, G., (1959): Charles DARWIN, Stuttgart.
[12] HERBERT, S., (1913): The First Principles of Evolution. London.
[13] KELLOG, V. L., (1907): Darwinism to-day. London and New York.
[14] LENIN, W. I., (1964): Philosophische Hefte. In: Werke, Bd. 38, Berlin.
[15] LÖTHER, R., (1972): Die Beherrschung der Mannigfaltigkeit. Jena.
[16] LÖTHER, R., (1982): DARWINsche Evolutionstheorie heute. Deutsche Zeitschrift für Philosophie, Vol. XXX, No. 3.
[17] MAY, E., (1947): Schöpfung und Entwicklung. Zeitschrift für philosophische Forschung, Vol. II.
[18] SUTT, T., (1977): The Problem of the Directedness of the Organic Evolution. Tallinn (in Russian).
[19] WICHLER, G., (1963): Charles DARWIN. München and Basel.
[20] ZAVADSKY, K. M., (1973): The development of the Theory of Evolution after DARWIN. Leningrad (in Russian).

Department of Zoology, University of Helsinki, P. Rautatiekatu 13, SF-00100 Helsinki 10, Finland

How to Conceptualize Macroevolution?

Y. HAILA

A distinction between the terms "microevolution" and "macroevolution" dates back to GOLDSCHMIDT [14], who argued that macroevolution, or evolution above the species level, should be studied separately from microevolution, or changes within a species. Furthermore, he argued that the two processes are based on different mechanisms (for a history of the concepts, see e.g. STANLEY [48]). GOLDSCHMIDT's hypothesis of a special mechanism for macroevolution ("macromutations") was soon rejected, but the terminological distinction has remained.

Should we accept this terminology? Indeed, e.g. SIMPSON ([46] 339) suggested that the distinction should be abandoned because organic evolution is a single process with "innumerable levels". Also SHMALGAUZEN ([39] 268; original in 1946), in his discussion of the concepts heavily emphasized the unity of the evolutionary process — the fact that the process is based on a single set of rules (or laws).

However, in recent discussions of evolutionary biology there has been a revival of the term macroevolution (see e.g. SHMALGAUZEN [40], STANLEY [48] PAAVER and SUTT [30]). It has even been claimed (LEWIN [26]) that the synthetic theory of evolution developed in the 1940s has not been able to explain evolutionary divergence above the species level.

New scientific concepts should be adopted whenever this improves our understanding of the underlying phenomena. In this paper I shall adopt the term macroevolution as a descriptive one. As a term it refers to the fact that organic evolution as a historical process comprises a totality including qualitative steps, the reaching of new levels of organization by organisms. Microevolution refers to processes going on in populations here and now, macroevolution to the divergence of various higher taxa above the species level in a historical perspective. I agree with GALL and KOLCHINSKY [11] that the concept of organic evolution, to be really "synthetic", should explain both micro- and macroevolution with the same set of basic mechanisms. I shall argue that evolution above the species level, i.e. macroevolution, should be studied and conceptualized separately from microevolution although both processes are based on the same mechanisms and rules. This is useful in order to indentify all important structural units in the process of organic evolution, studied as a totality (see also SHMALGAUZEN [40], TIMOFEEFF-RESSOVSKY et al. [51]).

A Conceptual Scheme for Macroevolution

When searching for a conceptual scheme for macroevolution the important question is: In the historical process of evolution, is it possible to discern any general, recurrent features that could be used in charcterizing and analyzing the process?

On the one hand the question concerns the problem of a general trend, or progress, in evolution. Without entering deeply in the discussion, it can be claimed that a general agreement exists about the existence of a general progressive trend in evolution, the most important characteristics of evolutionary progress being increasing complexity, integration, and relative independence from the physical environment realized at the organismic level (e.g. Rensch [32], Simpson [46], Shmalgauzen [39], Timofeeff-Ressovsky et al. [51]). The term progress does not, of course, imply any evaluations of the direction, or desirability, of the trend. Shmalgauzen [40] indentified the progressive trend of evolution as the decrease of entropy.

The mere existence of a general trend does not, however, provide a basis for a conceptualisation of the phases of macroevolution. The trend is realized as an overall result of the material mechanism of evolution (e.g. SHMALGAUZEN [39]). It is more interesting to ask whether it is possible to make any generalizations about the ways by which the evolutionary progress is realized. Several authors have discussed this problem; here I shall make use of the ideas of I. I. SHMALGAUZEN [39], taken over in a modified form from A. N. SEVERTSOV [34] (see also A. S. SEVERTSOV [37]). SHMALGAUZEN [39] distinguished between three main phases of progressive evolution: **aromorphosis,** leading to the invasion of a new adaptive zone by the taxon in question; **allomorphosis,** leading to adaptive radiation within the new adaptive zone; and **specialization,** leading to more strictly defined interrelations between organisms and the environment within the adaptive zone. SHMALGAUZEN also classified different types of specialization, but they are not vital to the general scheme (see SEVERTSOV [37]).

It is important to stress that these different phases in progressive evolution are based on the same material mechanism, natural selection. Natural selection is realized through the dialectical contradiction between the organism and the environment. In the different phases, the relative significance of directional and stabilizing selection varies (SHMALGAUZEN 39, SEVERTSOV 37). Through an analysis of the type of natural selection prevailing during the different phases, environment, or ecology, is incorporated into the description of macroevolution.

The SEVERTSOV-SHMALGAUZEN distinction between different phases of progressive evolution is convergent to several conceptions of evolution developed in the west. In the first place, the understanding or aromorphic evolution as an invasion into a new adaptive zone is analogous to the quantum evolution of SIMPSON [46] (see also SEVERTSOV [37]). On the other hand, the process which SHMALGAUZEN described as allomorphosis and specialization, i.e. evolution (adaptive radiation) within one and the same adaptive zone, is the domain of the "Red Queen" hypothesis of VAN VALEN (1973). Both SHMALGAUZEN and VAN VALEN maintain that evolution within a certain adaptive zone is greatly affected by interspecific interactions. There is only a limited amount of energy at hand, and therefore each new adaptation achieved by a certain species makes the environment relatively worse for the other species in the same zone. According to the Red Queen hypothesis, evolution is a zero-sum game for energy (VAN VALEN [55]) between species in any adaptive zone. VAN VALEN [55] explicitly restricts the domain of the hypothesis to a single adaptive zone. Also note that SHMALGAUZEN [39, 40] explicitly regards the acquisition of energy ("control of energy", VAN VALEN [54]) as a crucial criterion of progressive evolution.

The convergence of the conceptual schemes of SHMALGAUZEN and SIMPSON or VAN VALEN is caused by a common orientation towards organism-environment relations as an explanatory basis for evolution. This orientation towards ecology is, indeed, the fact that makes the scheme of SHMALGAUZEN fertile and usable in analysing evolutionary change at the level of macroevolution. After all, the original DARWINian approach is anchored in ecology; another matter is that, in the western tradition, evolutionary genetics has mainly developed separately from ecology (LEWONTIN [27]).

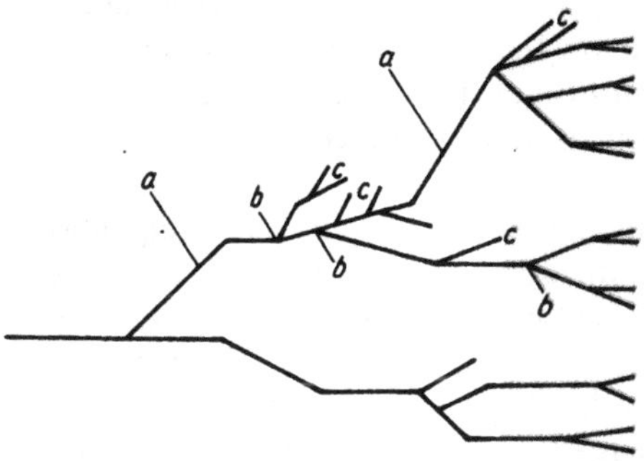

Fig. 1. The main phases of progressive evolution: Aromorphosis (a), leading to new levels of adaptation (adaptive zones); allomorphosis (b), leading to adaptive radiation; and specialization (c). Modified from SEVERTSOV [37].

SHMALGAUZEN [39] also discerned a **change of phases** in progressive evolution (see Fig. 1). The cycle starts with aromorphosis, continues as allomorphosis, and ends as specialization (see also SEVERTSOV [35, 36]). For specialized species there are three alternative fates in the evolutionary scale: Extinction as a result of drastic changes in environmental conditions, the development of a "living fossil" if the environment is very stable, or a new aromorphosis, giving rise to the invasion of a new adaptive zone and, consequently, to a new phase-cycle (see also SEVERTSOV [37, 269]). This change of phases in the progressive evolution of different taxa is connected with environmental conditions by SHMALGAUZEN ([39] 422—426) in his discussion of the factors that determine the phases. Especially SEVERTSOV [35] emphasizes that a new aromorphosis is attained by relatively specialized species; environmental conditions determine whether a certain specialization can be a basis for a new aromorphosis or not. Specialization is, of course, a relative concept. Strictly speaking, all species are "specialized" to certain conditions of the environment. Preadaptation is the concept often used to describe the possibility that a species adapted (specialized) to certain conditions may be able to invade a new adaptive zone when the conditions change. Thus, a perfectly materialistic interpretation, based on DARWINian natural selection, can be given to preadaptation (e.g. Bock 1).

Epistemological Status of the SEVERTSOV-SHMALGAUZEN Scheme

The explanatory power of the conceptual scheme presented above lies in its being based on DARWINian natural selection as the basic mechanism of evolution. This fact, however, creates a special problem concerning the epistemological status of the concepts. Natural selection acts on individual phenotypes, and leads to changes in the composition of populations. But it then follows that the analysis of macroevolution is methodologically essentially reductionist in the sense that it is based on a mechanism working at a lower level, at the level of populations (see also BOCK [2], LEVINTON and SIMON [25]. The phases of progressive evolution are realized as a consequence of microevolutionary processes, without any mechanism of their own.

I would characterize the SHMALGAUZENian conceptual scheme of evolutionary progress a **synthetic-descriptive generalization.** The scheme is, certainly, a descriptive generalization, while it is based on a reductionist analysis. But on the other hand, the scheme is not **only** descriptive, while important recurrent features of the process of evolution are synthetized by it. Consequently, macroevolution as a total process can not only be described, but also analyzed, by the scheme.

To make this point clearer, I shortly comment the model of "punctuated equilibria", developed in order to overcome some of the difficulties in uniting micro- and macroevolution on a common basis (see e.g. ELDREDGE and GOULD [7], GOULD and ELDREDGE [15], STANLEY [48]; for a critical overview, see LEVINTON and SIMON [25]). The idea of the model is simple: Speciation is the essential phase leading to macroevolution; speciation always takes place in small, isolated, peripheral populations; "phyletic evolution" following "punctuated" speciations is very slow and essentially insignificant to the course of evolution; what is significant is "species selection", realized through competition between different daughter species when they get into contact after allopatric speciation; the characteristics realized by the daughter species are random with respect to the future course of evolution — a trend can be realized only through species selection which can lead to an apparently unified trend that may be observed in the paleontological record. Some proponents of the model have explicitly characterized it as an alternative to the traditional DARWINian view (e.g. ELDREDGE [6]).

Methodologically characteristic of the punctuated equilibrium model is that it is based on extremely simple premises: speciations as essentially random events in small, peripheral populations, and species selection following speciations. Taking microevolutionary processes as a basis, one might say that instead of striving upwards, towards a synthetic scheme including all the consequences of microevolution, the proponents of the model descend downwards, by taking a thin abstraction of speciation as an explanatory basis for the whole process of macroevolution. The model can thus be regarded as an analytical one, striving for an explanation of macroevolution by starting from a few simple premises.

Now, as with all analytical models, the crucial question is the realism of the premises. It turns out that they are not realistic at all. In the first place, speciation is not necessarily always realized in peripheral isolates (for an excellent evaluation of the genetic processes involved, see LANDE [22]). As to "species selection", LEVINTON and SIMON [25] studied the possible evidence for a special mechanism of species selection, and found it nonexistent. The same result can be reached also through theoretical reasoning: Species selection cannot play any role when a certain taxon has reached a new adaptive zone through aromorphic evolution; on the contrary, the taxon has entered an unoccupied zone for adaptive radiation.

Consequently, the "punctuated equilibria" model certainly does not meet the demand of GALL and KOLCHINSKY [11] of creating a synthetic explanatory framework of micro- and macroevolution. This goal can be reached through a synthetic comprehension of all different features of evolution, based on the acknowledgement of natural selection as the basic mechanism (i.e. on the reductionist model, **sensu** LEVINTON and SIMON [25]).

Explanatory power of the SEVERTSOV-SHMALGAUZEN Scheme

What does it, strictly speaking, mean if we characterize the conceptual scheme of SHMALGAUZEN as a synthetic generalization? How far can we explain the processes of macroevolution by adopting this concept? Here I start from the premise that the explanatory of any conceptual scheme in science depends on whether the scheme can correctly reflect power the determinatedness of the corresponding object or not. It is possible to explain a phenomenon if we know the factors that determine it; then we can also make predictions about its occurrence. This is realized through the adoption of appropriate theoretical concepts.

When studying determinism in macroevolution, as described by the conceptual scheme of SEVERTSOV-SHMALGAUZEN, we must find a solution to an apparent contradiction: Evolutionary progress, in its different phases, is realized as a result of a lower-level causal mechanism, that of natural selection. Therefore, it is realized as a consequence, and can only be described **a posteriori**. Then, how can evolutionary progress, or its different phases, be determinate?

The point is that determinism in biological evolution needs a characterization different from classical causality (see e.g. BORZENKOV [4], HAILA [17]). Descriptively, I would characterize the kind of determinism observed in biological evolution as **pattern determinism** (analogously to the characterization of the determinism in community ecology by MAY [28], see HAILA [17]). Somewhat simplifying, it might be said that through the dialectical contradiction of the organism and the environment the organisms are posed by a certain set of problems which they must solve in order to survive and reproduce. Natural selection is the creative mechanism providing organisms with prerequisites for the solutions (which are expressed, of course, as the progressive adaptation of organisms to the environment). But each single problem can be answered in different ways — or maybe a certain species cannot solve a specific problem at all, but goes extinct. Consequently, a pattern results, determinate at a general level, but a detailed prediction of the **how**, and **where**, and **when**, is impossible.

At the level of the functional morphology of organisms, evolutionary convergence between different taxa is a good illustration of the patterns. When different organisms face essentially the same problem, like sensing light, or correlating bodily movements by an effective nervous system, the resulting solutions (functional adaptations) may be surprisingly similar. Any textbook on comparative anatomy and morphology abounds with examples.

At the level of phylogenic progressions, the change of phases of SHMALGAUZEN [39] is a similar kind of pattern. When ecology is incorporated into the picture the scheme of SHMALGAUZEN can be maintained — similarly as SIMPSON [46] maintened Quantum evolution, or VAN VALEN [53] maintained the Red Queen principle.

The situation is illustrated by Fig. 2. Natural selection is the material mechanism leading to progressive adaptation, i.e. evolutionary progress. However, adaptations are always expressed as specific results, and these are determined by a set of preconditions, comprising the genetic variability of populations, ontogenic (and evolutionary) canalization of the organisms (**sensu** SHMALGAUZEN [38], or WADDINGTON [58]), existing environment, and history. These preconditions act as chance factors relative to the process, although each of them is determined by a set of causal agents. Therefore, the details of the resulting patterns cannot be predicted, an **a posteriori** element is preserved in the explanation of evolution (see SEVERTSOV [37]), but general features, as patterns, are predictable.

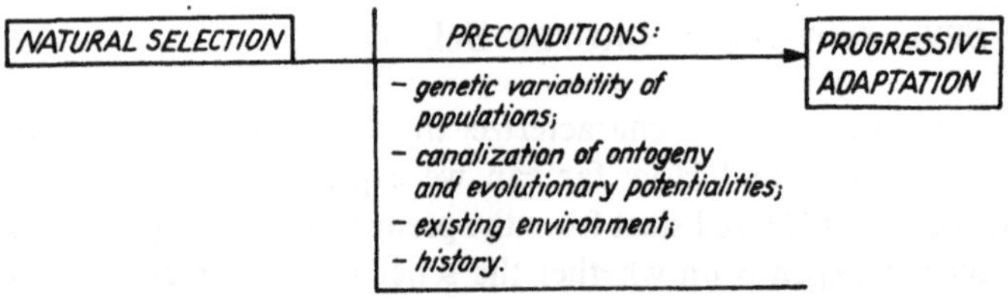

Fig. 2. Progressive adaptation as a consequence of natural selection, working under specific preconditions.

Evolutionary Patterns as a Heuristic Basis for Predictions

Starting from HUTCHINSON's [19] beautiful phrase "the ecological theatre and the evolutionary play", the patterns of evolutionary progress can be viewed at two levels. In the evolutionary time-scale we are dealing with phylogenic progressions, and in the ecological time-scale with populations, communities and ecosystems. At any moment, the totality of evolutionary lineages is expressed as the totality of ecological communities and ecosystems of the earth; and, **vice versa**, evolution is at any time going on in ecological systems.

There are analogous features in the evolutionary patterns observed at both levels: The patterns are results of material, causal mechanisms, but only results, with no mechanisms of their own that were independent of the lower level causal agents. Thus, TIMOFEEFF-RESSOVSKY et al. [52] define biogeocenoses (ecosystems) as "elementary biochorological units" (i.e. elementary biological units in space). Certain structural patterns of ecosystems as "biochorological units" are predictable and have a relatively independent character (e.g. circulations of vital elements, trophic structure, relations between the number of individuals and the number of species, etc.), but they are produced by organic evolution based on natural selection. What is needed is a dialectical view on ecological communities — a view acknowledging both the reality of communities as ecological units and the crucial significance of lower-level mechanisms in determining the structure of communities (LEVINS and LEWONTIN [24]).

It is obvious that ecological, synthetic-descriptive concepts (community, ecosystem) provide a heuristic basis for predictions; any ecological textbook is based on the idea, although the exact status of the concepts needs elucidation (e.g. EMLEN [8], ENGELBERG and BOYARSKY [9], LEVINS and LEWONTIN [24], SIMBERLOFF [44], JORDAN [20]).

Similarly, I argue that synthetic-descriptive concepts of the patterns in phylogeny also have a predictive element. As an example, consider the evolutionary perspectives of the environmental changes produced by man. It would certainly be a most desirable thing to be able to predict the consequences that environmental changes will have for the organic nature, especially to predict the evolutionary fate of organisms in a modified environment. I suggest that the SHMALGAUZENian change of phases of organic evolution may help us in making predictions about the extinction-proneness vs. invasion-proneness of different species.

Traditionally, the evolutionary consequences of human impact on nature have almost exclusively been discussed from the perspective of conservation and preservation of endangered species and genetic resources (see e.g. FRANKEL and SOULÉ [10]). The coin, however, has another side, i.e. question, which species will be most successful in man-modified habitats, and what their evolutionary future will be. We certainly know that evolutionary radiation may be rapid in these conditions as was already demonstrated by the pioneering works of VAVILOV [56] in the case of cultivated plants.

The characteristic feature of the environmental change today is its scale and speed, which are enormous compared with anything that has occurred in nature. However, environmental change is one of the fundamental characteristics of the milieu on the Earth (e.g. NITECKI [29]). Even in the environment considered the most stable of all terrestrial habitas, the tropical rain forest, ephemeral, early successional patches seem to be an essential element of the whole (FRANKEL and SOULÉ [10]). Consequently, the relative stability of any specific environment is a most important feature for the adaptations of the species living in it (see GHILAROV [12, 13], LEVINS [23]). Remind the basic criteria by which SHMALGAUZEN [39] distinguished between the different phases of progressive evolution: The criteria are, whether the adaptive zone of the species is widening (as in aromorphosis), or staying relatively stable (as in allomorphosis), or getting narrower (as in specialization). Using these criteria, we might predict that specialist species that have a strict and narrow adaptive zone are extinction-prone when the environment changes. On the contrary, species with a widening adaptive zone would be invasion-prone.

It should be clear that the classification cannot be applied directly, schematically. Even if we could identify a species that is facing a phase of aromorphic evolution, i.e. invasion of a new adaptive zone, that would refer to evolutionary time, with thousands of generations of the organism involved, but we can only observe ecological time. However, it is possible to study evolution even in ecological time, here and now, by studying the prevailing modes of natural selection (SEVERTSOV [36]). SHVARTS [41] suggested that directional selection can be observed as skewness of the distribution of morphological characteristics (see also SHVARTS [43]). In the same vein, SEVERTSOV [36] suggested that it might be possible to analyze prevailing modes of selection under specific circumstances, and make generalizing predictions. As he admitted, the suggesion is somewhat speculative because of scarcity of data, but it opens one possible approach even for experimental work. Colonizations of new habitats by species also offer interesting research problems (see also WADDINGTON [59]).

Another approach would be to try to define ecological correlates of extinction-prone vs. invasion-prone species. Here good autecological data about the exact adaptations of different species to their specific environment are crucial. So, GHILAROV [12] emphasized that the degree of stability of the environment is essential for the degree of dispersal ability of different species, and the ability to colonize isolated habitat patches, quite naturally, is a function of dispersal ability. An important generalization seems to be that good colonists are species living in ephemeral batches, adapted to a continuously and rapidly changing environment, the "fugitives" of HUTCHINSON [18] (see also SOUTHWOOD [47], WILSON [61], DEN BOER [3]. This also applies to plants; e.g. RAUP [31] gives a beautiful example from Greenland, where he observed the invasion of new roadside habitats by some members of the original flora of Greenland, living originally in occasionally flooded sand and gravels along rivers but acting as weeds in their new habitats.

However, a distinction should be made between "colonization" of an ephemeral habitat patch and "invasion" of a new area, or even adaptive zone. Good colonizing ability as an ecological strategy may be extremely specialized, without evolutionary perspectives (parthenogenetic species adapted to rapidly changing environments in the north represent an extreme, see VEPSÄLÄINEN and JÄRVINEN [57]). What we are interested in is the aptness of species to invade new habitats, which possibly is connected with the acquisition of new adaptations and may open new evolutionary pathways. Thus, e.g. SAFRIEL and RITTE [33] recently searched for ecological criteria for potential colonists of new environments.

The ability to colonize new habitat patches and proneness to invade new types of environment are probably correlated, although not strictly linearly. By the hypothesis of the "taxon cycle" WILSON [60, 61] tried to set the ecology of colonizations into an evolutionary frame. According to the hypothesis, good colonizing ability can also be regarded as an evolutionary strategy. Although good colonists are "fugitives", they may be able to invade new, previously unoccupied environments after having arrived for example on an oceanic island. As a basis of WILSON's hypothesis served the distributional patterns of the Melanesian ant fauna.

To conclude, I suggest that the Shmalgauzenian phase-cycle of macroevolution provides us with a heuristics that makes it possible to make predictions about the future course of evolution, albeit at a general, qualitative level.

However, two important qualifications should be emphasized:

1 The SEVERTSOV-SHMALGAUZEN scheme refers to evolutionary time but what we can observe are ecological patterns. Therefore, the predictions are **essentially** qualitative; the real situation is determined by too complicated a network of factors to allow strict predictions. This problem is magnified by the fact that ecological systems, as stages of the evolutionary future, also can be analyzed only in synthetic-descriptive terms. The ecology of invasions is a good example here: SIMBERLOFF [45] reviewed the available data on ecological consequences of invasions, and concluded that our present theories of ecological communities cannot predict the community level consequences of successful invasions or introductions (mostly no consequences could be detected, contrary to the expectations that can be derived from the theories of species packing, limiting similarity, competition etc.).
2 The predictions are made even more complicated by the fact that the organism as an integrated whole is the object of natural selection (**sensu** SHMALGAUZEN [38]; see also the recent review by WRIGHT [62] of his original ideas). This means that no single vector of the functional whole of an organism can be absolutized as "the" criterion for future evolution. The adaptationist programme, criticized by GOULD and LEWONTIN [16], is doomed to failure.

Ecology and Macroevolution: Concluding Remarks

I have presented the SEVERTSOV-SHMALGAUZEN conception of macroevolution as a synthetic view on evolutionary progression. The fertility of the approach lies in its being based on the organism-environment interaction. As DOBZHANSKY [5] emphasized, natural selection, realized through specific environmental conditions, is the antichance factor in evolution. This assertion should explicitly be widened to macroevolution, i.e. speciation, and evolution above the species level, as recently pointed out by e.g. SHVARTS [42, 43], LEVINTON and SIMON [25] and SEVERTSOV [37].

As the environmental conditions of the earth are constantly changing in a random fashion relative to evolution, there can be no final end in organic evolution. The same assertion can be made by emphasizing the opportunism of evolution: Adaptation is always realized here and now, and diversification is consequently another essential feature of evolution (e.g. KURTÉN [21]). But one might speculate a bit further: Maybe opportunism is exactly the factor that ensures that usable preadaptations are probably at hand when environmental conditions change. Geologic and climatic changes of the earth are, after all, so slow that all environmental conditions probably have been represented earth in small scale long before they became dominant. Thus, e.g. before the formation of present deserts, or boreal forests, or arctic tundras as major biomes, small areas of similar types of habitats have most probably been present on the earth, with organisms with suitable preadaptations to invade the new biomes and give rise to aromorphic evolution and adaptive radiation. STEGMAN [49, 50] explained the origin of the bird fauna of the Palearctic taiga by this mechanism. G. M. DLUSSKIJ (in press) has concluded on the basis of fossil evidence that the desert ant fauna originated from a few ancestral species adapted to sand dune habitats, preadapted to present deserts.

But does this assertion apply also to new, man-made habitats? And if not, what will be the future course of evolution — a really dramatic decrease of organic diversity?

I want to emphasize the role of ecology in evolutionary biology. This also means the realization of the original Darwinian approach; after all, DARWIN was an outstanding naturalist and ecologist. Consequently, the theme "DARWIN today" should also include a perspective of creating a synthesis between evolutionary biology and ecology.

Acknowledgements

O. JÄRVINEN, E. RANTA and K. VEPSÄLÄINEN read through a draft of the manuscript and gave many useful comments.

References

[1] BOCK, W. J.: Preadaptation and multiple evolutionary pathways. Evolution **13** (1959), 194—211.

[2] BOCK, W. J.: The synthetic explanation of macroevolutionary change — a reductionist approach. Bull. Carnegie Mus. Nat. Hist. **13** (1979), 20—69.

[3] DEN BOER, P. J.: On the survival of populations in a heterogeneous and variable environment. Oecologia **50** (1981), 39—53.

[4] BORZENKOV, V. G.: Printsip determinizma i sovremennaya biologiya. Izd. Mosk. Univ., Moskva (1980).

[5] DOBZHANSKY, T.: Chance and creativity in evolution. In: AYALA, F. J. and T. DOBZHANSKY (eds.): Studies in the philosophy of biology. Macmillan, London; and University of California Press (1974), 307—338.

[6] ELDREDGE, N.: Alternative approaches to evolutionary theory. Bull. Carnegie Mus. Nat. Hist. **13** (1979), 7—19.

[7] ELDREDGE, N. and S. J. GOULD: Punctuated equilibria: an alternative to phyletic gradualism. In: SCHOPF, T. J. M. (ed.): Models in paleobiology. Freeman, Cooper and Co., San Francisco, California (1972), 82—115.

[8] EMLEN, J. M.; Ecology: An evolutionary approach. Addison-Wesley Publ. Company, Reading, Massachusetts (1977).

[9] ENGELBERG, J. and L. L. BOYARSKY: The noncybernetic nature of ecosystems. Amer. Natur. **114** (1979), 317—324.

[10] FRANKEL, O. H. and M. E. SOULÉ: Conservation and evolution. Cambridge Univ. Press, Cambridge (1981).

[11] GALL, Y. M. and E. I. KOLCHINSKY: Synthetic conceptions of evolution and problems of macroevolution. In: PAAVER, K. L. and T. SUTT (eds.) 1980), 12—17 (in Russian with an English summary).
[12] GHILAROV, M. S.: Influence of the character of dispersal on the ontogenesis in insects. Zhurnal obshchei biologii **6** (1945), 26—37 (in Russian with an English summary).
[13] GHILAROV, M. S.: Trends in the evolution of passively dispersing insects and the feed-back control in phylogenesis. Z. f. zool. Systematik und Evolutionsforschung **7** (1969), 1—18.
[14] GOLDSCHMIDT, R.: The material basis of evolution. Yale Univ. Press, New Haven, Connecticut (1940).
[15] GOULD, S. J. and N. ELDREDGE: Punctuated equilibria: The tempo and mode of evolution reconsidered. Paleobiology **3** (1977), 115—151.
[16] GOULD, S. J. and R. C. LEWONTIN: The spandrels of San Marco and the Panglossian paradigm: a critique of the adaptationist programme. Proc. R. Soc. Lond. B **205** (1979), 581—598.
[17] HAILA, Y.: Determinism in biological evolution. In: HAILA, Y., I. TUOMINEN and O. VILHU (eds.): Evolution — different aspects, common problems. Proc. 1st Tutkijaliitto Symp., Tutkijliitto, Helsinki (1982), 115—135.
[18] HUTCHINSON, G. E.: Copepodology for the ornithologist. Ecology **32** (1951), 571—577.
[19] HUTCHINSON, G. E.: The ecological theatre and the evolutionary play. Yale Univ. Press, New Haven, Connecticut (1965).
[20] JORDAN, C. F.: Do ecosystems exist? Amer. Natur. **118** (1981), 284—287.
[21] KURTÉN, B.: The „gestalt" of human evolution. In: HAILA, Y., I. TUOMINEN and O. VILHU (eds.): Evolution — different aspects, common problems. Proc. 1st Turkijaliitto Symp. Tutkijaliitto, Helsinki (1982), 135—140.
[22] LANDE, R.: Genetic variation and phenotypic evolution during allopatric speciation. Amer. Natur. **116** (1980), 463—479.
[23] LEVINS, R.: Evolution in changing environments. Princeton Univ. Press, Princeton, N. J. (1968).
[24] LEVINS, R. and R. C. LEWONTIN: Dialectics and reductionism in ecology. Synthese **43** (1980), 47—78.
[25] LEVINTON, J. S. and C. M. SIMON: A critique of the punctuated equilibria model and implications for the detection of speciation in the fossil record. Syst. Zool. **29** (1980), 130—142.
[26] LEWIN, R.: Evolutionary theory under fire. Science **210** (1980), 883—887.
[27] LEWONTIN, R. C.: Fitness, survival, and optimality. In: HORN, D. J., R. G. MITCHEL and G. R. STAIRS (eds.): Analysis of ecological systems. Ohio State Univ. Press, Columbus, Ohio (1978), 3—21.
[28] MAY, R. M.: The evolution of ecological systems. Sci. American **239(3)** (1978), 118—135.
[29] NITECKI, M. H. (ed.): Biotic crises in ecological and evolutionary time. Academic Press, N. Y. (1981).
[30] PAAVER, K. L. and T. SUTT (eds.): Mikro- i makroevolyutsiya. Tartu, Est. S. S. R. (1980).
[31] RAUP, H. M.: Physical disturbance in the life of plants. In: NITECKI, M. H. (ed.) (1981), 39—52.
[32] RENSCH, B.: Evolution above the species level. Columbia Univ. Press, N. Y. (1959).
[33] SAFRIEL, U. N. and U. RITTE: Criteria for the identification of potential colonizers. Biol. J. Linn. Soc. **13** (1980), 287—297.
[34] SEVERTSOV, A. N.: Morfologicheskie zakonomernosti evolyutsii. Izd. AN SSSR, Moskva-Leningrad (1939).
[35] SEVERTSOV, A. S.: Stanovlenie aromorfoza. In: Organizatsiya i evolyutsiya zhivogo. Izd. Hauka, Leningrad (1972), 164—167.
[36] SEVERTSOV, A. S.: O nekotorikh podkhodakh k izucheniyu estestvennogo otbora. In: Fiziologicheskaya i populyatsionnaya ekologiya zhivotnykh. Izd. Saratovskogo universiteta, Saratov (1980), 5—13.
[37] SEVERTSOV, A. S.: Vvedenie v teoriyu evolyutsii. Izd. Mosk. univ., Moskva (1981).
[38] SHMALGAUZEN, I. I.: Organizm kak tseloe v individual'nom i istoricheskom razvitii. Izd. AN SSSR, Moskva—Leningrad (1938).
[39] SHMALGAUZEN, I. I.: Problemy Darwinizma. Izd. Nauka, Leningrad (1969).
[40] SHMALGAUZEN, I. I.: Faktory progressivnoi (aromorfnoi) evolyutsii = snizheniya entropii. In: ZAVADSKIJ, K. M. (ed.): Zakonomernosti progressivnoi evolyutsii. Izd. Nauka, Leningrad (1972), 5—24.

[41] SHVARTS, S. S.: Opyt izucheniya napravleniya izmenchivosti v prirodnykh populyatsiyakh zhivotnykh. Dokl. AN SSSR **166** (1966).
[42] SVARTS, S. S.: The evolutionary ecology of animals. Plenum, N. Y. (1977).
[43] SHVARTS, S. S.: Ekologicheskie zakonomernosti evolyutsii. Izd. Nauka, Moskva (1980).
[44] SIMBERLOFF, D. S.: A succession of paradigms in ecology: essentialism to materialism and probabilism. Synthese **43** (1980), 3—39.
[45] SIMBERLOFF, D. S.: Community effects of introduced species. In: NITECKI, M. H. (ed. (1981), 53—81.
[46] SIMPSON, G. G.: The major features of evolution. Columbia Univ. Press, N. Y. (1953).
[47] SOUTHWOOD, T. R. E.: Migration of terrestrial arthoropods in relation to habitat. Biol. Rev. **37** (1962), 171—214.
[48] STANLEY, S. M.: Macroevolution. H. Freeman and Co., San Francisco, California (1978).
[49] STEGMAN, B. K.: O proiskhozhdenii ornitofauny taigi. Dokl. AN SSSR **13** (1931), 350—357.
[50] STEGMAN, B. K.: Psnovy ornitogeograficheskogo deleniya Palearktiki. Trudy ZIN AN SSSR, Fauna SSSR, No 19, **1(2)** (1938).
[51] TIMOFEEFF-RESSOVSKY, N. V., N. N. VORONCOV and A. N. JABLOKOV: Kurzer Grundriss der Evolutionstheorie Gustav Fischer Verlag, Jena (1975).
[52] TIMOFEEFF-RESSOVSKY, N. V., A. N. JABLOKOV and N. V. GLOTOV: Grundriss der Populationslehre. Gustav Fischer Verlag, Jena (1977).
[53] VAN VALEN, L. M.: A new evolutionary law. Evol. Theory **1** (1973), 31—49.
[54] VAN VALEN, L. M.: Energy and evolution. Evol. Theory **1** (1976), 179—229.
[55] VAN VALEN, L. M.: Evolution as a zero-sum game for energy. Evol. Theory **4** (1980), 289—300.
[56] VAVILOV, N. I.: The origin, variation, immunity, and breeding of cultivated plants. Chron. Bot. **13** (1949—50), 1—366.
[57] VEPSÄLÄINEN, K. and O. JÄRVINEN: Apomictic parthenogenesis and the pattern of the environment. Amer. Zoologist **19** (1979), 739—751.
[58] WADDINGTON, C. H.: The strategy of the genes. Allen & Unwin, London (1957).
[59] WADDINGTON, C. H.: Introduction to the symposium. In: BAKER, G. H. and G. L. STEBBINS (eds.): The genetics of colonizing species. Academic Press, N. Y. (1965), 1—6.
[60] WILSON, E. O.: The nature of the taxon cycle in the Melanesian ant fauna. Amer. Natur. **95** (1961), 169—193.
[61] WILSON, E. O.: The challenge from related species. In: BAKER, G. H. and G. L. STEBBINS (eds.): The genetics of colonizing species. Academic Press, N. Y. (1965), 7—24.
[62] WRIGHT, S.: Genic and organismic selection. Evolution **34** (1980), 825—843.

Discussion

H. A. ROSENTHAL: What, in your opinion, is the role of an organism of species within the environment? Do organisms behave passively, being merely objects of the environment, or are they actively involved in evolution as brought about by an interaction between them and the environment? Of course, environment includes here also those organisms whose evolution is not being specifically considered.

HAILA: I would use the concept preadaptation in this connection: the species favoured by environmental modification caused by human culture are adapted (specifically adapted) to corresponding condition in nature. From the point of these species it is pure chance that human culture produces similar habitats in a much wider scale.

BRETSCHNEIDER: I would like to make a few remarks about the new quality of the synthetic theory of evolution. My starting point is ENGELS Opinion of DARWIN's "On the origin of species by means of natural selection, or the preservation of favoured races in the struggle for live" expressed in a letter to MARX. In 1859 ENGELS wrote:
„Übrigens ist der DARWIN, den ich jetzt gerade lese, ganz famos. Die Technologie war nach einer Seite hin noch nicht kaputt gemacht, das ist jetzt geschehn. Dazu ist bisher noch nie ein so großartiger Versuch gemacht

worden, historische Entwicklung nachzuweisen, und am wenigsten mit solchem Glück. Die plumpe englische Methode muß man natürlich in Kauf nehmen.“ (MEW, 29, p. 524)
(“By the way, DARWIN, who I am now reading, is a wonderful man. In one respect teleology had not been destroyed; now it has. And in addition there has never been such a fabulous attempt to furnish proof of historical development, and much less has there been such a successful one. The clumsey English method, of course, has to be put up with.”)
The synthetic theory of evolution can be considered the present form of Darwinism. Compared with DARWIN's theory of evolution, it has the new quality of a biological theory. What are the characteristics of this new quality?
1. The idea of development no longer refers almost exclusively to the organism and species but to all steps of the encaptive hierarchy of biotic phenomena;
2. The synthetic theory of evolution reflects increasingly precisely the objective dialectics of biotic evolutionary processes. This includes the laws of biotic evolution, and role of necessity and chance, and the existence and mode of action of contradictions in the evolution of biotic phenomena.
3. The synthetic theory of evolution discusses the problem of the elementary unit of biotic evolution. This elementary unit is the population, which is the primary reproductive community of individuals belonging to a species.
4. The synthetic theory of evolution considers the causes of biotic evolution to be a multifactoral interactive structure or formation. In this connection it is of fundamental importance to know that the factors of evolution and the mechanisms by which it acts are also subject to evolution.
5. Knowledge from many different fields of biology has been integrated into the synthetic theory of biotic evolution. The synthesis of biotheories has become of decisive importance in this process.
According to WUKETITS the theory of evolution can be regarded as the central theory of the biological sciences and as the basis of biology because of its character as a theoretically based science. It can thus be considered a fundamental biological theory. All biological theories which can be immediately synthesized into a general theory of living matter can be considered fundamental.

RUSE: I think that Dr. LÖTHER overestimates DARWIN's materialism. In fact, when he developed his theory of evolution, and right up to the time when he wrote the “Origin”, DARWIN was a that is to say, he believed in God as an unmoved mover. Only in later life, probably under the influence of HUXLEY, did DARWIN become an agnostic.

LÖTHER: I see no contradiction between “western” and “eastern” histories of science with regard to the philosophy of DARWIN, to DARWIN's materialism. This is why my paper is based essentially on DARWIN's own statements and the research work of authors from the United States, especially by GRUBER and BARRETT's fundamental book which contains DARWIN's notebooks on man, mind and materialism and was published in 1974 in New York.

NOVAK: To the communication by Dr. JENSSEN: The book by DAWKINS is bad and misleading in my opinion, both for scientific reasons and regarding its social consequences. His conclusions are quite one-sided on the level of the earlier social Darwinism. He accepts from DARWIN only his wrong conception of interspecific struggle, which DARWIN himself used only formally to understand the natural reaction correctly (cf. NOVAK, 1978). When speaking of the selfish gene, he so to say projects into it his own character's feature. His antisocial approach is manifested by his opinions being unrelated to the subject of the book, for instance his claim that all the economic evils of Great Britain are caused by the British working class because they are not ready to work for their capitalists sufficiently hard.
To Dr. HAILA's paper: I should just like to add that SCHMALGAUSEN later supplemented the SEVERTROV conception by distinguishing three categories: **aromorphosis**, i.e. rising to a higher organisation level, **adaptiomorphosis**, i.e. adaptation at one level of organisation, with two subcategories: allomorphosis, or adaptation without limiting further adaptive capacity and telomorphosis, and specialisation, which means such extreme adaptation to a given environment that, when the environment changes, further adaptation is impossible (special consideration of telomorphosis and hypermorphosis, like with the dinosauriers ostitionotheyia). The third category is **phylogenetical degeneration**, in which the general

organisation falls to a lower level. Here there are two types: hypomorphosis (like in Apterygot insects) and kotamorphosis (like in most parasites).

JENSSEN: I agree with Dr. NOVAK that there is a lot to criticise in DAWKIN's book. This was actually a main point in my presentation. DAWKIN forwards an unacceptable philosophical interpretation of the concept of gene.
But I do not agree that we can criticise his views by relating them to a supposed conservative ideology. We must analyze his arguments, and I have suggested how to criticize them.

ZIRNSTEIN: DARWIN wrote one year before his death, I remember, in a letter that also the problem of the origin of life will be solved by the naturalists. The "Origin of Life" was the point which DARWIN saw in an idealistic manner.

Institut für Anthropologie und Humangenetik,
Universität Heidelberg, Im Neuenheimer Feld 328, D-6900 Heidelberg

Effect of Cultural Evolution on Biological Evolution in Man

F. VOGEL

Hephaistos was the god of fire. He was a blacksmith, and his forge was located in the core of volcanos. HESIODOS describes him as the most skillful of the sons of heaven in all industries, — the teacher of mankind in skills such as processing metal, bronze casting and many handicrafts. He hammered, among others, the sceptre and shield of Zeus, Poseidon's trident and the armour of Achilles. He was born limping, and his mother Hera, being ashamed of this malformed child, threw him down into the ocean, where he was rescued by the daughters of Okeanos.

Here, Greek mythology establishes a connection between occupational specialization and the aquisition of specialized skills, on the one hand, and a physical impediment, on the other. This is not the only example in mythology for such a connection, and we may conclude with confidence that the thorough change in many aspects of daily life that was brought about by the advent of the neolithic period — the gradual replacement of the age-old hunter-gatherer way of life by a more sedentary life style with agricultural animal breeding — has also created ecological niches for disabled individuals who would have had but little chance for survival under the much harsher conditions of a hunter-gatherer society.

Standing on DARWIN's shoulders, we naturally look at this change in life patterns from the viewpoint of natural selection. Those of us who are geneticists will immediately ask for the ensuring genetic consequences. Which have been the influences of the changing life styles due to introduction of the neolithic ways of production — and, more generally, the process of cultural evolution as a whole —, on the genetic composition of human populations?

Naturally, direct evidence for such changes is scarce. We are left with the analysis of sceletal relics. They lend themselves for a small number of very general conclusions, — for example that some individuals were now achieving a somewhat higher age than before, — or that individuals with bodily impediments such as fractures, rheumatic disease or tuberculosis or even milder congential malformations had a better chance for survival. An interesting problem is posed by the phenomenon of gracilization: It has been observed that, during the millenia since the beginning of the neolithic period, the bones, especially of the skull, have become, on the average, less and less heavy and clumsy. There appears to be some evidence, on the other hand, for a correlation between gracilization of the skeleton and mental or intellctual refinement in the presentday human population. It has therefore been speculated that the observed skeletal changes might reflect corresponding behavioral alterations (SCHWIDETZKY 1962). This is a very interesting hypothesis, because it connects aspects of paleoanthropology with those of constitution biology and psychology. But direct evidence is difficult to find. We have to look for harder data. Such data have come mainly from comparisons of genetic parameters between present-day human populations.

Such a comparison makes use of the fact that the transition of agriculture did not occur at the same time all over the inhabited world but spread from a few centers such as the Near East or Egypt to other areas (see TACKENBERG 1953); this process needed several millenia and was not finished at all when, in recent centuries, conquest of the world by white nations together with the industrial revolution changed the life styles of mankind entirely. Therefore, in the present-day world population there are groups with a history of sedentary life and agriculture going back for many thousands of years side by side with others who have been living under neolithic and post-neolithic conditions for relatively short times. Changes in the genetic composition of populations due to relaxed selection need many generations to become visible. Hence, the former groups are expected to show genetic changes not present in the latter.

One example is the frequency of color vision defects. It is reasonable to assume that the hunter in jungle or savannah needed intact color vision in order to protect himself from snakes and leopards, while the same ability helped his gatherer-wife to spot wild berries and edible insects. For the agriculturist, on the other hand, color vision might have offered very little in terms of selective advantage. Consequently, one expects a much higher frequency of color vision defects in populations with an ancient history of agriculture as compared with those who have lived as hunter-gatherers until more recently.

Such differences have, indeed, been described (POST 1962; 1965; 1971). There are two X-linked gene loci, — protan and deutan, — for each of which exist, in addition to the normal allele, at least two mutant alleles, one causing a complete defect, and another one leading to a milder anomaly. All these alleles together have a gene frequency of about 7—8% in European populations, and the same phenotype incidence among males, because the gene is X-linked. An only slightly lower frequency has been observed among Chinese whose aquisition of agriculture dates back almost as long as ours (see EBERHARD 1953). Japanese frequencies are lower than those of the Chinese, — in accordance with the cultural lag between China and Japan in early times. On the other hand, gene frequencies among recent hunter-gatherers have been found to be much lower — about two percent on the average (Fig. 1). Populations somewhat removed from hunters in time or in habitat showed intermediary values. Considering the relatively short time of about 5000 to 6000 years (in Europe) since the onset of the neolithic age and the introduction of agriculture, the increase of the frequency of color vision anomalies is suprisingly high, suggesting either an unusually high mutation rate of up to $\approx 3\times 10^{-4}$, or a transient selective advantage of color defectives (POST 1971). Possibly, some of them, not acting very efficiently as hunters or warriors, were delegated for gardening or special handicrafts. This might have increased their chances for surviving, having contact with women and transmitting their genes to the next generations. Does the myth of Hephaistos, the limping blacksmith, tell us precisely this story?

Similar population comparisons, as for color vision defects, have been performed for visual acuity (Myopia and hyperopia, see Fig. 2), hearing acuity, deformed nasal septa, small tear ducts, breast cancer and spina bifida occulta, — always with the result that slight deficiences are more common among civilized than among primitive populations. While some of the data are open to criticism, the overall argument that civilized populations differ from primitive ones in the genetic variability of functional systems with

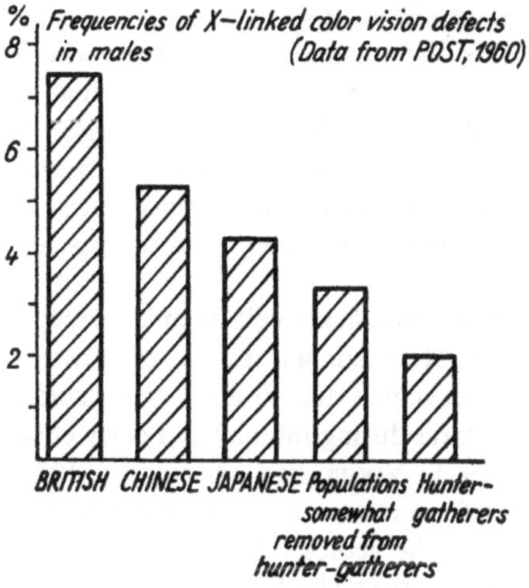

Fig. 1. Frequencies of X-linked color vision defects in males of various populations (Protanopia, protanomaly, deuteranopia, and deuteranomaly together) (Data from POST, 1960).

adaptive value, and that the differences are due to relaxed selection, can hardly be disputed.

These results remind us of the well-known observations on fish and salamanders who have lived for many generations in complete darkness: First, variability of their eyes increases; later, they are reduced more and more, until the species finally becomes completely eyeless. Obviously, maintenance of the genetic basis for a complex functional system such as the eye requires permanent selection against deviant mutations.

However, aquisition of the neolithic way of life has not only led to relaxation of selection — it has also created new selection pressures. For example, the emerging pattern of continuous settlement has created ideal conditions for spreading of infective agents. This does not mean that hunter-gatherer groups were entirely free of infectious diseases; recent studies on primitive Indians in Brazil (NEEL 1970) have shown surprisingly high blood levels for antibodies against certain *Salmonella* strains. But widespread epi-

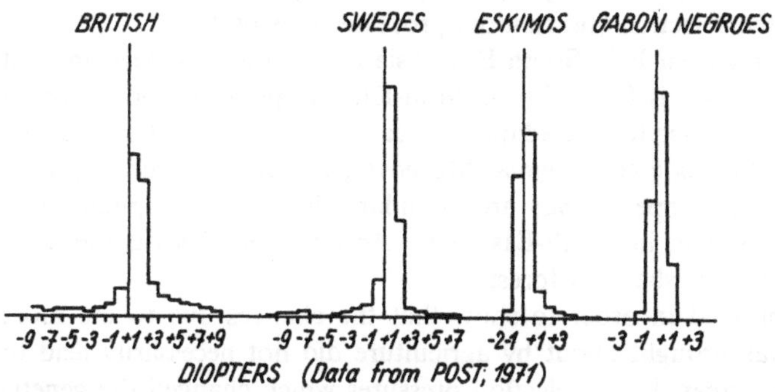

Fig. 2. Distribution of refraction anomalies (myopia and hyperopia) in four population groups (Data from POST, 1971).

demics require easy diffusion of infective germs. Considering the enormous mortalities from infections before and during reproductive age as documented for the 18th and even 19th century in Europe, (see for example SÜSSMILCH, 1786), it must be concluded that the present-day genetic composition of the human population has been changed largely by interaction with various infective agents in the few thousand years of human civilization. This problem has been analyzed in great detail for abnormal hemoglobins, especially sickle cell hemoglobin in Africa and hemoglobin E in Thailand, in their relationship with *Falciparum* malaria.

The sickle cell homozygote suffers from a severe disease, sickle cell anemia, which is almost invariably fatal under primitive living conditions. The ensuing loss of genes, however, is compensated for by a selective advantage of heterozygotes due to a lower morbidity and reduced death rate from tropical (*Falciparum*) malaria during infancy and early childhood. (For a detailed discussion see LIVINGSTONE, 1971; VOGEL and MOTULSKY, 1979). Hence, the increase in gene frequency of the sickle cell gene was caused by the spread of *Plasmodium falciparum*, and this, in turn, depended on the ecological conditions for the vector, the malaria mosquito, *Anopheles gambiae*. LIVINGSTONE (1958) has shown that the West African black populations could not exploit the tropical rain forest before they learned to work on iron, because only iron axes enabled them to clear the forest. Iron working was introduced in West Africa during the first centuries A. D. The mosquito, on the other hand, thrives best in ponds of stagnant water that are exposed to bright sunlight. Such ponds do not exist in the rain forest. Therefore, spreading of malaria and, as a consequence, increase of the sickle cell gene required clearing of the forest for agriculture, because only in such an environment, the mosquito could survive. Hence, the high frequency of sickle anemia — and of other hemoglobinopathies — is a fairly recent development, — only thousand years old at best. Up to this date, the sickle cell gene has its lowest frequency in some West African tribes — such as the pygmies — who have lived until recently as forest hunters.

The area in which tropical malaria was hyperendemic is not confined to the parts of the world in which sickling is common. Malaria also occurred in South East Asia, and here, other hemoglobin mutants have established themselves. Obviously, a number of different mutations could, in principle, meet the requirements of the selection pressure almost equally well. Those which happened to be present by chance at the time at which selection started could then increase in frequency. By the way, this principle of evolutionary "sufficiency" (ZUCKERKANDL 1976) is, in my opinion, very important for our understanding of interaction between selection and random processes in evolution.

The high frequency of hemoglobin E in South-East Asia (Fig. 3) is of special interest (FLATZ 1967). Its highest frequency is found in the Mon-Khmer speaking population of Cambodia. This hemoglobin is a much luckier adaption because the homozygotes are much less disabled than patients with sickle cell anemia. Moreover, its highest frequencies are observed not in the plains, where open ponds are abundant, but in the population of hills and forests. The reason is that, in South-East Asia, *Anopheles minimus* is the main vector, — and this mosquito prefers shadowy forests.

These examples were mentioned in order to show that the changes due to the more sedentary way of life that was brought about by agriculture did not necessarily lead to selection relaxation; they also created new selection pressures which changed the genetic composition and health status of present-day population thoroughly. We have good reason to assume that malaria selection is not the only such example. The distribution of ABO

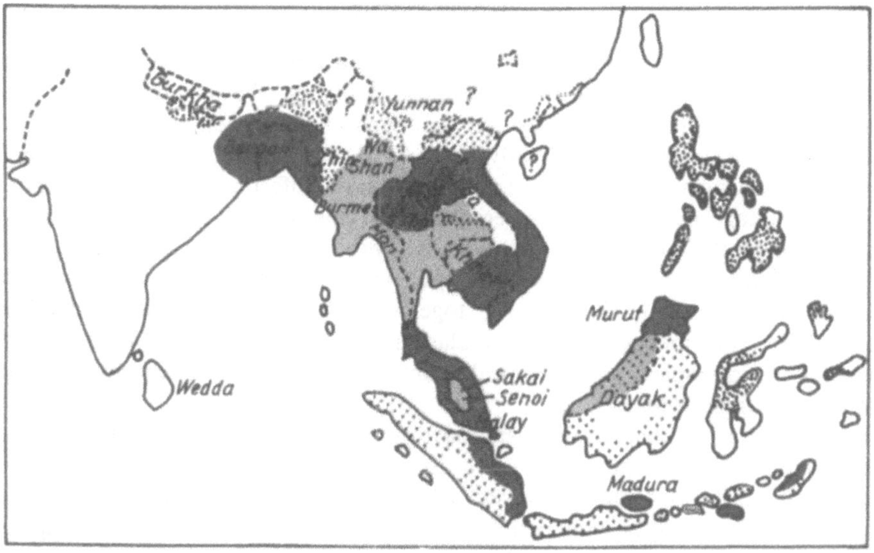

Fig. 3. Populations in which HbE has been observed (FLATZ, 1967). Shaded areas: High frequency; stippled areas: Occasional observations (from VOGEL and MOTULSKY, 1979).

blood group genes in the populations of the world is highly suggestive of an influence of the great epidemics, especially plague, smallpox and syphilis, — and there is some direct evidence for influences on the ABO blood group genes on susceptibility and outcome of infections (for discussion see VOGEL and MOTULSKY, 1979). It is reasonable to assume that also the extreme polymorphism of the Major Histocompatibility Complex, especially the HLA-System, has something to do with selection against infectious diseases. However, both polymorphisms, — ABO and HLA, — are important components of life of most of us.

Meanwhile, a second change in life style has occurred. It started only in the last century and is spreading right now almost into the entire world population. It has been brought about by what we may call "modern civilization". You know we have experienced — mainly since the end of the 18th century — a through change of production that has often been called the "Industrial revolution". It was accompanied by changes in the reproduction patterns (MACKENROTH 1953), which must also have genetic effects. Much more important genetically, however, is a side-line of this revolution and the growth of natural sciences concomitant with it: The development of modern medicine and hygiene.

As mentioned, only two hundred years ago, more than 50% of all newborns died up to the age of 10. At present, mortality during the first year of life has dropped to less than 2%, and mortality up to reproductive age is almost negligible — with the result of a steep increase of life expectation. This must alter the pattern of natural selection, — probably still more thoroughly than the advent of sedentary life and agriculture some thousand years ago.

The sickling-malaria example may serve to discuss some aspects of the influence of modern medicine on selection. Two achievements are important: The eradication of *Falciparum* malaria, on the one hand, and preventive measures for sickle cell anemia, such as population screening, genetic counseling and prenatal diagnosis, on the other. What are the genetic consequences of malaria eradication? In a non-malaria environment, the sickle cell heterozygote has no selective advantage. However, the disadvantage of the homozygote remains. With each homozygote, two such alleles disappear from the gene pool. Hence, gene frequency will gradually be *reduced*. The compromise which nature has undergone — in which the advantage in coping with the infection has been traded off against the disadvantage of having many severely disabled children with sickle cell anemia — is not necessary any more, and is resolved gradually. This is undoubtedly an advantageous effect of modern medicine. Until a few years ago, one could confidently hope that the sickle gene would disappear in due time, — together with other abnormal hemoglobin genes such as HbE, HbC and especially the thalassemias. Everybody who has ever visited medical institutions in tropical or subtropical countries can imagine the enormous relief this would mean to these countries.

More recently, however, other developments have occurred: Patients suffering from sickle cell anemia and other hemoglobinopathies can be treated fairly succesfully and some of them reproduce. Much more important, however, are the screening programs and prenatal diagnosis. Sickle cell screening is being performed expressedly with the goal of discouraging marriages between heterozygotes. They should rather marry normal homozygotes (see MOTULSKY 1975). If successful, such programs will necessarily slow down the process of gene elimination. This is, of course, an advantage for individual present-day families; but in the long run, it will *increase* the genetic burden for the population (VOGEL 1981a). This trend will be reinforced if the recent developments in prenatal diagnosis will lead to world-wide application. Sickle-cell anemia is the first hereditary disease that can be diagnosed directly at the gene level: It is closely linked with a polymorphic DNA site in a nontranscribing DNA sequence. The sickle cell mutation is always transmitted together with one of two DNA variants that can be recognized by digesting the DNA of the patient with certain restriction enzyme (KAN and DOZY 1978; see also DAVIES 1981). DNA is available in all cells including those from the amniotic fluid; hence, prenatal diagnosis is possible.

Families will now use this service, and homozygous fetuses will be aborted. They will be replaced, however, by healthy children, until the preplanned number of children is reached. Two thirds of these children will be heterozygous. This would actually lead to an *increase* of gene frequency — certainly a disadvantage for the population as a whole.

The hemoglobin diseases are the best-analyzed examples of genetic variants in humans from every viewpoint, — molecular genetics; clinics; and population genetics. Therefore, they are so wellsuited for demonstrating the difficulty in predicting genetic consequences of cultural evolution.

We observed a three-step development:

1 Sedentary life style with agriculture and iron working → Spread of tropical malaria → Establishment of a balanced polymorphism with heterozygote advantage.

2 Modern medicine → Eradication of malaria → No heterozygote advantage → *Decrease* of gene frequency.

3 Preventive measures in human genetics → Abortion and replacement of homozygote fetuses → *Increase* of gene frequency.

Keeping the complexity of the problem in mind, we may now turn to some more general questions:

1. Where else can we observe changes — especially in medical cane — that will probably influence natural selection; and
2. how do preventive measures — genetic counscling and prenatal diagnosis — modify the "natural" course of events? Both problems will be discussed together.

We have learned to treat an increasing number of genetic diseases more and more successfully. This has undoubtedly enhanced the changes of these patients to survive and to transmit their defective alleles to future generations. A classical example is X-linked hemophilia which used to kill many patients by bleeding whereas at present, substitution therapy with factor VIII helps them to lead a nearly normal life, — and to have children. This is a great achievement, but it would necessarily lead to an increase of this disease unless we replace relaxed selection by artificial selection. Knowledge of the mode of inheritance together with the techniques of prenatal diagnosis will encourage more and more families to prevent hemophilia genes from being transmitted. The same is true for many other genetic diseases: The "natural" trend toward increase is counteracted by the effects of genetic counseling. Counseling is carried out exclusively in the interest of the individual family; but its side effects are in most cases advantageous for the population as a whole (see VOGEL 1982). We may be confident that this development will continue; probably, prenatal diagnosis will, in due course, become a routine procedure in the care for all pregnant women. Then, occurrences such as DOWN's syndrome and other severe chromosome anomalies will become very rare, indeed. Already now, these breathtaking developments of our techniques are creating ethical rather than genetic problems: Where are the limits of deviations from normality that are considered tolerable? And how far do we have the moral right to interfere with the course of nature? Recently, somebody told me a story: Two Neandertal men sit in front of their cave, frying their mammoth steak. Suddenly a jet plane crosses the sky above. One of the two points up, saying: "They know, how". Asks the other one: "Do they also know, why?" —

But even, if we shall, indeed, develop our technical skills much further and shall therefore succeed in reducing the burden of genetic disease and malformation appreciably, — this will not save us at all from the problems created by selection relaxation due to progress in medicine. As mentioned, the most important single factor for selective adaptation in the past has been infective disease. Infections, however, have now become less and less dangerous, — not only because .of improved hygiene, — as in the malaria case, — but also, because we have learned to treat them successfully with antibiotics. The sickle cell-malaria case has shown that selection relaxation might even have a beneficial effect for the genetic composition of our population. This, however, does not mean that the entire *effect* of relaxed selection due to improved management of infections *must* be beneficial. There is one other aspect that has always concerned me: Evolution provided us with an intricately complex and highly efficient biological system for coping with foreign material, especially infections: *The immune system.* It consists of many components, — cellular and humoral. For all these components, genetic defects are known (see STANBURY et al., 1978). In former times, they have usually led to early death of their bearers. Still more important: Even milder anomalies would, in many cases, diminish the chances of survival appreciably; and it is a general experience in genetics that wherever defect mutants are known, milder ones with less extreme effects do also occur. Moreover, many of these defect mutants are recessive, and human geneticists are now learning that heterozygotes of recessive diseases will often show slight anomalies (for disc. see VOGEL and MOTULSKY, 1979, Sect. 4.2.2.; 8.2.3.). Relaxed selection due to infective agents might easily lead to gradual accumulation of such mutants and, in consequence, to a slow decomposition of our immune system, until, in the end, we have to be protected carefully from all germs — and, despite all this care, will die early from cancer. The situation of our progeny will then be the same as that of the present-day species of fish and salamanders who have lost their eyes because they have lived for millenia in lightless caves (see VOGEL 1973).

Our considerations on effects of cultural evolution to the biological evolution of the human species are now approaching the end. For demonstrating such effects, I confined myself to examples from medical genetics. This does not mean that I would regard possible genetic influences on genes determining variability in the normal range as unimportant. In the contrary, I believe that selection for genetic determinants of behavior might be influenced quite as much by the altered living conditions in our modern world. Too little is known, however, on the genetic mechanism underlying behavioral variability in humans to permit more than very vague speculations. Here, research is just beginning (VOGEL 1981).

Probably, our progeny will be much like the Greek god Hephaistos: They will be superbly skilled in many techniques, scientific, industrial and social. They certainly will know how. Let us hope that they will also know, why — better than we do at present. But many of them will, more or less, limp.

Summary

Two major steps in cultural evolution have influenced the pattern of natural selection, and hence, the genetic composition of human populations:

1 The development of a sedentary life style with agriculture and animal breeding in the in the neolithic period.
2 The industrial revolution together with the development of modern science and medicine in the last ≈ 200 years.

The first step led to selection relaxation for hereditary abilities such as color vision or visual acuity. On the other hand, selection due to infectious diseases was intensified. The second step — and especially modern medicine — led to selection relaxation mainly due to improved management of infections. This may have advantageous (hemoglobinopathies and tropical malaria) or disadvantageous (gradual loss of the immune system?) consequences for the human population.

References

[1] DAVIES, K. E.: The application of DNA recombinant technology to the analysis of the human genome and genetic disease. Hum. Genet. **58**, (1981), 351—157.

[2] EBERHARD, W.: Geschichte Chinas bis zum Ende der Han-Zeit. In: Historia Mundi, (F. VALJAVEC, Ed.) Vol. X. München: Leo Lehnen Verlag 1953.

[3] FLATZ, G.: Haemoglobin E: Distribution and Population dynamics. Hum. Genet. **3**, (1967), 189—234.

[4] KAN, Y. W., A. M. DOZY: Polymorphism of DNA sequence adjacent to the human β globin structural gene. Its relation to the sickle mutation. Proc. Natl. Acad. Sci. USA **75**, (1978), 5631—5635.

[5] LIVINGSTONE, F. B.: Anthropological implications of sickel cell gene distributions in West Africa. Am. Anthropologist **60**, (1958), 533—562.
[6] LIVINGSTONE, F. B.: Malaria and human polymorphisms. Annu. Rev. Genet. **5**, (1971), 33—64.
[7] MACKENROTH, G.: Bevölkerungslehre. Berlin—Göttingen—Heidelberg: Springer Verlag (1953).
[8] MOTULSKY, A. G.: Problems of screening for genetic disease. In: Early diagnosis and prevention of genetic disease. L. WENT, C. VERMEIJ-KEERS, A. G. J. M. Van der LINDEN (eds.), 132—140. Leiden: University of Leiden Press, (1975).
[9] NEEL, J. V.: Lessons from a "primitive" people. Science **170**, (1962), 815—822.
[10] POST, R. H.: Population differences in vision acuity. Eugen. Q. **9**, (1962), 189—212.
[11] POST, R. H.: Notes on relaxed selection in man. Anthropol. Anz. **29**, (1965), 186—195
[12] POST, R. H.: Possible cases of relaxed selection in civilized populations. Hum. Genet. **13**, (1971), 253—284.
[13] SCHWIDETZKY, A.: Das Grazilierungsproblem. Homo 13, (1962), 188—195.
[14] STANBURY, J. B., J. B. WYNGAARDEN and D. S. FREDERICKSON: The metabolic basis of inherited disease. 4th ed., New York, Toronto, London: McGraw Hill (1978).
[15] SÜSSMILCH: Die göttliche Ordnung. 9th ed. Teil II. Berlin (1786).
[16] TACKENBERG, K.: Die jüngere Steinzeit Europas. Das Neolithicum. Historia Mundi (F. VALJAVEC, ed.). Vol. **2**, München: Leo Lehnen Verlag (1953).
[17] VOGEL, F.: Der Fortschritt als Gefahr für die genetische Beschaffenheit des Menschen. Klin. Wochenschr. **51**, (1973), 575—585.
[18] VOGEL, F.: Populationsgenetische Folgen der genetischen Familienberatung und der vorgeburtlichen Diagnostik. Biol. Zentralblatt, **101**, (1982) 73—79.
[19] VOGEL, F.: Neurobiological Approaches in Human Behaviour Genetics. Behaviour Genetics 11, (1981), 87—102.
[20] VOGEL, F., MOTULSKY, A. G.: Human Genetics — Problems and Approaches. Berlin—Heidelberg—New York: Springer Verlag (1979).
[21] ZUCKERKANDL, E.: Evolutionary process and evolutionary noise at the molecular level. II. A lectional model for random fixations in proteins. J. Mol. Evol. 7, (1976), 269—311.

Discussion

GRANTHAM: Two points, the first merely humorous is regarding certain fish losing eyesight through living in dark places. Let us not forget that LAMARCK was ridiculized for using this example on the mole (or gopher).

Second, it seems to me difficult to document that selection is more relaxed now than in neolithic and and paleolithic times. But what we can be sure of is that the selection is of a different nature, perhaps more tuned to the future of our species. The artificial selection we are establishing may be more efficient than natural selection. In any case it will often oppose natural selection.

VOGEL: There are certainly good as well as bad tendencies, but as the whole, I am not visite as optimistic as Dr. GRANTHAM.

H. A. ROSENTHAL: You have given a clear picture of the advantages and disadvantages of high and low selection pressures and of the influence of modern medicine on the genetics of mankind. We have discussed these problems in our country too and came to the conclusion that genetic counselling will provide us with the most powerful means to avoid the disadvantage of lowering the natural selection pressure. What is your opinion? Do you believe that an increase of better and improved knowledge and methods genetic counselling eventually might solve these problems? When discussing the balance between exploioting nature by man and the progress of medicine (mankind has to pay for each progress in exploiting nature) one might take into account also the conclusions drawn by F. ENGELS 100 years ago, namely that mankind will not exist for ever. However, these times are be so far in the future that we should not regard them in this this context. Therefore, my question touches upon the real future.

VOGEL: I share H. ROSENTHAL's opinion that, in the immediate future, genetic counselling and possibly — in same rare diseases — genetic engineering techniques will certainly help single families. Long-range problems might be left for future generations of scientists.

SCHOLZ: Would you advocate genetic engineering to counteract a possible genetic deterioration of mankind ("positive eugenics")?

VOGEL: I agree that the *main* effect of genetic counseling will be advantageous. However, there are also some disadvantageous side effects (sickle cell anemia; Tay-Sachs-disease). Being interested in the well being of the individual family, we should promote genetic counselling including prenatal diagnosis with all our powers.

Zentralinstitut für Alte Geschichte und Archäologie der Akademie der Wissenschaften der DDR, Berlin

Evolution of Man and Development of Labour

J. Herrmann

Darwin saw the essential driving force of evolution in living beings coming to grips with nature. Coming to grips with nature, in other words the activity of living, puts living beings under the pressure of adaptation and natural selection through which the evolution of new species takes place. This basic idea of Darwin leads logically to the question: which specific features under the given circumstances determine this coming-to-grips with nature in the evolution of man. The diversified and changing nature of the conditions to which man has been exposed are responsible for considerable variability. Darwin notes: "It cannot be denied that changed conditions produce some, and occassionally a considerable effect on organisms of all kinds; and it seems at first probable that if sufficient time were allowed this would be the invariable result. But I have failed to obtain clear evidence in favour of this conclusion; and valid reasons may be urged on the other side, at least as far as the innumerable structures are concerned which are adapted for special ends. There can, however, be no doubt that changed conditions induce an almost indefinite amount of fluctuating variability, by which the whole organization is rendered in some degree plastic" [Darwin 1874, p. 29].

In this connection the increased use and disuse of parts (p. 31) operates as a law which is also valid for lower animals (p. 34): "Although man may not have been much modified during the latter stages of his existence through the increased or decreased use of parts, the facts now given show that his liability in this respect has not been lost; and we positively know that the same law holds good with the lower animals. Consequently we may infer that when at a remote epoch the progenitors of man were in a transitional state, and were changing from quadrupeds into bipeds, natural selection would probably have been greatly aided by the inherited effects of the increased or diminished use of the different parts of the body" [Darwin 1874, p. 34].

Through continual migration and distribution over several climatic zones, evolving man repeatedly exposed himself to changed living habits (p. 46). Multiplication at times went beyond the means for preservation. "The early progenitors of man ... must, therefore occassionally have been exposed to a struggle for existence, and consequently to the rigid law of natural selection" (p. 46).

These conditions take effect on the organism as a whole via correlated variation (p. 42).

These observations and ideas which Darwin had already expounded in his work "On the origin of species" (1859) were assessed by Karl Marx as follows:

„Darwin has interested us in the history of Nature's Technology, i.e. in the formation of the organs of plants and animals, which organs serve as instruments of production for sustaining life. Does not the history of the productive organs of man, of organs that are the material basis of all social organisation, deserve equal attention? ... Technology discloses man's mode of dealing with Nature, the process of production by which he sustains his life, and thereby also lays bare the mode of formation of his social relations..." [MARX 1958, p. 372].

In "The descent of man" (1871) DARWIN goes far beyond this interest in the "history of natural technology". He is concerned with describing the specific characteristics of the conditions of life and the activity of life in the evolution of man. He notes that "Man in the rudest state . . . is the most dominant animal that has ever appeared on this earth" (p. 46). He continues: "He manifestly owes this immense superiority to his intellectual faculties, to his social habits, which lead him to aid and defend his fellows, and to his corporeal structure. The supreme importance of these characters has been proved by the final arbitrament of the battle for life. Through his powers of intellect, articulate language has been evolved; and on this his wonderful advancement has mainly depended." (p. 47). DARWIN goes on to write: "He has invented and is able to use various weapons, tools, traps, &c., with which he defends himself, kills or catches prey, and otherwise obtains food. He has made rafts or canoes for fishing or crossing over to neighboring fertile islands. He has discovered the art of making fire, by which hard and stringy roots can be rendered digestible, and poisonous roots or herbs innocuous. This discovery of fire, probably the greatest ever made by man, excepting language, dates from before the dawn of history. These several inventions, by which man in the rudest state has become so preeminent, are the direct results of the development of his powers of observation, memory, curiosity, imagination, and reason" (p. 47).

This primaeval man practised a division of labour (p. 48). In these activities the hand of the ape became the human hand, and, DARWIN continues, "the vocal organs, which in the apes are used for uttering various signal-cries . . . have become adapted through the inherited effects of use for the utterance of articulate language" (p. 48).

According to DARWIN, then, the formation and development both of natural and of artificial organs characterize equally the specific nature of man's coming-to-grips with the natural conditions of life. In the case of man too it was via this coming-to-grips with nature that natural selection (p. 46) and the battle of life (p. 47) acted.

Thus DARWIN mentioned essential elements and peculiarities of the evolution of man, but did not see them as the expression or as a component of a new quality of evolution compared with the epochs of prehuman evolution.

Such elements as the making of tools, the development of language and the discovery of fire are seen as ranking equal with the development of organs in prehuman evolution, and are thus regarded as being subject to the same laws, such as natural selection and the battle of life. As a consequence of this evaluation of specific features in the framework of one and the same quality, DARWIN failed to grasp the correlation between specific human activity and specific features of human evolution. Thus he assesses activity as the result of an intelligence gained in the general evolutionary process [DARWIN 1874, p. 607]; the development of intelligence, that is of the brain, is not retarded as a result of the interactions of the specific human activity of life. The complicated interlocking network of interrelations resulting from arts and the production of artifacts in the bio-psychological and intellectual development of man thus remained inaccessible to DARWIN. On the other hand he attempted to discover the dialectical interrelation in the development of language and thought (p. 607). From today's point of view we can say that DARWIN's insight only got half way. In fact, of course, things were far more complicated. In political economy the effectiveness of man as a tool-maker had been identified since Benjamin FRANKLIN. In MARX and ENGELS' materialist conception of history and nature the specific nature of human existence and human development was similarly well identified by DARWIN's time. In 1846 Karl MARX and Friedrich ENGELS, in the "German Ideology", which was not printed until 1932, had already recognized the role of labour as the driving force and the essential feature in the evolution of man. In 1867 Karl MARX in his work „Das Kapital" wrote: "Labour is, in the first place, a process in which both man and Nature participate, and in which man of his own accord starts, regulates, and controls the material re-actions between own forces, arms and legs . . . By thus acting on the external world and changing it, he at the same time changes his own nature." [MARX 1958, p. 177]. While MARX and ENGELS of course knew DARWIN's work, the converse was by no means the case[1].

In DARWIN's time, archaeological research had begun to discover facts and describe connections, such as Neanderthal man and the stone implements of Abbeville. DARWIN assessed these finds from the point of view of evolution. Unlike Rudolf VIRCHOW, he classified Neanderthal man within human evolution [DARWIN 1874, p. 53].

However, there were too few historical-archaeological facts available to be able to base a theory of labour on them. Thus DARWIN, on the border between natural history and social history, was unable to sharpen his theoretical insights to any significant new degree. Instead, he attempted to rediscover the workings of the evolutionary laws of nature in social life too, via the stages of savagery and barbarism right up to the capitalist competitive society of his time [ENGELS 1962, p. 565].

Thus a peculiar contradiction arises in his findings: while citing that, in addition to intellectual abilities and physical development, man dominates also ". . . by his social qualities which lead him to give and receive help from his fellows" [DARWIN 1874, p. 61] as an important principle of human evolution he sees the basis of society in the "struggle for existence" between individuals. And he assumes: "if he is to advance still higher, it is to be feared that he must remain subject to a severe struggle" (p. 612). He immediately weakens this statement by pointing to moral characteristics of man which developed less out of natural selection than from other causes.

The complicated interrelation between the activity of life and the biological development of man in the course of human evolution has not yet been fully explained even today, a hundred years after DARWIN.

The coming-to-grips with nature or the metabolistic relation to nature during the history of human evolution took on the character of labour. As we have seen, DARWIN failed to recognize this quality and consequently had no term for it. Even today views on this question are not unanimous; religious and ideological reservations or social-DARWINist interpretations avoid the elaboration of the peculiarities of man's acquisition of nature or polemicize against the description of such a quality. Often such polemics are based on an incomplete conception of what human labour represents. Christian VOGEL, for

1 In 1873, Karl MARX dedicated DARWIN a copy of the second edition of "Das Kapital". DARWIN answered with a kind letter to MARX on October 1th 1873. The letter has been repeatedly reprinted — see for instance R. REIPISCH 1969, S. 105; S. KIRSCHKE, in this volume. It seems, that the authenticity of the letter cannot be seriously questioned, as Lewis S. FEUER tries (Is the 'DARWIN-MARX-Correspondence' authentic? In: Annals of Science 32-1, 1975, p. 1—12).

example, has very emphatically expressed the view that the MARXist conception of the role of labour is today no longer valid. Permit me, though, to point out an error in his argument. VOGEL supposed that MARX and ENGELS — and Marxism — ascribe the emergence of man to "technology in hominid evolution" [VOGEL 1975, 1977, 248 pp.]. This is of course not true. Labour is more and includes considerably more. Labour is conscious, directed activity by man, in which, with the aid of artificial means and methods, he changes the object of labour according to a preconceived purpose in productive acquisition. This process does not take place in isolation, but in society. The process of labour is thus a social process. It is clear — if we are talking about labour in a Marxist sense — that language and thought are at the same time included as a precondition and as a component of the process of labour.

The problem is not that of coming to an understanding on the term "labour" or "activity of life". If, like DARWIN, we regard coming to grips with nature as the basis of evolution, we cannot avoid examining and describing the specific form of this activity in human evolution. The problem is, rather, to discover the development of this activity, labour, to examine the stages by which labour became the basis of human existence and social development. DARWIN saw this question only vaguely and touched on individual aspects in the framework of general evolutionary concepts of the edevelopment of mental, social and moral faculties (DARWIN 1874, p. 57) and the inventions which resulted from them.

MARX attempted a philosophical-logical reconstruction of the development of labour, distinguishing three stages:

— ". . . those primitive instinctive forms of labour . . ."
— ". . . in which human labour was still in its first instinctive stage"
— ". . . labour in a form that stamps it as a exclusively human . . .

At the end of every labour-process, we get a result that already existed in the imagination of the labourer at its commencement." [MARX 1958, p. 178]
It was not possible at the time to relate these stages to concrete historical facts and events. The first attempts which MARX nevertheless made to do so have since been superseded through the discovery of new sources. On the other hand the basic idea expressed for the first time by MARX, that labour has undergone a development, the stages of which can be clearly defined, is of topical importance, though it still remains to be done. In connection with the evolution of man we often find a very static conception of labour. Even ENGELS in "The part played by labour in the transition from ape to man" did not go into MARX's proposal or its consequences in any great detail. He merely wrote: "With each generation labour itself became different, more perfect, more diversified. Agriculture was added to hunting and cattle raising" [ENGELS 1876/1953, p. 20].

Owing to the large number of facts and questions resulting among other things from interdisciplinary cooperation, the question of the development of labour can no longer be avoided. The basic facts are known in outline. For 1.8 or 2 million years implements have been made. The oldest of these are found in encampments and work places, in other words in places where hunted game was divided up. The biological evolution of the beings involved can also be described in outline. Yet there are considerable gaps owing to the sometimes meagre and variously interpretable fossil finds. Likewise the great changes in the conditions of life, the environment, can be recognized today. Added to this there are the observations of ethologists on higher primates, discoveries by psychologists on the development of thought, and so on.

This state of our knowledge enables us to discuss the difficult problem of the development of labour.

Although implements and tools are the expression and the means of labour, they nevertheless reflect the course of the development of labour itself. In the history of tools and technology it is possible to judge the effectiveness of labour. If these various traces of activity are related to the knowledge available from the analysis of fossil finds concerning biological development, in particular that of the brain and the hand, then we have a good chance of being able to trace the development of labour, at least in outline, and to describe some basic elements and patterns of activity.

The following stages can be recognized:

Pre-labour stage (fig. 1)

Objects given in nature are occasionally used by recent anthropoid apes. In the period prior to man we can conclude from the "biological division of labour" between foot and hand that the use of such objects from nature became normal behaviour. The signs of right-handedness in the approximately 3-million-year-old Australopithecus skeleton AL 288-1 ("Lucy") from the Hadar region in Ethiopia could hardly be explained otherwise. If we leave out of account the very uncertain find of Fort Ternan in Kenya, there are no finds of such natural implements. Thus we can conclude that there was a pre-labour stage which was characterized by a simple activity on one level of time and in one visual field. This stage developed and lasted some millions of years.

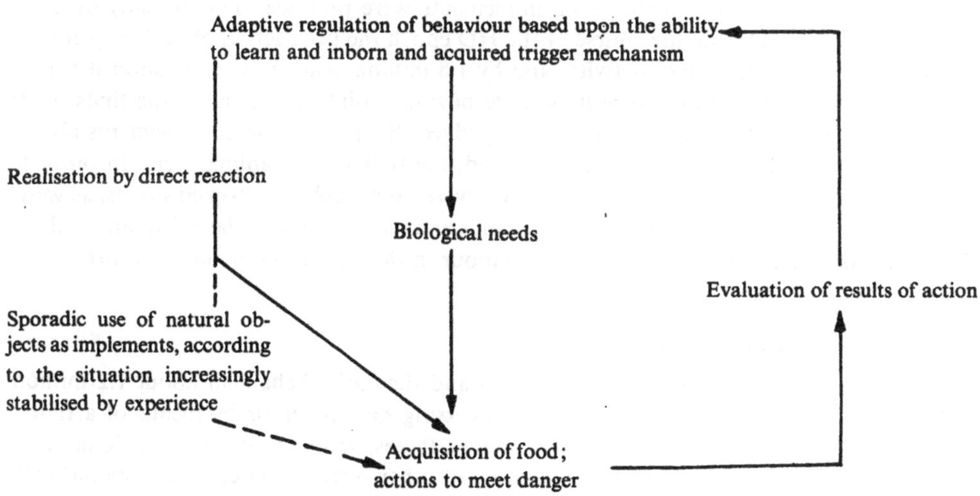

Fig. 1. Diagram: Pre-stage of labour (Author).

First stage of labour (fig. 2)

In the period of homo-deviation (HD epoch) about 2 million years ago, the use of implements reached a new quality. There was production of artifacts by breaking up and preparing things existing in nature. This was the way in which the oldest rude artifacts of the Oldovan type emerged. The artifacts required little time and effort, and after serving their

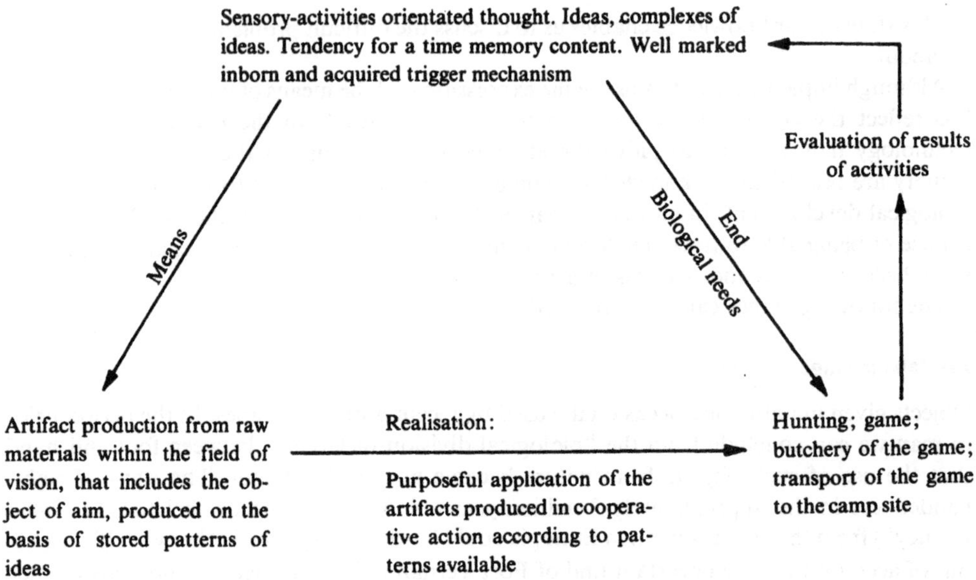

Fig. 2. Diagram: First stage of labour in the epoch of *Homo*-deviation (Author).

purpose they were left at the site [LEAKEY 1971; CLARC and KURASHINA 1979]. In many cases it is difficult to tell whether such implements were produced intentionally or as an accident of nature. During the course of the HD epoch the production of such implements became the normal pattern of activity, and by 1.8 million years ago at the latest it can be assessed as the expression of something quite new in evolution. As far as the finds made and the connections recognized so far can be judged, the production of implements always had the purpose of directly appropriating food, the butchery of animals, etc. In order to make implements, unprocessed natural objects were used, such as fist-sized stones, as were, in the case of wooden implements, the bare hands and teeth. This stage lasted about 1 million years. It was replaced by a second stage of labour in the transition to *homo erectus*.

Second stage of labour (fig. 3)

This stage began with the developed Oldovan and the early Acheulian about 1.2 million years ago. Artifacts began to be made by processing raw materials by means of artifacts already made. Suitable stone implements, bifaces, flakes, scrapers, etc. were made in order to fashion stones, bones, horns, tusks, wood and other organic material into special artifacts, such as spears, bone awls, bone gouges, digging implements or hide containers for transport. At the same time, of course, the habit of producing short-lived implements for the purpose of acquiring food continued. On the other hand, what was essentially a new form of behaviour was the production of artifacts in order to produce other artifacts or what, in my view, characterizes more clearly this new and qualitatively higher plateau: the production of tools in order to produce artifacts and tools. Thus a new link was added to the purpose-means-ends chain. The purpose and the end were divided by a longer, more abstract and more general activity.

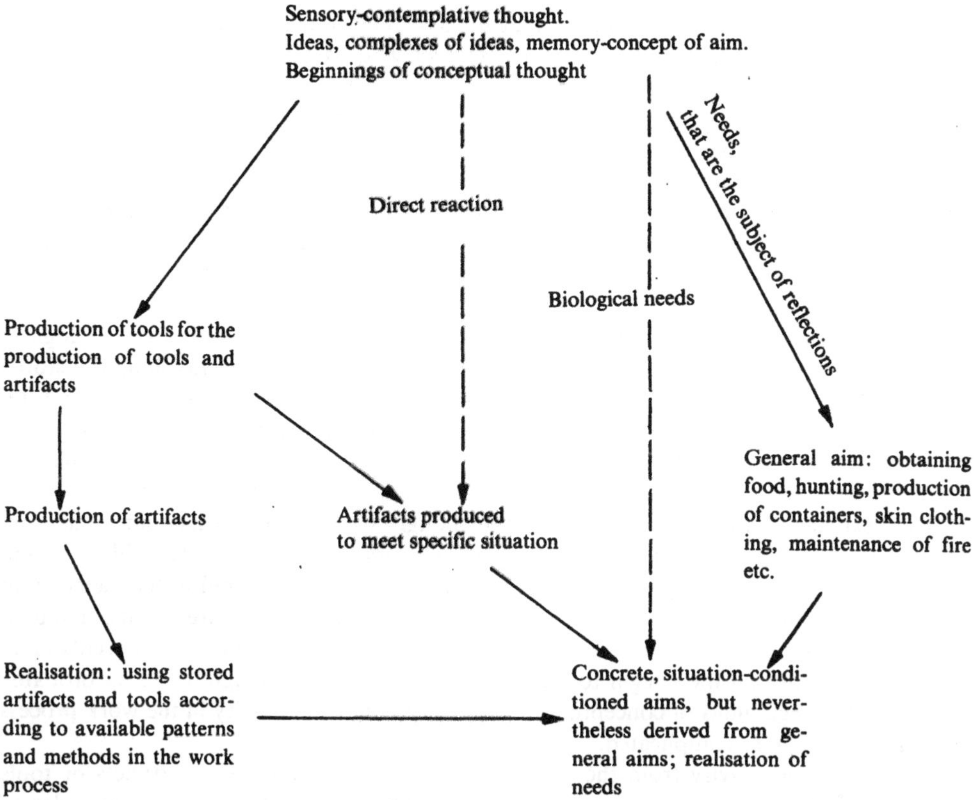

Fig. 3. Diagram: Second stage of labour in the epoch of anthroposociogenesis (Author).

The distinction between the individual stages become clear as soon as the labour process as a whole is taken into account. In the first stage of direct, purposeful artifact production, purpose-means-end lay on one time level and in one field of vision. The end arose from such unreflected needs as hunger, thirst, elimination of danger and other stress situations. As a means of achieving this end that array of artifacts was chiefly used which actually lay in the field of vision; experiences stored in the memory, the results of learning from similar situations, but also inborn and acquired trigger mechanism, i.e. instinctive reactions and conditioned reflexes, were activated. Game is to be hunted and cut up. To this end raw material directly available from nature is collected and fashioned into artifacts. Sensory and activity-oriented thinking going back to ideas and complexes of ideas, the beginnings of which can also be observed in the higher primates, guided both the production of artifacts and the use of these artifacts for a purpose already envisaged when they were produced: hunting, obtaining meat and sharing the prey. Certain patterns of behaviour and execution must have formed which guaranteed a quick and coordinated course of activity. These reaction patterns were far above the conditioned reflexes; they included the latter, but were on the whole determined and controlled by sensory and active thought (Darwin did in fact point out the decline of the instincts and reflexes).

To return to the hunting example: a pattern existed according to which large game was driven by a group of men, by means of suitable deployment, movements and cries, into a bog hole, where it was restricted in its movements and they were able to kill it. The general procedure had been tried and tested for ages and was, as a rule, successful. It was the hunting pattern. To put the pattern into practice however, required not only the reaction of each individual involved but the coordinated reaction of the hunters as a group. The production of implements was similar, whether they were to kill the animal or to carve it up. The type of artifacts, whether heavy, hand-axe-like implements for striking and breaking open, or smaller but sharper splinters or edges etc., were present as an idea and their production was present as a pattern of activity in the memory. However, variations corresponding to the specific need and the material available had to be "worked out" in thought.

The work process that took place in this way thus emerged essentially from biological, unreflected needs, and to put it into effect in a given situation meant invoking the stored memory and stimulating the patterned reaction. There is no indication at all for any foresighted adaptation to future actions beyond immediate needs. The level of activity and thought of the first stage of the labour process was thus two-dimensional. It was stimulated by the present and directed to the present by using stored experiences and patterns.

The next stage required a higher degree of thought capacity and behaviour regulation. The level of reaction is no longer confined to one time plateau and one field of vision. Rather, it is based on a large range of ideas and idea complexes stored in the memory, of stored behavioural experiences, and extrapolates from these to future events. Thought capacity and consciousness become three-dimensional and begin to grasp past events more clearly, and above all they begin to grasp future ones. Such a thought capacity probably antedates the beginnings of conceptual thought. The course of activity in the work process was thus much more complicated.

The models of activity from the point of view of the production of artifacts or tools make the qualitative changes clear that began with the production of tools in order to produce artifacts and tools. To grasp the work process in several stages was a condition and a consequence of this relation. In terms of political economy we are encountering the first beginnings of Department I of production: the production of tools and the three-dimensionality in which the elements of the work process interlocked with each other came the emergence of firmer, and to a large degree normative, ideas on the shape of tools and artifacts, their proportions and their processing. Certain work traditions emerged, the surfaces of tools and artifacts were shaped, sometimes regularly, without serving the actual purpose, but simply to make them more pleasant to handle. There have been various attempts to understand this process quantitatively, for example by finding the number of sometimes very distinct strokes uses to sharpe such artifacts. Qualitatively, as a consequence of the growing role of traditional technical processing that had no linear relationship to the function of artifacts on account of varying geographical milieu, there emerged different "archaeological cultures".

Corresponding to the three-dimensional character of the course of activity in the work process another essential factor was added: at this stage fire was used and its use cultivated. Thus for the first time a "force of nature" was directly involved in the work process and in man's way of life.

These are the statements we can make so far on the basis of archeological finds and findings directly concerning the general history of the first stages of labour.

The development of labour in the first stage determined the way of life and the biological development of man in the homodeviation field and in the transition to the epoch of

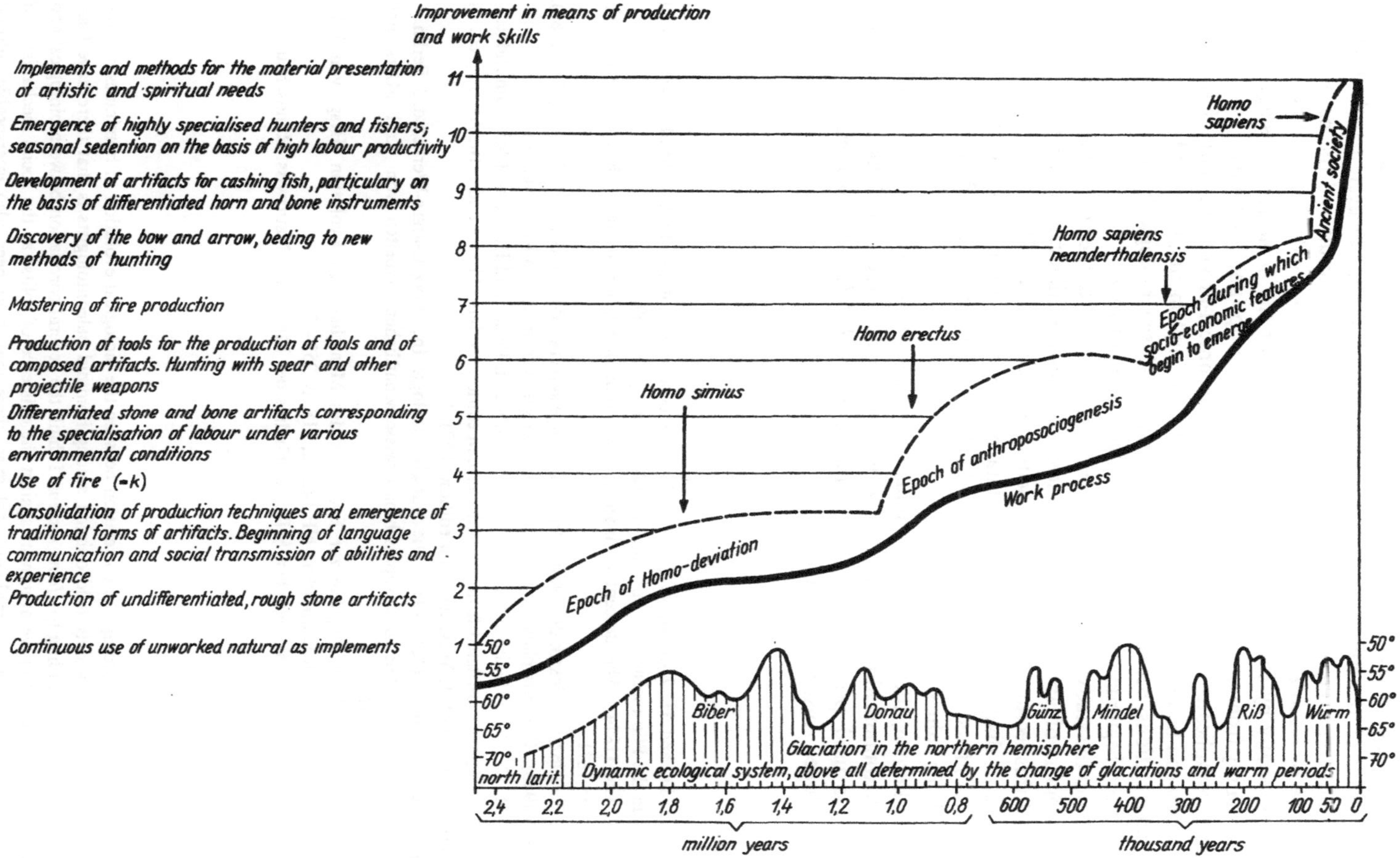

Fig. 4. The growth of productive abilities and the main epochs of human evolution (Author).

anthroposociogenesis. The activities became consolidated, more highly developed and richer. The new quality of the model described in the second stage, however, hardly became distinctive until some 0.6 to 0.8 million years ago. It was the result and the basis for the full emergence of *homo erectus* or primaeval man. While the model for the first stage is supported especially by findings in Africa, the model for the *homo erectus* stage can be recognized mainly from archaeological excavations in Europe and East Asia.

Far more complicated than the changes in the technical-productive part of the work process, which can be reconstructed thanks to finds of implements, tools and work places, are the interrelations between this technical part and the biological and psychosocial changes and the beginnings of social forms of communal living.

Research and the study of material in the past years has led to the entry of essential steps in the development of labour and of fossil finds in a time-space diagram which will be shown at the end of this lecture and on which discussion is invited (fig. 4).

It is a growth diagram. The essential data on the work process are in the nature of estimated values fixed within the points 1—11. The values for the growth of the brain are the direct volume values.

As is well known, there is no unanimous agreement on the naming of the individual *homo* types taxonomically. For the *homo* deviation epoch, therefore, the neutral term "ape man" is used provisionally, which means *homo habilis* or *homo modjokertensis* or other developed Australopithecines.

References

[1] CLARK, J. D. and H. KURASHINA (1979): Hominid occupation of the East Central Highlands of Ethiopia in the Plio-Pleistocene. In: Nature 282, 1. Nov. p. 33—39.

[2] DARWIN, Ch. (1874): The descent of man and selection in relation to sex. Chicago, New York. First edition 1871.

[3] ENGELS, F. (1962): Dialektik der Natur. In: Marx-Engels-Werke Bd. 20, Berlin. (Written 1873—1886).

[4] — (1876/1953): The part played by labour in the transition from ape to Man. Moscow.

[5] LEAKEY, M. D. (1971): Olduvai Gorge, Vol. 3 Excavation in Beds I and II, 1960—1963. Cambridge.

[6] MARX, K.: Capital, translated from the 3rd German edition by Sam. Moore and Edw. Aveling, ed. by Frederik ENGELS, Vol. I, Moscow (First edition 1867).

[7] MARX, K. and F. ENGELS (1846/1958): Die deutsche Ideologie. In: MARX-ENGELS-Werke Bd. 3, Berlin.

[8] REIPISCH, K. (1969): Die philosophisch-naturwissenschaftlichen Arbeiten von Karl MARX und Friedrich ENGELS. Berlin.

[9] VOGEL, Ch. (1975): Neue Aspekte zur Evolution des Menschen. In: Evolution. Hrsg. von J.-H. SCHARF, Halle (Nova Acta Leopoldina NF Bd. 42, 218, p. 253—269.

[10] VOGEL, Ch. (1977): Zum biologischen Selbstverständnis des Menschen. Naturwiss. Rundschau 30-7, p. 248.

Discussion

HAILA: I have a question regarding the correlation between the biological evolution of hominides and the evolution of labour. A very widely held opinion among hominid paleontologists is that there have been several splits in the lineage leading to *Homo sapiens*. I think the general correlation between biological and cultural evolution is not a decisive argument against the existence of splits; as the social mechanisms of inheritance of capacity of labour gradually tool over, there may have been also interspecific cultural exchange among the ancestors of modern man, e.g. between *Homo neanderthalensis* (if it was an independent species), and his contemporaries who where the ancestors of *H. sapiens*.

HERRMANN: It was not possible to touch upon the problem of splits in the development of mankind. Generally I agree with you. There were, in my understanding, direct relations between the development of the work process and the integration and disintegration of splits.

GRANTHAM: One of the fascinating aspects of human evolution is the rapid replacement of *Homo sapiens neanderthalensis* by *Homo sapiens sapiens* (HSS). The oldest specimens of HSS are only about 40,000 years old. So we may wonder where HSS came from. Your work suggests an African origin for modern man. However a very recent study on the distribution of 6 morphes of mitochondria in 235 individuals from 5 ethnic groups is compatible with an Asian origin. Please comment.

HERRMANN: At present fossils and artefacts from this period of mankind are only known from Africa. It may be, of course, that there will in future also be some record from East Asia showing that development to a species like Australopithecus also took place there. Nevertheless I myself assume that there was only one region in which the species *Homo* came out, spreading from there and, of course, reaching Asia already at a very early time (about 1,5 million years ago). These interrelations may be responsible for the phenomena you mentioned.

ROSENTHAL: Do fossils exist which permit the brain volume of human neonatals to be measured? They would permit estimation of the influence of postnatal conditions on brain development.

HERRMANN: There is no fossil or archaeological evidence. The question is touched upon for the period of Australopithecus on a general level by v. KOENIGSWALD. He pointed out that there must have been an important prolongation of the infancy period in which the brain developed and education to social behaviour became possible.

FÖRSTER: I want to make three short remarks on a philosophical problem of anthropogenesis: 1. In the past days we repeatedly touched upon and discussed the problem of the relationship between continuity and discontinuity, quantitative and qualitative changes and qualitative jumps in biotic evolution. This problem was treated comprehensively with regard to various evolutionary processes at the 5th Kühlungsborn Colloquium. 2. I want to stress that from the standpoint of Marxist philosophy every evolution is a materialistic determinated dialectical unity of continuity and discontinuity, quantitative and qualitative change. This conception of evolution is contradictory to the metaphysical overemphasis of either continuity or discontinuity in evolutionary process. Only against this background can one understand the process of the qualitative jump in evolution as well. It is necessary to take into account that the qualitative jump surpasses and includes quantitative changes. Otherwise the qualitative jump would be mystifying. The overemphasis of the quantitative aspect of evolution leads to a mechanical materialistic position, i.e. evolutionism. 3. Even if these philosophical theoretical reflections are necessary prerequisites to understanding all concrete evolutionary processes, they are no substitute for specific research into chemical, pre-biotic and biotic evolution, for instance. In this sense Anthropogenesis was a unity of quantitative and qualitative change as well. On the whole it was a qualitative jump from the biotic motion form of matter to the social one. It embraced the liquidation, preservation and elevation of essential characteristics of the lower motion form of matter into the higher one. This dialectical negation involved morpho-physiological, mental and behavioural properties. However, one has to take into account that evolution of *Homo sapiens sapiens* and his society did not consist of only a single qualitative jump but included several qualitative jumps of different classes, importance, etc. The deeper understanding of them requires further studies in natural and social sciences, including philosophy, and the synthesis of their results.

Museum für Naturkunde der Humboldt-Universität zu Berlin

Some Methodological Remarks on Current Studies in Phylogeny of Hominids

G. PETERS

Interest in a better understanding of our own biological past has led — among many other results in anthropology — to the proposal of numerous hypotheses on the phylogeny of hominids and of mankind in particular. Some of them have been presented as testable theories in popular journals and even in textbooks. Every palaeoanthropologist knows that such presentations are far from the actual truth. I think that the very different opinions published in scientific papers on this subject are sufficient evidence of the still highly hypothetical nature of all current phylogenetic models of the genus *Homo* and its nearest relatives. Quite recently CRONIN, BOAZ, STRINGER and RAK (1981) have reviewed the most widely propagated models. Ongoing skilled examination of the fossils already excavated and studies of a new material in the future will not only lead to further hypotheses but also enhance or detract from the probable validity of existing models (fig. 1).

I shall not try to propose a new scheme for the phylogeny of man and related taxa nor to review the schemes now at hand. A justifiable phylogram, or so called "dendrogram", of the hominids which is free of the methodological defects of present sketches so that the probability of its correctness would be greater would depend on a better quality of fossils and their palaeoanthropological description and analysis. From the position of a phylogenist I would like to make some more general remarks, which should not be ignored in future discussions.

Whenever new species come into being by the splitting of existing ones — and this is the common way of speciation in vertebrates and most other animals — evolution becomes phylogeny. The phylogeny of man or — more correctly — the phylogeny of the genus Homo in the current version of its conception, poses two essential questions; the first is the question of the initial or founder species of the genus, and the second is that of the number of species which existed up to the recent *Homo sapiens*. Phylogenetic systematics poses these questions, but it should be emphasized that they will never be completely answered by studying fossil remnants.

The definition of the biological species concept states clearly that the problem of whether or not two or more morphologically similar forms are distinct species can be solved only by examination of living taxa. But palaeontologists are at a disadvantage: at best they deal with more or less distinct "morpho-" or "palaeospecies".

Concurrent and sympatric occurrence of similar but morphologically distinguishable fossil forms can increase the probability that the fossils being studied are remnants of distinct species. As an example I can cite the possible coexistence of *Homo habilis* and/or *Homo erectus* respectively, *Australopithecus robustus* and *Australopithecus africanus* in some parts of the Koobi Fora Formation (WALKER and R. E. F. LEAKEY 1978).

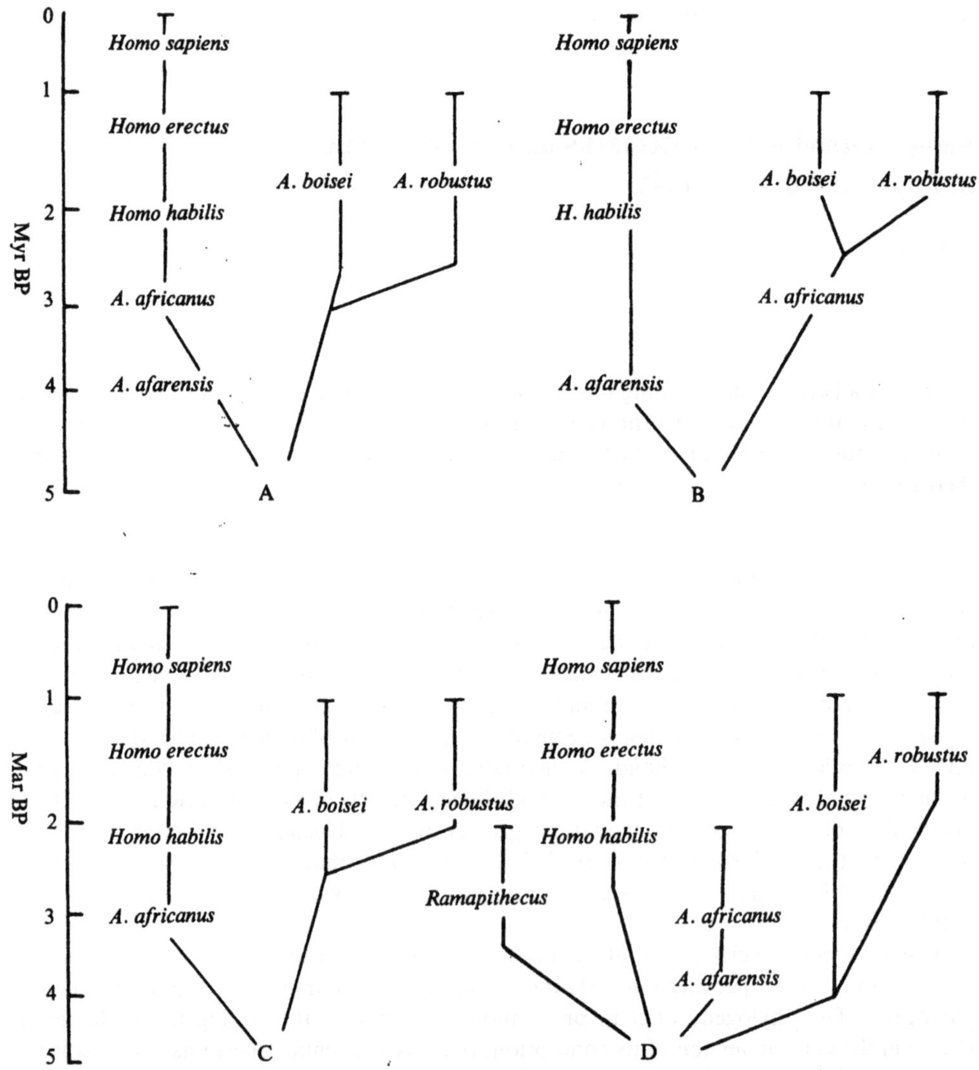

Fig. 1. Some typological concepts of hominid phylogeny (after CRONIN et al. 1981). The sketches A, B and C are monophyletic hypotheses, D shows a polyphyletic one favoured by R. E. F. LEAKEY.

As is well known to all specialists, diagnostic difficulties in fossils result from the scarcity of specimens, their fragmentary appearence, usually poor state of preservation and — last but not least — the inherent uncertainties of stratigraphic dating. Problems in the estimation of the traits exaggerated by inadequate premises for weighting their variability have to be regarded as an expression of the very nature of fossils (BOY 1981).

If palaeoanthropologists cautiously were to agree that there are strong indications for *Homo*-like fossils much older than the hominid ones of Koobi Fora (HENKE 1981), there would be some reason to conclude that in Koobi Fora remnants of reproductively isolated species have been found, at least from one species of each genus. But the palaeoanthropological discussions on the number and distinctness of upright walking hominid species that lived in the surroundings of Koobi Fora some two to one and a half million years ago continue and have resulted in thorough investigation of the material so far available. The results, including the discovery of new and better preserved fossils, may produce more likely hypotheses. But, nonetheless, all possible conclusions will remain hypotheses!

Cronin and coauthors (op. cit.) recently once more demonstrated the postulates of population genetics when they pointed out the probability, "that a few bottlenecks in effective breeding size, some resulting in speciation, without major effects on the phenotype may have occurred along the hominid lineage." The fission of a given species leads to the origin of two new species. The parental one becomes extinct by incorporation into its daughter species. This type of species origin is usually not accompanied by morphological changes of diagnostic value to palaeontologists. Visible structural transformations usually gradual in appearance, occur later in the lifetime of the new species, and often in only one of the "daughter lines". This is what was expressed by Hennig in 1950 in the rule of deviation. The misunderstanding of this rule is frequently expressed in the common figure of speech that "species b split off from species a" or, for instance, that the lineage of gracile australopithecines gave rise subsequently to the lineages *Homo* and robust australopithecines (fig. 2).

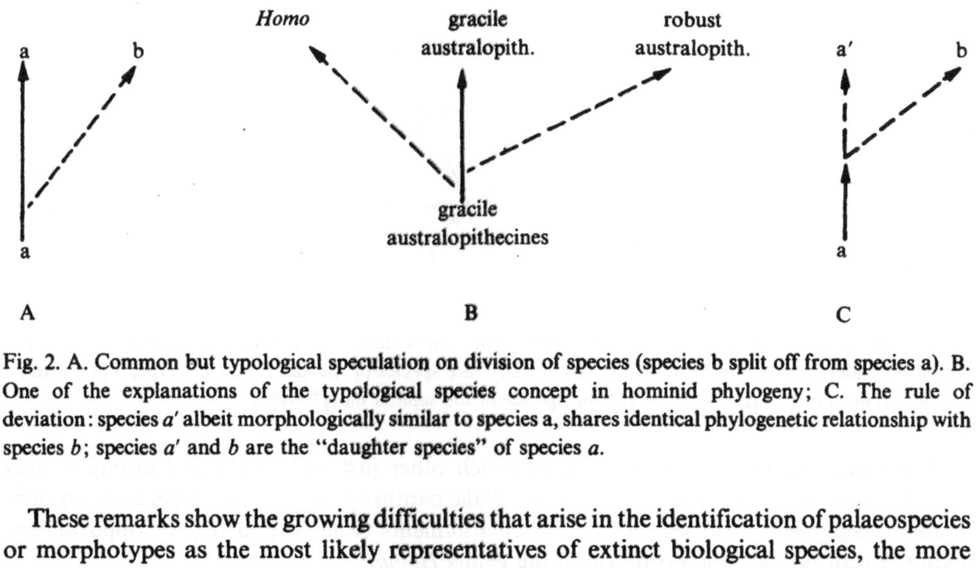

Fig. 2. A. Common but typological speculation on division of species (species b split off from species a). B. One of the explanations of the typological species concept in hominid phylogeny; C. The rule of deviation: species *a'* albeit morphologically similar to species a, shares identical phylogenetic relationship with species *b*; species *a'* and *b* are the "daughter species" of species *a*.

These remarks show the growing difficulties that arise in the identification of palaeospecies or morphotypes as the most likely representatives of extinct biological species, the more one reaches down to the roots of monophyletic groups. As one goes back in time the morphological traits of the genus *Homo* tend to become fused with the character set of ancestral australopithecines. Diagnostic evidence of the earliest ancestral remnants of man will always be dated later than the origin of the true initial *Homo* species.

The question of whether or not the essential characters of mankind are applicable to the earliest *Homo*-like creatures does not belong here because it is not a subject of phylogenetic consideration (Peters. 1982).

The limitations of palaeontological diagnosis mentioned above lead to yet another somewhat surprising conclusion: at any time in the late Pliocene, early and even mid-Pleistocene the actual number of existent hominid species will always be different from the number proposed by the results of palaeomorphological investigations. Quite a few anthropologists believe in the idea that the genus *Homo* at any time consisted of only one species: *Homo habilis* was followed by *Homo erectus* and the latter by *Homo sapiens*. According to this peculiar picture, to which, as we know, there has been frequent opposition, there have occured — apart from all other important features of development — only morphological transformations from the *habilis* morphotype to the *erectus* type and then to the *sapiens* type. Thus, it describes the morphological evolution of man by the typological fixation of three morphospecies in the *Homo* lineage. If this idea were in agreement with reality one would have to conclude that the progressive evolution of mankind took place without phylogenetic events. In other words: the genus *Homo* has undergone anagenetic evolution but not phylogeny (fig. 3).

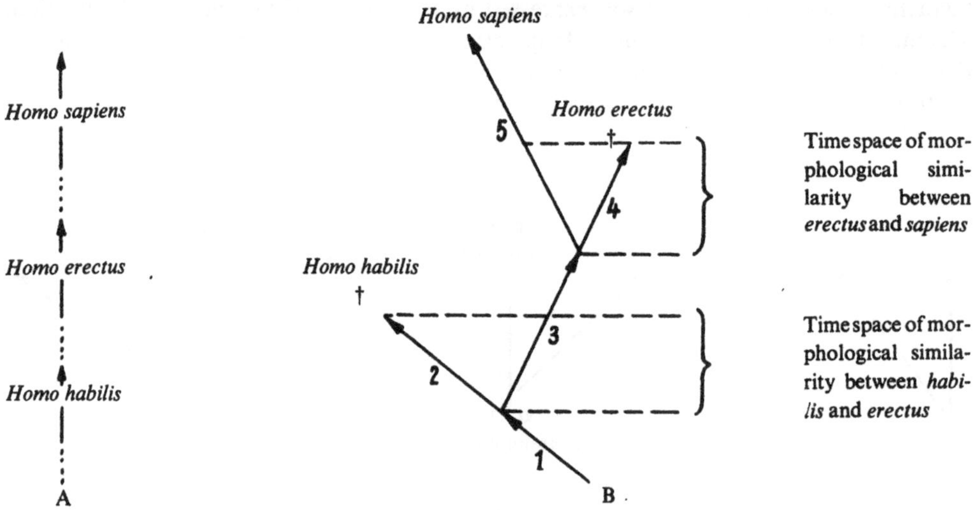

Fig. 3. The common hypothesis on the phylogeny of the genus *Homo* (A) and its correct phylogenetic interpretation under supposition of its reality (**B**); 1 and 2 are species of the *habilis* morphotype, 3 and 4 those of the *erectus* morphotype and 5 means the species of the *sapiens* morphotype.

This would be an unusual process, of which other just as hypothetical examples may exist in other sexually reproducing species. Palaeoanthropologists are, depending on their opinion, occupied in the search for further arguments either supporting or opposing the hypothesis of nonphyletic evolution in the genus *Homo*.

From the phylogenetic point of view only a small but important methodological detail can be contributed to the controversy between the specialists: the concurrent and sympatric existence of forms identified with good reason as palaeospecies supports the hypothesis that phylogenetic events have occurred in the group under study. For the genus *Homo* this would mean that the presence of different morphotypes of similar geological age and spatial neighbourhood can be interpreted as the result of an earlier origin of species in this genus.

But, on the other hand, fruitless efforts to find such morphotypes or uncertainty in their distinctness cannot be used as arguments against the occurrence of phylogenetic events in the history of mankind. An uninterrupted preservation of fossils in time and space to such an extent that would lead to this proposition being refuted cannot be expected. After all, the palaeoanthropologists are facced with the same theoretical and practical difficulties that restrict palaeontological studies in general.

As fas as I know, attempts to obtain evidence of the very moment of speciation in the fossil record have so far been successful only in the case of marine protozoa, the foraminifers, which are accessible as millions of well preserved specimens in undisturbed and continous layers of thinly packed sediments deposited over long intervals of time (GRABERT 1958, BRTTENSTAEPT 1968). The possibility that palaeoanthropology can find exact periods of species deviation in hominids therefore seems rather remote.

References

[1] BOY, J. A.: Zur Anwendung der Hennigschen Methode in der Wirbeltierpaläontologie. Paläont. Z., **55** (1) (1981), 87—107.

[2] BETTENSTAEDT, F.: Wechselbeziehungen zwischen angewandter Mikropaläontologie und Evolutionsforschung. Beih. Ber. Naturh. Ges., 5 (Keller-Festschrift) (1968), 337—391.

[3] CRONIN, J. E., N. T. BOAZ, C. B. STRINGER & Y. RAK: Tempo and mode in hominid evolution. Nature (London), **292**, (1981), 113—122.

[4] GRABERT, B.: Phylogenetische Untersuchungen an *Gaudryina* und *Spiroplectinata* (Foram.), besonders aus dem nordwestdeutschen Apt und Alb. Abh. senckenb. naturf. Ges., **498** (1959), 1—72.

[5] HENKE, W.: Zum Ursprung der Hominidae. Naturwissenschaften (Berlin/W), **68** (1981), 407—417.

[6] HENNIG, W.: Grundzüge einer Theorie der phylogenetischen Systematik. Dtsch. Zentralverl. Berlin 1950.

[7] PETERS, G.: Die ‚Menschwerdung des Affen' und die biotischen Voraussetzungen der menschlichen Arbeit. In: R. LÖTHER (ed.) „Biologische Evolutionstheorie und Medizin" (Reihe: Medizin und Gesellschaft", **28**). Verl. Volk und Gesundheit Berlin 1982, 109—130.

[8] WALKER, A. & R. E. F. LEAKEY: The Hominids of East Turkana. Sci. American (New York), **239** (2) (1978), 44—56.

Discussion

SEIFERT: Was there any connection between the stages of ice-age and the splitting of *Homo*?

PETERS: Not at all! This means that nothing is known in fact about splitting in the *Homo* genus.

HERRMANN: Dr. PETERS informed us about different diagrams given by palaeanthropologists on the evolution of *Homo*. I share Dr. PETERS' scepticism and think it is hardly possible to describe the different species etc. But of course, we have recorded evidence and all theories must be proved against this background. It is, for example, impossible to place *Ramapithecus* at the same level as *Homo habilis*, as was done in one diagram.

ANTONOV: I share your scepticism concerning the potencies of paleontology; moreover, I would like to make the situation even more dramatic by asking you: do you know **objective** criteria which may be used to differentiate the genus *Homo* from related genera?

PETERS: All presented schemes on phylogeny of mankind are speculations; their probability cannot be proved by fossils and the results of their investigation.

Zentralinstitut für Alte Geschichte und Archäologie der Akademie der Wissenschaften der DDR, Berlin

More Questions Than Answers in Hominid Taxonomy: Homo sapiens as an Example

H. ULLRICH

Charles DARWIN never used the term *Homo sapiens* in his work on the descent of man. He stated that the question of whether mankind is represented by one or several species had often been discussed by anthropologists but had not yet been answered because of absence of an agreed definition of "species" (Ch. DARWIN 1871, p. 183).

In the last hundred years very considerable progress has been made in the field of hominid evolution. Large numbers of fossil remains have been discovered and much scientific research has been done, so that many of the questions on the origin and evolution of man have been answered. The process of hominid evolution is now known in broad terms, but in detail more questions have been posed by new fossils than can be answered. This leads to new expeditions, excavations and stimulates scientific research.

In 1758 Carolus LINNEAEUS introduced the term "*Homo sapiens*" and used it for recent mankind. The great discovery of human fossils can be conveniently dated as 1856 with the Neanderthal fossils, named *Homo neanderthalensis* KING 1864. *Homo sapiens* has been used not only for living but also for fossil anatomically modern man. Besides these two valid taxonomic subspecies of the genus *Homo* many other terms were given to human fossils. Most of the specimens were named as new species or genera. Extreme confusion in taxonomy became typical for palaeoanthropology. According to B. G. CAMPBELL (1965) 60 fossil hominid specimens in 19 genera had been described, and 55 species/subspecies, together with a further 12 genera and 32 species were given as synonyms.

Revision of hominid taxonomy was very urgent and was carried out by B. G. CAMPBELL (1962, 1963, 1965) in the early sixties. He divided the hominids into 3 genera (*Australopithecus, Pithecanthropus, Homo*) and several species: *A. africanus* and *A. robustus, P. erectus, H. sapiens* (with the subspecies *H. s. sapiens, H. s. neanderthalensis, H. s. rhodesiensis, H. s. soloensis, H. s. krapinensis*). As a result of a Wenner-Gren symposium in 1962 the term *Pithecanthropus erectus* was replaced by *Homo erectus*. An active discussion on the revision of the hominid taxonomy started. The subspecies division of *Homo erectus* has very often been discussed, but further subspecies of *Homo sapiens* have also been used, e.g. *H. s. palestinus* (B. G. CAMPBELL 1978), *H. s. heidelbergensis* (E. MAYR (1963), *H. s. praesapiens* and *H. s. praeneanderthalensis* (H. SCHMIDT 1979), *H. s. capensis* and *H. s. anteneanderthalensis* (R. PROTSCH 1978).

One of the main results of the revision of hominid taxonomy has been that neanderthal and sapiens populations should not be classified as two independent species but only distinguished at the subspecies level as *H. s. neanderthalensis* and *H. s. sapiens*. Consequences and problems resulting from this taxonomic revision are directly relevant to sapienisation. The term sapienisation was used before the taxonomic revision in connection with the origin of anatomically modern man, who was named *Homo sapiens*. Since the revision sapienisation can be understood to refer to the origin either of the species *Homo sapiens* (that means neanderthal and sapiens populations together) or of the subspecies *H. s. sapiens*, i.e. anatomically modern man.

In recent years some scholars, such as J. JELÍNEK (1978, 1980, 1981), have discussed whether *Homo erectus* was not already *Homo sapiens*. New finds of Middle and Upper Pleistocene fossil man in Africa, Europe and Asia have shown that there are only few morphological characters are typical of *Homo sapiens* and that the boundary between *Homo erectus* and *Homo sapiens* cannot be precisely fixed. The subspecies division of *Homo erectus* has been doubted by some scholars because we do not know the functional meaning of such variable characters or distinguish geographical and temporal variants. On the other hand some palaeoanthropologists (C. B. STRINGER 1974, 1978, 1980, 1981; C. B. STRINGER, F. C. HOWELL, J. K. MELENTIS 1979; M. H. WOLPOFF 1980) have argued that the European Middle Pleistocene finds of Mauer, Vértesszölös, Bilzingsleben, Arago and Petralona cannot be grouped with *Homo erectus* but should be considered as early forms of *Homo sapiens*. That means that *Homo erectus* would neither have existed nor coexisted with *Homo sapiens* in Europe. But it is possible to separate the European from the African and Asian *Homo erectus* finds in the sense that the former are *Homo sapiens* and the latter *Homo erectus*? The Middle Pleistocene finds show marked differences, but also a complex of common morphological characters. Morphologically and culturally they exhibit mosaic structures consisting of both archaic and progressive (modern) characters which varying in form, intensity and number. The evolutionary trend of rapidly increasing cranial capacity found in the majority of *H. s. neanderthalensis* is already indicated in some European (Vértesszölös), Asian (Sambungmatja, Sangiran 17) and African *Homo erectus* finds.

All these facts point — according to the hypothesis of J. JELÍNEK — to the conclusion that there was no genetical barrier between *Homo erectus* and *Homo sapiens*. The morphological differences between these two types should therefore no longer be considered on the species level but only on the subspecies level. "The logical consequence of such a situation is to lead us to consider the different African, European and Asian finds of *H. erectus* type as *Homo sapiens erectus*" (J. JELÍNEK 1978, p. 428).

Palaeoanthropological research has been influenced for decades by the first finds of *Homo erectus* in Asia on the one hand and by the classical West European neanderthals on the other. It had been expected that European and African Middle Pleistocene hominids would show the same *erectus*-complex of characters as those in Asia. In the same way Europe has been considered to be typical for the Upper Pleistocene evolution of man, and neanderthals should have existed in Africa (so-called tropical neanderthals) and Asia, too. Recent finds and comparative studies (G. P. RIGHTMIRE 1980, 1981 and others) have shown that *H. s. sapiens* in Africa should be much older than in Europe (about 40,000 years). The skull Klasies River and Border Cave (80,000 to 100,000 years) in South Africa and Omo 1—3 in East Africa (130,000 years) represent obviously anatomically modern man. It has been shown by new fossils and re-examinations of previous finds that *H. s. neanderthalensis* has never existed in Africa and Asia and that this form is the result of a special evolution under extreme climatic conditions. In Africa and Asia *H. s. sapiens* could have originated directly from *Homo erectus* or *H. s. erectus*. Such a hypothesis should be taken into consideration when discussing the origin of anatomically modern man.

New finds and new hypothesis stimulate the discussion on the origin and evolution of man and shed light on the evolutionary process. Although the discussion about *Homo erectus* and *Homo sapiens* is continuing, we should recognize that the delineation of *Homo sapiens* in the fossil hominid material is more problematical and difficult the older *Homo sapiens* is regarded. *Homo sapiens* as anatomically modern man once had an age of approximately 40,000 years, but following the revision of hominid taxonomy and his association with the neanderthals his age increased to 250,000 years and would date back still further to 1.5 Mio years, if *Homo erectus* had to be included as *H. s. erectus*. The oldest *H. erectus* finds have recently been discovered in East Africa (Olduvai, East Turkana, Omo, Melka Kunture) and range from 1 to 1.5 Mio years old. Following another hypothesis, all or some of the *Homo habilis* finds from East Africa (Olduvai, East Turkana, Omo) with an age of 1 to 2 Mio years should also be classified as *Homo erectus* or *pre-erectus* forms of the genus *Homo* (J. E. Cronin, N. T. Boaz, C. B. Stringer, Y. Rak 1981). Can *Homo sapiens* be regarded as reaching an age of 2 Mio years? We would hardly get accustomed to the idea that *Homo habilis* — represented for example by the specimen ER 1470 from East Turkana — with an age of 1.8 Mio years and the recent anatomically modern man should represent the same species *Homo sapiens*. We would never get accustomed to this idea if we understand man not only as a biotic but also as a cultural-social being! We should bear this in mind when applying biological taxonomy to man.

One hundred years ago the question of whether recent mankind represents one or several species was under discussion. Nowadays the question of whether fossil man represents one or several species is a subject of heated discussion. Nobody can tell us now exactly when *Homo sapiens* emerged. There are more questions than answers in hominid taxonomy.

References

[1] Campbell, B. G. (1962): The systematics of man. Nature **194**, 225—232.

[2] — (1963): Quantitative taxonomy and human evolution. In: S. L. Washburn (ed.), Classification and human evolution, 50—74. New York.

[3] — (1965): The nomenclature of the Hominidae including a definitive list of hominid taxa. London (Royal Anthropological Institute, Occasional Papers No. 22).

[4] — (1978): Some problems in hominid classification and nomenclature. In: C. J. Jolly (ed.), Early hominids of Africa, 567—581. London.

[5] Cronin, J. E., T. Boaz, C. B. Stringer and Y. Rak (1981): Tempo and mode in hominid evolution. Nature **292**, 113—122.

[6] Darwin, Ch. (1871): Die Abstammung des Menschen und die geschlechtliche Zuchtwahl. Leipzig o. J.

[7] Jelínek, J. (1978): *Homo erectus* or *Homo sapiens*? In: D. C. Chivers, K. A. Joysey (eds.), Recent advances in primatology. Vol. 3. Evolution, 419—429. London.

[8] — (1980): Variability and geography. Contribution to our knowledge of European and North African Middle Pleistocene hominids. Anthropologie (Brno) 18, 109—114.

[9] — (1981): Was *Homo erectus* already *Homo sapiens*? In: D. Ferembach (dir.), Les processus de l'hominisation, 85—89. Paris.

[10] Linnaeus, C. (1758): Systema Natura per Regna tria naturae secundum classes, ordines, genre, species cum characteristibus, differentiis, synonymis, locis. Editio decima, reformatia. Stockholm.

[11] Mayr, E. (1963): The taxonomic evaluation of fossil hominids. In: S. L. Washburn (ed.), Classification and human evolution, 332—346. New York.

[12] Protsch, R. (1978): Der Mensch stammt aus Afrika. Umschau in Wissenschaft und Technik **78**, 554—562.

[13] RIGHTMIRE, G. P. (1980): *Homo erectus* and human evolution in the African Middle Pleistocene. In: L.-K. KÖNIGSSON (ed.), Current argument on early man, 70—85. Oxford.
[14] — (1981): Later Pleistocene hominids of eastern and southern Africa. Anthropologie (Brno) **19**, 15—26.
[15] SCHMIDT, H. (1979): Materialien zur Fossilgeschichte des Menschen. Köln.
[16] STRINGER, C. B. (1974): A multivariate study of Petralona skull. Journal of Human Evolution **3**, 397—404.
[17] — (1978): Some problems in Middle and Upper Pleistocene hominid relationships. In: D. J. CHIVERS, K. A. JOYSEY (eds.), Recent advances in primatology. Vol. 3. Evolution, 395—418. London.
[18] — (1980): The phylogenetic position of the Petralona cranium. Anthropos (Athens) **7**, 81—95.
[19] — (1981): The dating of Europe Middle Pleistocene hominids and the existence of Homo erectus in Europe. Anthropologie (Brno) 19, 3—14.
[20] STRINGER, C. B., F. C. HOWELL and J. K. MELENTIS (1979): The significance of the fossil hominid skull from Petralona, Greece. Journal of Archaeological Science **6**, 235—253.
[21] WOLPOFF, M. H. (1980): Cranial remains of Middle Pleistocene European hominids. Journal. of Human Evolution **9**, 339—358.

Discussion

OESER: Is it still possible to differentiate between *Homo sapiens praeneanderthalensis* and *Homo sapiens praesapiens*, for instance in respect to *Homo steinheimensis* or the fossils of Swanscombe?

ULLRICH: Some scholars have tried to differentiate between pre-sapiens and pre-neandertal lines in the European human fossil remains. But it is very difficult, especially if you keep in mind that hominid evolution has been mosaic evolution, that means that fossils show archaic as well as progressive (modern) characters, in different shape, number and intensity.

Leningrad Department of Philosophy of the Academy of Sciences of the USSR

Ch. DARWIN and Philosophic Problems of Theory of Anthropogenesis

Y. I. YEFIMOV

The principle of historism which has, on the basis of DARWIN's theory, brought together the advance of various branches of biology materilized most completely in the doctrine of rise and development of mankind. Highly appreciating the DARWINian doctrine as one of the most significant natural-scientific prerogatives of the formation of the dialectic-materialistic concept of development, F. ENGELS wrote that Ch. DARWIN "dealt a tremendous blow to the metaphysical view on nature by having proved that the whole modern organic world, plants, animals, and, consequently, man, are products of the process of development which had lasted millions of years" [1]. The greatest merit of Ch. DARWIN was that he "discovered the law of development of the organic world" [2] having revealed the true causes and mechanisms of the many-sided, contradictory process of organic evolution and thereby "for the first time placed biology on a scientific soil" [3].

The evolutionary theory of Ch. DARWIN not only brought to ruin the idealistic dogmas on the origin of man, which isolated man from the whole organic world, but also gave a powerful impetus to the birth of new hypotheses and development of new generalizations.

The founders of Marxism vigorously insisted on the synthesis of the evidence of biological and social sciences in elucidating the problem of anthropogenesis. Giving a dialectico-materialistic substantiation to the doctrine of the origin of man, F. ENGELS rested mainly upon the evolutionary ideas of Ch. DARWIN. Darwinism, stressed ENGELS, is extremely important not only as an original biological concept, but also a key to the materialistic explanation of the prehistory of consciousness.

The obvious priority of the evolutionary theory in the development of problems of the natural pre-history of mankind stems from the fact that DARWINism is not only a theoretical basis of biological sciences, but it also has been, and remains to be one of the most important natural-scientific foundations of the dialectico-materialistic theory of development.

The present-day synthetic theory of development has been formed during the recent decades as a result of wide synthesis of data on the regularities of functioning and development of living systems and of the accompanying reconstruction of the very logic of biological thinking. By this is meant the transition from the typological (organism-centred) to a new, populationistic, thinking. This aspect of the evolutionary theory has assumed an outstanding role in the study of primitive history and, in particular, of its onset, as well as in solving the key-problems of the origin and evolution of consciousness, language, social means of communication as the basis of primitive collectivism, and of the problem of genesis of labour.

In this field there has been a marked rise of interest in the problem of progressive development of living nature. The classical ideas of the doctrine of the aromorphic (progressive) evolution by A. N. SEVERTSOV and I. I. SCHMALHAUSEN have been supplemented by some new ones formulated by their proponents and followers: "the trunk line of progress" (K. M. ZAVADSKY); "unlimited progress" (D. HUXLEY — K. M. ZAVADSKY); "noogenesis" (M. M. KAMSHILOV). This system of concepts worked out together with the advance in the fields of paleo-anthropology, primatology, ethology, ecology and genetics of present-day human populations along with the results of current investigations in archeology, ethnography, psychology, neurophysiology and physiology of the higher systems, has greatly stimulated the philosophic elaboration of the evolutionary-biological aspects of the theory of anthropogenesis.

The last decade has been marked by noticeable advance in the development of the problem of regularities of anthroposociogenesis including the achievements of modern evolutionism in biology. Contributions have appeared which use as their material for study the findings and generalizations of modern evolutionary theory on the rates of development and particular models of speciation, on the character and various forms of evolution of populations, evidence for regularities of formation and integration of individual living systems.

All this underlies certain theoretical principles of analysis of the specific role and significance of these factors in anthroposociogenesis which do not exclude from their field of study the peculiarities of reconstruction of mechanisms of biological control under conditions of formed human society, as well as the tendencies of their likely changes in future.

In the recent times some development gained the study of evolutionary-ecological and ethological aspects of anthroposociogenesis. Quite a few works appeared in our country which reflect the results of these investigations.

It is typical that it is the problem of group organization, its evolutionary-biological aspect, which has become a point of special attention of philosophers, experts in the field of prehistory.

One of the major advances of recent years in the manysided complex investigations on the problem of anthroposociogenesis has been a philosophic and natural-scientific development of F. ENGELS' idea of the necessity of regarding the process of hominization as a unique type of transitional state. In this connection of great importance has become the study of specific regularities of anthroposociogenesis as an intermediate state. The formulation and analysis of this problem have not only noticeably advanced the philosophic study of the theory of anthroposociogenesis, but also have led to an essential re-interpretation and enrichment of the contents of a number of philosophic study of the theory of anthroposociogenesis, but also have led to an essential re-interpretation and enrichment of the contents of a number of philosophic categories (leap, dialectic removal).

At the same time, the philosophical-methodological evaluation of modern advances in evolutionary thought in the area of problems of natural pre-history of mankind is, at present, the task of future research. A further improvement is necessary in the conceptual stare of the theory of anthropogenesis. Such notions as "animal-like labour", "instinctive labour", "reflectory labour", "biosocial selection", "zoological individualism" etc. are of such conventional and metaphoric character as to call for a well-grounded critical analysis from the viewpoint of dialectico-materialistic methodology. Even so, this does not rule out the possibility of further search and substantiation of new concepts fixed in terms which adequately reflect the peculiarity of anthroposociogenesis as an object of philosophic analysis.

A special importance is gaining the further development of natural selection as a factor of humanization. Up to recent time this problem has not been systematically studied. To a large measure this is to be explained by an absence of fairly close contact between philosophy and particular branches of science in the development of methodological, philosophic-gnoseological aspects of the theory of anthroposociogenesis. This, undoubtedly, has created certain difficulties for a successful realization of the dialectic principle of a comprehensive consideration of this problem.

Thus, philosophical analysis of the evolutionary-biological interpretation of anthroposociogenesis stimulates the search and contributes to a concrete realization of the heuristic function of the evolutionary theory as applied to the solution of some top-problems in the study of human prehistory, to the creation of an integral concept of birth of humanity, revealing the dialectics of the commencement of human history.

At the same time, the study of evolutionary history of mankind not only constitutes the general outlook of the present day, but also makes for the development of ideas on the prospects of organic evolution and ecological future of humanity.

References

[1] MARX, K., ENGELS F., Collected works, v. 20, p. 666.
[2] IBID, v. 19, p. 350.
[3] LENIN, V. I., Complete collection of works, v. 1, p. 139.

Humboldt-Universität zu Berlin, Sektion Biologie, Bereich Verhaltenswissenschaften

Biosocial Aspects of Human Evolution

G. Tembrock

Biosocial behaviour of primates has evolved to become a special form of behavioural adaptation with an inherent motivational content. Darwin's general idea of selection was discussed by Dobzhansky [2] with respect to principle environmental properties. MacArthur and Wilson [6] coined the terms K-selection and r-selection for natural selection in the tropics and temperate zones, K referring to carrying capacity and r to the maximum intrinsic rate of natural increase. Fisher [3] formulated the idea of the budgeting of time, matter and energy into the circumstances of life history and environment, which "would render profitable the diversion of a greater or lesser share of the available resources toward reproduction".

In respect of these ideas, Pianka [7] states general features of the two endpoints in the continuum of selection. Biosociogenesis of primates will correlate in this sense with K-selection, which is characterized by the following criteria:

- fairly constant and/or predictable climate,
- more directed, density-dependent mortality than in r-selection,
- population size fairly constant in time, equilibrium at or near carrying capacity of the environment — saturated communities — no recolonization necessary,
- intra- and interspecific competition usually keen,
- selection favours:
  - relatively slow development,
  - delayed reproduction,
  - relatively large body size,
  - iteroparity: repeated reproduction
- length of life usually longer than one year,
- K-selection leads to efficiency (r-selection to productivity).

Parental investment will be high in the case of K-selection, and the quality of the individual produced is a constitutional part of this strategy of evolution. The state of being individual is a positive asset in the process of natural selection. We will return to this problem later. First we wish to consider biosocial structures that have evolved in this context to become non-anonymous entities comprising less than one hundred individuals. Substructures tend to develop in such biosocial systems. Biological prerequisites for this special form of biosocial interaction will be seen in the following elementary characters of animal behaviour:

1. Animal behaviour consists of organic interactions with the environment, including the exchange of information. This takes place via three vectors:
   - the input vector, corresponding to the assimilation of information,

— the internal vector, corresponding to motivation, emotional state, or the general state of activity or arousal,
— the outpout vector, which includes the whole efferent activity of the organism.

This definition involves two pairs of conceptions:

- organism and environment, the latter being defined by those properties of the surroundings which directly or indirectly affect a particular organisms,
- interaction and exchange of information; this dyad forms the behavioural processes.

2. Behaviour as defined by these terms includes three classes of properties:
   — constitutional properties given by structural and functional attributes of the organism,
   — energetic properties, and
   — informational properties in the sense of structural information.
3. With respect to the origin of information and the goal of the interaction with the environment, we distinguish between three classes of environment:
   — the own body,
   — the ecosystem, given by the context of interactions controlled by information, and
   — the population-system, the context of interactions, controlled by biocommunication (a signal is communicative if it is intentended by the sender to infirm the receiver, LYONS [5]).
4. The interactions with the environment will be based on special needs derived from elementary functions of living systems. We distinguish between the following classes of needs:
   — need for space,
   — need for time,
   — metabolic needs,
   — need for protection,
   — need for information about the environment, and
   — need for partnership, especially sexual partnership but also in several species for age-specific partnership.
5. To satisfy these needs for environmental properties, organisms evolved three functional types of interaction:
   — first order needs related to the own body,
   — second order needs related to the ecosystem and
   — third order needs related to the population.

   The higher order needs can modify the parameters of lower order needs (enkapsis).
6. Needs for special environmental properties will induce motivations which are materialized in the readiness to perform behavioural acts.
7. The elementary form of realization of a motivational state is the consummatory act with contact to the goal-object of the particular form of behaviour. This field of interactions with the environment thus designates the field of contact. The implementation of this form of interaction, for instance the uptake of food, establishes a negative feedback to the motivational state, the typical effect of the consummatory behaviour, triggered by releasing stimuli. Evolution leads to a further phase of motivated behaviour as a result of the development of telereceptors permitting identification of distant goal objects. This resulted in adjusting stimuli leading to located behaviour, which is sometimes also called orientated appetitive behaviour. This behaviour established

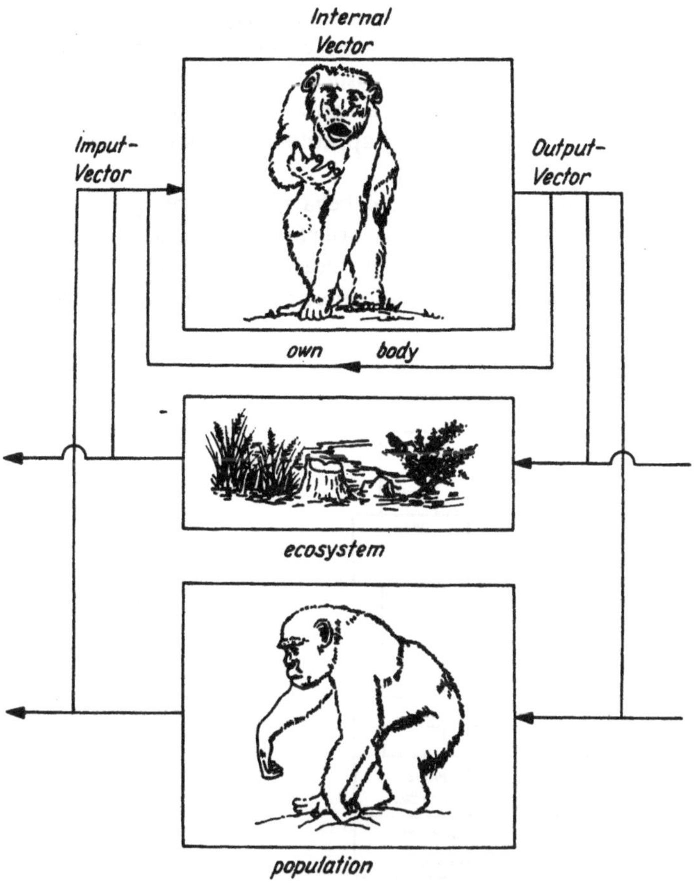

Fig. 1. The three classes of environment correlated with behavioural interactions, determined by constitutional, energetic, and informational properties.

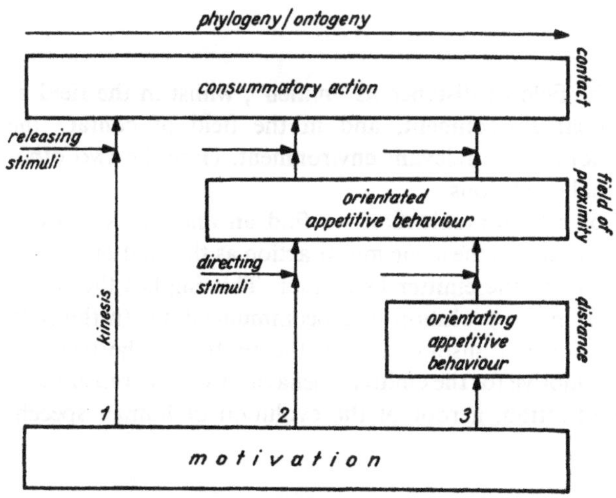

Fig. 2. The three phases of realization of motivated behaviour with respect to phylogeny and ontogeny.

the field of proximity. Certain properties of the environment may be implicated in this aspect of behaviour and act as functional parameters of the environment. Finally, a third field of motivated behaviour will be established. This, the field of distance, appears if no information of the goal-object is present so that special search strategies must be executed. This is the orienting appetitive behaviour or search behaviour.

Now we have three typical phases of motivated behaviour: orienting or location appetitive behaviour, induced directing stimuli, orientated or located appetitive behaviour, and consummatory behaviour aroused by releasing stimuli. The consummatory phase has a weighting function, whereas the preceeding phases optimize this process of motivated behaviour.

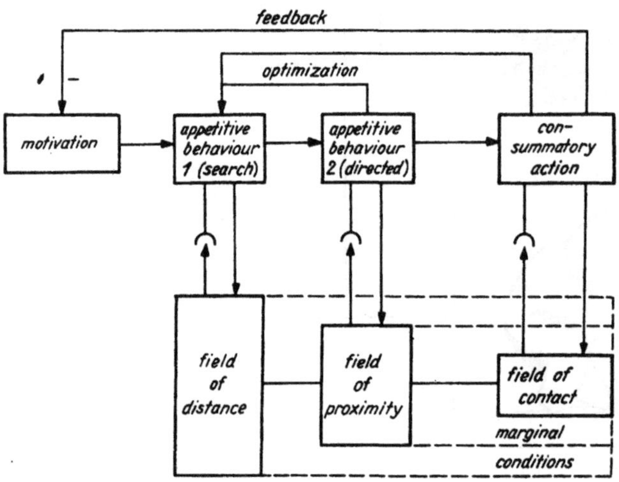

Fig. 3. The three "fields" of organism-environment interactions of motivated behaviour; selection processes will be realized by optimalization and feedback.

8. We define the environment in the field of distance as "milieu", whilst in the field of proximity we have the functional environment, and in the field of contact the goal-object of the special behaviour is the relevant environment. Here the two other environmental classes are marginal conditions.
9. With respect to the evolution of biocommunication we find an analogous process. In the field of contact the receiver executes the same motor action as the emitter. This is triggered by "mimic exaggeration" of the emitter behaviour. This implies the transmission of function, the first step in the evolution of biocommunication. In the field of proximity the state of orientation is transmitted from the emitter to the receiver. In the field of distance, finally, the motive for the emitter's behaviour will be transmitted. This is the transmission of signification, a root of the evolution of human speech. We give an example:
   First step:
   The emitter shows flight behaviour — the receiver also.

Second step:
The emitter signalizes fear — the receiver shows fear and thereafter flight.
Third step:
The emitter signalizes "enemy!" — the receiver accepts this information, which will be followed by fear and flight behaviour.
In the third step, the receiver finds himself in the same state as the emitter, but without actually perceiving the primary cause. Now, selection is directed towards an increasing ability to discriminate between environmental events or things, their significance and communicative transmission by special signs.

10. Biocommunication leads via evolution to a fourth quality of needs regarding environmental parameters: These needs involve learning processes, building up traditions of behavioural traits, a basis of biohistorical components concerned in the evolution of biosocieties. This quality will be transmitted during anthroposociogeny with the development of human societies to historical processes: Parabiological evolutionary processes are superposed upon biological evolution.

This very concentrated survey of general behavioural traits may be helpful in finding ways to understand the process of human evolution also in the sense meant by DARWIN and his adherents. In this context another aspect — mentioned earlier — should be taken into account. Life processes are performed only by organismic systems: life is restricted to individual organisms, and this is the level at which selection acts with respect to the gene pool of populations in the course of evolution. Therefore we may ask about the relation between the personality of the human individual and such elementary entities. We distinguish in this context between the following:

1. The state of organisms in the form of individualized beings.
2. levels of individuation, i.e. unicellular organisms, multicellular organisms, units of organisms either physically connected or in the form of eubiosocial entities.
3. levels of individuality,
   - first order individuality resulting from constitutional traits and differences, i.e. castes,
   - second order individuality determined by informational relations to the environment, especially the ecosystem,
   - third order individuality determined by communication with the environment, especially the population,
   - fourth order individuality based upon tradition and individual learning processes, and
   - fifth order individuality resulting from individual cultures and norm systems: the personality of human beings.

Castes are determined by natural traits. Elementary castes are the two sexes and differences resulting from age. In eubiosocial insects there are in addition morpho-types, the so called "workers" and "soldiers". In higher vertebrates selection promotes individual differences. We find special types of individual expression described not very successfully by words like "temperament" or "nerval types". These traits are independent of age and sex, and are connected with evolutionary processes which led to anthroposociogeny. We think that individual differences promote the development of interaction by communicative signs: in individuals with the same behavioural repertoire the probability of being able to make receiver aware of something increase with the degree of individual differences. A special function is associated with differences in age due to the connection between age and ex-

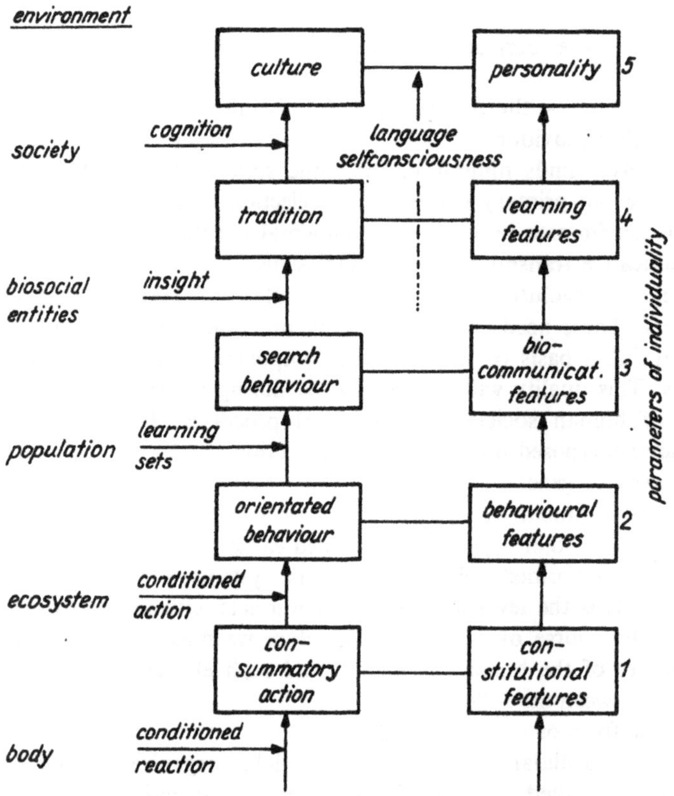

Fig. 4. The levels of realization needs for environmental properties with respect to individual adaptation (learning in general) and levels of individuality (individual differences) leading also to special features of human personality.

perience. In primates this is an important fact related to special roles in the biosociety, and sex and age become secondary functions in this context. Many behavioural studies in recent years have provided examples of individual traits of behaviour and acoustic signals. Individuality on the fourth level in biosocieties promotes innovation, exploration and inventive interactions which are assessed by the group concerned. This leads to new relationships between conservative transmission, which is necessary to species-specific behaviour, and innovative transmission. The latter leads to group-specific behavioural traits and promotes intergroup activities and selection mechanisms of a new quality. Kin-selection may be discussed in this context.

Sexual dimorphism and mating patterns like monogamy and polygamy also are related to biosocial processes. The states described by FLEAGLE et al. [4] in oligocene anthropoids confirm sexual dimorphisms which in recent primates correlate with polygynous mating types; monogamous forms show insignificant sexual dimorphism. In gibbons the relation between the mass of the male body divided by the mass of the female body is only 1,03. In primitive anthropoids sexual dimorphism seems to be relatively pronounced or distinct. According to ALEKSEEV neandertaloid humans show a higher degree of sexual dimorphism than later hominids.

In general, individual traits in primitive biosocial entities will be promoted by evolution, and identification by signals also becomes of great value to the sender; this may be one of the ways that led to human selfconsciousness. We think that there are four levels of needs for biosocial partnership:

— first order biosocial partnership: community with one or more individuals of the own species, regardless of sex and age,
— second order biosocial partnership: community with one or more special individuals of the own species of a certain sex, age or individual state,
— third order biosocial partnership: community with one or more groups of special individuals of the own species,
— fourth order biosocial partnership: community with one or more special groups of special individuals of the own species.

We have discussed the mechanisms promoting the specifity of individuality, and of groups also. Biosocial systems embody individual needs in the third order framework: distribution in space and time will be no longer determined only by the motivation of the individual, but becomes integrated into the complex interaction of the biosocial unit with its surroundings. This applies also for metabolic needs and the need for protection, the need for information and the need for partnership. Some of the essentials of first order needs relating to the constitutional properties of the own body will be transformed or inhibited. Differentiation between the individuals of groups and between the groups themselves leads to an increase in adaptation and innovation, and these in turn imply new states in the population analogous to the rates of mutation at the genetic level. Innovations in the behavioural states of individuals will be transmitted from generation to generation as tradition. In this way parabiological evolution takes place and biohistorical traits in the primate groups lead to new forms of interaction with the environment. The prospective value of interactions in the field of distance becomes recognized as a consequence of insight-learing processes, these leading to new forms of adapted behaviour which in the biosocial context are promoted by communicative mechanisms. This is preadaptation in the direction followed by the descent of man and transformed into a historical process. Facultative learning abilities gain a new value in the arrangement with the surroundings, and play behaviour promotes not only biosocial integration but also the appearence of new behavioural patterns in the interaction with the environment and with respect to group processes. Biocommunication assumes the function of a consummatory act, and the exchange of information now takes place mainly by means of behavioural interactions between individuals that are realized by physical contact. But physical contact is still essential for the group interactions in primates, where it is necessary also to comprehension (in the double sense of this word) of the facts of the environment. The possibility of transforming visual inputs of the three dimensions of space into signals gives biocommunication a new quality and paves the way for the evolution of speech. The unique combination of these different principles of behaviour opens the way for anthroposociogeny and provides a wide spectrum of behavioural determinants also in the human actions of today. Nonetheless, evolutionary traits that we have explained here do not permit us to reduce the biological mechanisms to genetic processes or innate behavioural patterns. But they do open the windows giving a view of the new modes of interaction with the environment which have been realized and developed by human beings.

Summary

Biosocial behaviour in primates has evolved to become a special form of behavioural adaptation based on K-selection. This process was determined by general features of animal behaviour, defined as interactions of the animal with the environment. Special needs for environmental properties such as space, time, nutritional resources, information, features for protection and partnership produce functional processes promoted by motivation. These are realized via three levels: orientating appetitive behaviour, orientated appetitive behaviour, and the consummatory action. In biosocial evolution special forms of learning lead to traditions and promote the development of specific levels of individuality which in humans lead to the features of personality. Some general traits of the evolution are demonstrated.

References

[1] ALEKSEEV, V. P.: Morphologische Aspekte des Sexualdimorphismus beim paläolitischen Menschen. Tag. Ber. „Anthroposoziogenese — biotischer und gesellschaftlicher Entwicklungsprozeß der Menschheit. Akad. Verl. Berlin. (in press).

[2] DOBZHANSKY, T.: Evolution in the tropics. Amer. Sci. **38** (1950), 209—221.

[3] FISHER, R. A.: The genetical theory of natural selection. University Press, Oxford (1930).

[4] FLEAGLE, J. G., R. F. KAY and E. L. SIMONS: Sexual dimorphism in early anthropoids. Nature **287** (1980), 328—330.

[5] LYONS, J.: Human language. In: "Non-verbal communication" (Ed. R. A. HINDE), 49—85. University Press, Oxford (1972).

[6] MACARTHUR, R. H. and E. O. WILSON: The theory of island biogeography. University Press, Princeton (1967).

[7] PIANKA, E. R.: On r- and K-selection. Amer. Natural. **104** (1970), 592—597.

Discussion

HAILA: I have a comment on the concepts of r- and K-selection you made use of in the beginning of your paper. I think the scheme is too narrow to be used in the context of biosocial behaviour. Comparisons of life cycle properties of species, supposed to be consequences of r- or K-selection, can be made within certain families of genera, i.e. within morphologically reasonably unified taxa. But I cannot see the usefulness of the scheme in the broad context of biosocial behaviour. In your paper you say that "herd" as a biosocial phenomenon is a result of r-selection; but take the elephant. It certainly is a K-selected animal but it moves in herds.

TEMBROCK: The value of concepts of selection strategies as K-selection and r-selection is in discussion. We can consider them perhaps as endpoints of a continuum of selection; many combinations are possible. In eubiosocial insects we also find r-K-selection, "K" with respect to the castes, r with regard to the number of individuals, the rate of reproduction.

NAUMANN: You derived your explanation of animal behaviour from the interaction between groups of animals and the environment. You regarded biosocial problems as involved in this approach of yours. But modern sociobiology is guided by the idea of preservation and the greatest possible propagation of the genes of a single representative of a species. How do you consider at present the relation of your theory to the sociobiology being developed by DAWKINS, WICKLER and SEIBT, WILSON and others?

TEMBROCK: Animal behaviour, in our sense, **on the third level** of realization of needs for environmental properties is a function of population structures. Our theory of animal behaviour does not require essentials of the concept of sociobiology, but they are not excluded. Further studies will be necessary to prove the value of the general concepts. We try to derive conceptions which are not dependent upon special mechanisms of selection.

SCHUSTER: I would like a comment on the way selection is acting in an ensemble of individuals. It may be helpful to consider an important difference in phenomena: Some are properties of the individual, some are related to an entire population. Behaviour is related to the individual and, eventually, partially as completely determined by the individual's genes. Then, selection seems to act genetically on the level of the individual as it is described by population genetics. There are phenomena related to entire populations. The age distribution in a population is an example. It is not a property of an individual, it is a property of a population. Selection of age distributions has to act at the level of whole populations. This idea might help, eventually, to bring order into the present debate on the role of individual, kinship or groups and species selection in sociobiology.

GRANTHAM: I do not believe in kin selection. I cannot see how favoritism of relatives in local groups where family members are frequent can have anything to do with the genetic composition of the species or its evolution. Can you reply to the argument of P. J. DARLINGTON (Proc. Natl. Acad. Sci. 78, 4440, (1981), if not perhaps you should not use this concept as support for your interpretations.

Humboldt University of Berlin, Department of Psychology, Berlin, GDR

DARWIN TODAY in Learning Theory (Are learning processes evolutionary-invariant?)

F. KLIX

I am not professionally engaged in evolution theory. As an experimental psychologist I am especially interested in the laws of thought and the elementary mechanisms which shape these laws. These elementary mechanisms which we are trying to resolve underly the intellectual capability of the human being. But this also gives rise to interest in the conditions which led to such laws. Exactly this point links our experimental work with learning processes, their basic mechanisms and their modes of performance during different evolutionary periods. So I took a second look at some principles of psychological learning theory[1]. And I have found that, from the beginning up to the present day, they have included a very general but unproven assumption. And this assumption is related to the topic of our conference. It is the assumption that the laws of learning are basically the same in all organisms. Because this assumption refers to organisms of very different levels it implies that laws of learning are evolutionarily invariant.

I would like to explain this assumption and show why it is incorrect and must be changed in favour of an evolutionarily dependent one; i.e. the DARWINian theory must be taken into account in psychobiological learning theory.

1. On the Assumption of Evolutionarily-Independent Processes in Psychology and Biology

With the discovery of conditioning the main paradigm of all learning processes seemed to have become clear. PAVLOV [7] assumed that conditioned-reaction learning performance in lower organisms can be explained by the same principles as the acquisition of language in human beings. (His so-called second signal system functions in the same way as the receiving of food after a light stimulus with regard to salivary reaction.) PAVLOV himself identified his learning mechanism with the one analysed by E. L. THORNDIKE [10]. THORNDIKE also favoured the same basic assumption regarding the evolutionary independence of this phenomena. He mentioned "that the underlying process also might be the same . . . in all animals, including man". This opinion is shared by all behavioristic learning theorists. It is well known that SKINNER's [9] position is the same. H. F. HARLOW [4] wrote: "... it is not unreasonable to hypothesize that the fundamental process of learning has probably remained essentially unchanged for nearly countless millions of years". What may change with the evolutionary level is the complexity of associations

1 This idea was stimulated by an article of BITTERMAN [1].

underlying the learning performance instead of its quality: "... the conditioned reflex is a universal mechanism of activity ... in the course of evolution of the animal world there took place only a quantitative growth or complication of higher nervous activity" (VORONIN [12]). What does it mean, this universal paradigm called condition reaction? It may be formulated in the following three statements:

(1) If two stimuli S_1 and S_2 respectively are perceptually related, and one of them is neutral and the other is related to an unconditioned response, then the neutral stimulus acquires the priming function for that response, for its becoming "conditioned".
(2) The main factor of this (and therewith all) learning procedures is reward (or reinforcement). Reward (reinforcement) links reactions to stimuli. This is valid for simple reactions such as salivary response as well as for complex activity like approach or avoidance-learning.
(3) The stronger the reinforcement the stronger will be the connection between stimulus and reaction and the faster will be the learning process.

These three statements are the essentials of the classical learning theory elaborated by biologists and psychologists alike.

In other words, learning is an organismic process which came into being outside the plan of the DARWINian evolutionary factors. This might be possible for two reasons, namely:

(1) Learning processes are universally organismic invariants because of their being optimal adaptivity procedures. They allow a priori and in all cases the best adaptive strategy. Evolutionary factors cannot make them better, but can produce only worse procedures and then eliminate them due to their disadvantages in terms of adaptivity.

Or:

(2) Learning processes have no biological significance so that they have been dragged unchanged through the selection filters.

I would like to begin with sentence 2. It seems trivially refutable. Notwithstanding, its detailed analysis leads us to some surprising consequences.

We shift the point of view and ask: Why did learning processes during evolution come into being and why do they become more marked with successive evolutionary levels in the development of species? In other words, why did the historically long-lasting stable, instinctive or innate behavioural patterns not carry through in phylogenesis as the best possible adaptive principle? One can find the answer in considering organisms in which behaviour control is mainly due to innate behaviour patterns. Lower vertebrates or insects, bees or spiders are good examples. The following rule seem to hold: The more stable and homogeneous the biotope over succeeding generations is, the more will instictive behaviour patterns prevail in behaviour control and vice versa:

The larger the degree of variation in any life space, the more will individually adaptive, i.e. learning-dependent, behavioural patterns dominate. The reason seems to be clear: a consequence of evolutionary laws is that innate patterns of animal behaviour may originate and survive only in environments that remain unchanged for long periods. The lay-out of instinctive behavioural patterns is related to a mean of environmental properties, i.e. to an environment which may be expected with sufficient degree of probability. Its adaptive value is the higher, the more stable long-lasting environmental conditions remain unchanged. Nonstationary environmental conditions put innate behavioural patterns at a disadvantage and, furthermore, favour learning-dependent behavioural patterns.

Learning-dependent and innate behavioural patterns have one essential point in common: they exploit regularities in the environment to predict expected events. They both are predictive in their function. However, while innate patterns receive their predictive value during long-lasting successions of generations, learning mechanisms use short term regularities, regularities which may shift even within only a short life span. Here we become aware of one essential difference between the two behaviour-adjusting principles. It is forgetting. While loss of the species' memory is maladaptive (and therefore eliminated by evolution), the forgetting of learned patterns may be highly adaptive and favoured in evolution wherever the regularities in the environment change or vanish.

Considering the connection between the evolution of species and the share of learned patterns, one is forced to the conclusion that the variation of the environment depends not only on the environment itself but, in the same sense, on the evolutionary level of the organism. Running and climbing terrestrians are confronted with the greatest variety of possible biotops upon our blue planet. Their share of learning depending patterns is by far greater than that of bird or fish species. Essentially it is the mammals which dominate these biotopes. Lower animals survive in local stable niches where a high proportion of instinctive patterns are appropriate; security patterns such as mimicry are often encountered.

Thus, consideration of sentence 2 leads us to the conclusion that learning processes are a result of DARWINian evolutionary factors. Factors which influence the development to higher organisms, especially terrestrian mammals, lead to increasing variability of life spaces which makes their calculability more difficult and more risky. Learning mechanisms are appropriate to reduce that risk by using temporarily valid information for predicting events. The rise of learning performance is a consequence of evolutionary conditions. So much for sentence 2.

Things are somewhat more difficult with regard to sentence 1: Are the laws of learning evolutionarily invariant? Doubtless they arise as a result of evolutionary conditions (sentence 2) but, once come into being, are they optimal and not changed by these conditions? Instead of engaging in comparative considerations of learning behaviour I would like to cite only one example which might be able to support some doubts regarding this postulate. It makes us doubt that it would be correct to assume more than quantitative growth in learning capabilities.

Fig. 1 (after BITTERMAN [1]) shows the choice reaction behaviour of goldfishes (left). The probability of reward is 70:30 and 50:50 respectively for left and right decisions.

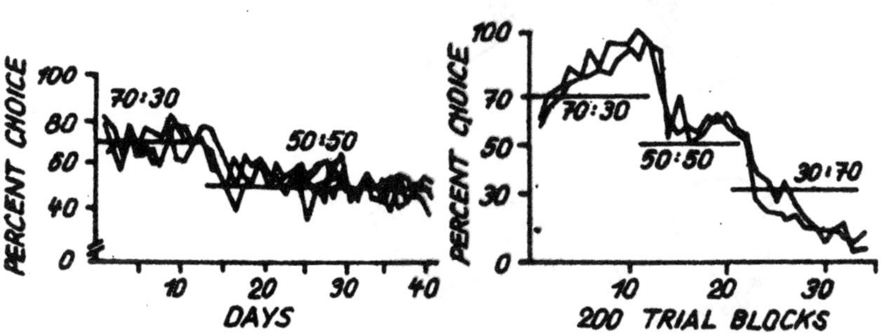

Fig. 1. Choice reaction behavior of goldfishes (left) and rhesus monkeys (right) trained with a 70:30 reward ratio and shifted to a 50:50 and (right side) 30:70 ratio (after BITTERMAN, 1975).

On the right side we have choice reaction behaviour under comparable conditions performed by rhesus monkeys. Fig. 1 indicates that the effectiveness in probability-learning is quite different. Rhesus monkeys are much faster, they respond more rapidly to a change, and they learn the inversion very easily. What is the reason? Is the amount of reward different? This has been checked and was not valid. Are there different sensoric capabilities with regard to the requirement? Surely not. Or is the capability to predict among statistically distributed events faster and more correct? Probably it is. But what does this mean? Is it due to a larger number of associations between the events? If one used comparable results with human subjects to explain that difference, the assumption of counting events must be eliminated. What happens seems to be more likely an inductive inference, i.e. a generalization across frequency proportions linked with an (unconscious) decision threshold. It is learning, but prediction seems to be result from (unconscious) inferences in a time series analysis and not from step-by-step reward evaluation. We have a hypothesis to explain why rhesus monkeys might be so much better: They use a more efficient procedure for calculating proportions. Now we see that there might be qualitative discontinuities in the learning-performances of organisms belonging to different evolutionary levels. Processing of information, mediated by reinforcement, seems to be changed by cognitive capabilities of the kind, found in pre-human primates. Now I would like to consider sentence (1) little more in detail.

2. On qualitative diversity in learning performance

Following W. H. Thorpe [11], J. Konorski and other systematizers of learning phenomena one can conclude that habituation of an inborn reaction is the most primitive kind of learning. It is a universally adaptive phenomenon found in almost all organisms, including protozoa. In experiments with *Aplysia californica* E. Kendal [6] analysed the biochemical background. He found that under habituation conditions the calcium influx is reduced, and this in turn inhibits the release of the transmitter norepinephrin. Thus the sensory linkage is broken so that the motor reaction vanishes.

Conditioned reactions are found in almost all metazoans. Fig. 2 indicates the conditions (left), the learning state (middle) as well as the learning result of these two kinds of individual adaptivity (right).

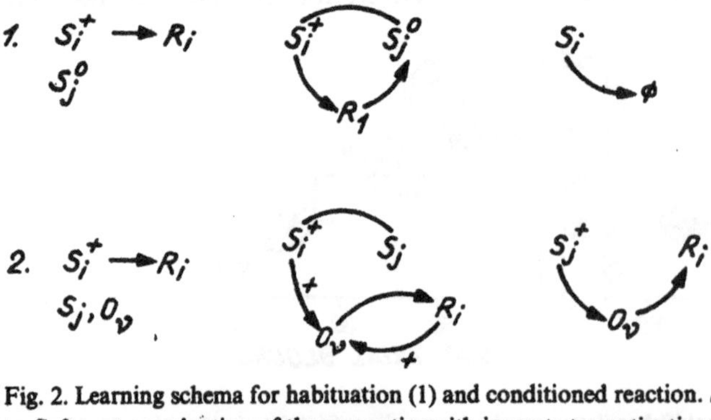

Fig. 2. Learning schema for habituation (1) and conditioned reaction. $S_{i,j}$: Situation properties, R_i: reaction to S_i;0_v—,+ : evaluation of the properties with impacts to motivational forces 0_v · →0 implies or leads to∅: Zero property. Left side: general conditions; middle: the learning situation; right: the learning result.

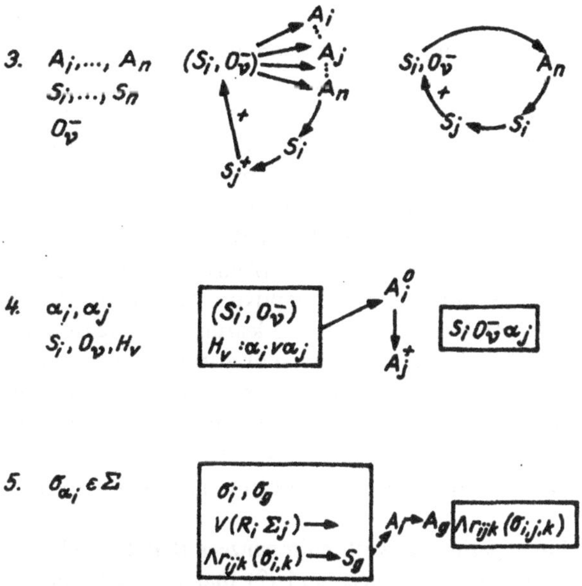

Fig. 3. Learning schema for (3) instrumental conditioning (4) trial and error learning and (5) insight learning. (3): A_i, ..., A_n: activities related to situational properties, shifting them from one state to another one until one shift relaxes the motive state 0_v^- to 0_v^0 (4): Hypotheses formation on s_i, checked by actions $\alpha_i \bullet \alpha_i$ respectively. The outcome controlles the consecutive behavior. (5) Insight: $V(R_i\Sigma_j)$: action programs in memory, related to situational properties $\sigma_i \cdot \sigma_g$: goal state. V $(R_i\Sigma_j)$: List of relations between properties and actions: ${}_ir_{ijk}(\sigma_{ijk})$: concatenation of an action sequence, due to situational (σ_i) and goal properties (σ_g). The difference guides the selection of actions until the goal is realized.

More complex levels of learning, behaviour instrumental conditioning, trial and error learning and insight are discussed in the literature. They are explained in fig. 3. These last three kinds are linked with self-acting, spontaneous movements driven by actual needs or motives. They predominate in higher animals, especially mammals. But it is impossible to attach different evolutionary levels to these three different kinds of learning levels.

The diversity of evolutionary courses produces a large variety of adaptive phenomena which are interlinked with specific needs and innate behavioural patterns. Our goldfish-rhesus example reminds us that behind superficially similar performance there may be quite different capabilities. Instrumental conditioning and trial and error learning seem to be very similar. A given motive O_v produces behaviour patterns A_{ij} of which one reduces the given need by $A_i(S_i \rightarrow S_j) \Rightarrow (O^- \rightarrow O^+)$. This effect is fixed in the memory and is applicable under similar internal and external cirumstances. But there is also the well-known phenomena of alternative choice reactions combined with a delayed response situation (Fischel [2]; Hunt [5]). A dog looks back during the baiting of two food places (1 and 2). Then the animal is turned around itself so that it looses its orientation. Food is „in about" one direction. The animal starts and finds place 1 empty. Instead of being punished, immediately after this experience the dog turns to place 2. Place 1 was not at all reinforcing itself, rather it bore the information that place 2 should be the remaining baited place. A non rewarded decision favours an alternative. Such a procedure is exactly what happens during hypothesis-checking. The decision itself is analogous to what we may call an uncounscious inference. H. v. HELMHOLTZ was the first who drew such a conclusion from learning dependent behaviour. The evolutionary power of such a learning procedure is obvious: hypotheses-driven decision-making is more effective because it uses preconfirmed information as knowledge from memory. The learning expenditure is reduced, the procedure is shortened and achieves the same goal more effectively than the more stupid instrumental conditioning. This links learning behaviour with the phenomenon of animal intelligence. It is related to the effectiveness in the use of individual memory contents (i.e. learning results) for new adaptivity. Hypothesis-driven learning bears the transition to the different kinds of cognitive learning and reasoning. I am able only to outline some general characteristics of this type of learning behaviour.

Cognitive learning is not only the use of memory contents for decisions in learning situations. It includes inference mechanisms in a definite sense, namely in deriving implicitly stored knowledge. The early modes of inferential (or cognitive) learning can be found among pre-human primates. I would like to demonstrate the main point by two examples.

D. J. GILLAN [3] demonstrated in his experiments with chimpanzees that they (1) are in a position to draw something like transitive inferences and (2) that they are able to perform some kind of deductive reasoning in reaching their behaviour decisions. Referring to (1) the author rewarded the pairs

$$F + E -, \quad E + D -, \quad D + C -, \quad C + B -, \quad B + A -$$

in the sense of the signs + or — (+ meaning reward, and — meaning no reward). In test trials the animals receive the *EB* as well as the *EC* condition. They choose *E* correctly with 88% under the first and 83% under the second condition. This is just the same performance as we bring when we draw from $A > B$, $B > C$ the conclusion $A > C$. It is the derivation of implicitly stored knowledge by inference.

Referring now to (2): a kind of deductive reasoning was recognized by GILLAN in the following context: Behind a wall, there are two hiding places, invisible to the animals, for food, say *p* and *q* (see fig. 4). Two young chimpanzees, say *A* and *B*, observe how food seems to be carried to these places. *A* is allowed to go in one direction. It prefers direction *p*. Having seen this direction, chimpanzee *B* is removed from the scene and is not returned until *A* is coming back with a fruit in its hand. If allowed, *B* immediately starts in the direction of *q*. The cognitive background of such a behaviour decision is like a disjunctive syllogism:

$$(p \vee q) \rightarrow (q \rightarrow p) .$$

It is use of knowledge, but in a revised, indirect mode. Inferences allow implicitly stored knowledge to be derived. Cognitive procedures allow the right decision to be predicted and derived without any overt false trial. It is exactly what Wolfgang KÖHLER denominated "insight learning". Insight is not entirely mediated by perceptual information but based rather on mental or cognitive operations which are applied to stored knowledge. This kind of learning is not reactive: it is not primarily guided by reward steps, but is instead generative in nature. The generative power comes from the recognition of rules, for which inductive or deductive reasoning are two examples. Construction of problem solving strategies, top-down or bottom-up techniques of

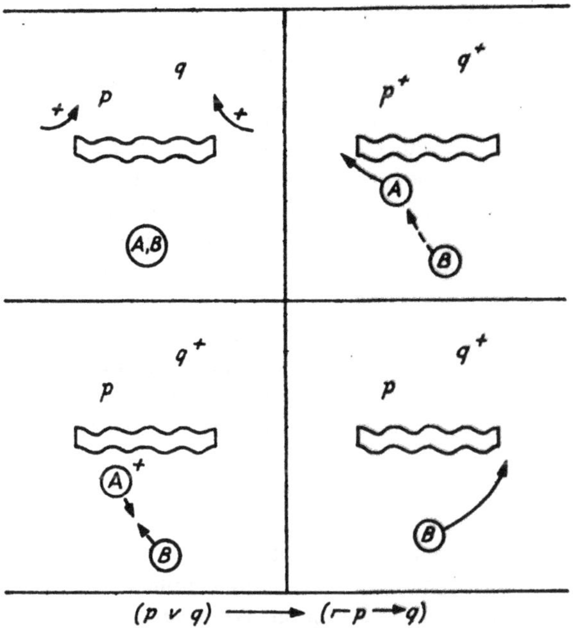

Fig. 4. Experiment with chimpanzees, performed by D. Y. GILLAN. *p,q*: hided baiting places; *A,B*: the animals; further description see text.

information processing, heuristic techniques — all of them are examples for the learning-dependent use of procedural knowledge.

We do not know definitely how procedural knowledge is acquired. We do know that there are mutual influences between it and language development in the human being. Pertinent investigations reveal that local reward of sentence understanding is by far insufficient to explain the development of language comprehension and generation. The derivation of completely new composite nouns, the use of individual morphological derivations and the transfer of semantic structures from one area of knowledge to another indicates that the acquisition of rules for both processing and producing information is the crucial point in understanding the cognitive development of the human being.

What we more definitely know is what and how cognitive procedures are used for the construction of abstract concepts or abstract knowledge in general. With the help of cognitive procedures we are also able to construct the meta-level in our cognitive world, for instance, how to establish knowledge about our knowledge, mental skills or beliefs.

In my book on the origin of human intelligence (Erwachendes Denken) I have demonstrated that the different historical figures or number systems reveal some information on the function modes of procedural knowledge. They can be used for a taxonomy of cognitive functions. In the book I gave reasons showing why these results of learning-dependent performance are not reducible to evolutionary laws. Basic motives are derivable from social contexts and different modes in the division of labour that came into being with the development of human society. Nevertheless it is amusing to look for a moment at the psychobiological value of the more abstract kinds of human thinking and the acquisition of its accomplishments. As the biotope diversifies, evolution again places increasing constraints on adaptive patterns in the given environments. Increasing specialisation for appropriate adaptivity is the consequence. The formation of intelligence as a result of (pre-dominantly) learning capability creats kinds of adaptivity which are not restricted to special kinds of biotops. Just look at our two examples: The employment of transitive relations in the information offer is independent of any special life space. Deductive reasoning remains valid in all cases, such as the detection of causality. With the cognitive background of learning an increasing independence of any specific environment comes into being. (The active change of environmental properties according to the own need is another consequence.)

The results of cognitive learning in comparison with those of conditioning reminds me to the difference between natural and computer languages. Computer languages are precisely differentiated and adaptable to a well defined, closed class of problem requirements (taken as a well-defined environment). Computer languages are optimal for that purpose. Natural languages are not so precise, they are more fuzzy, but optimal with regard to the infinite requirements of a non-specific environment. The analogy holds for learning: cognitive learning is nonspecialized adaptive behaviour. It is thereby applicable to an infinite variety of different environmental conditions.

We had asked if learning processes are derivable from any unique principle such as conditioning or if qualitative changes took place during evolution. We have to summarize that there are qualitative jumps in the functional principles of learning, even among animals. As to how such jumps might originate, this, I believe, is the subject of applied systems theory in biology. System theory knows solutions to that problem.

I would like to restrict myself in this context to the statement that there are different levels of learning capability: one linked to the conditioning principle (habituation, conditioned reaction, operant or instrumental conditioning might be subclasses) and another is linked to cognitive procedures in the use of internal information (hypotheses-driven behaviour, reasoning or inferences), and constructive rules for behaviour patterns (such as heuristic or analogue reasoning) might be subclasses (fig. 5).

Between these kinds of learning there are probably differences in quality. It remains to be explained how such quality changes may occur. Probably not as macromutations. I mentioned systemtheoretical solutions in that context.

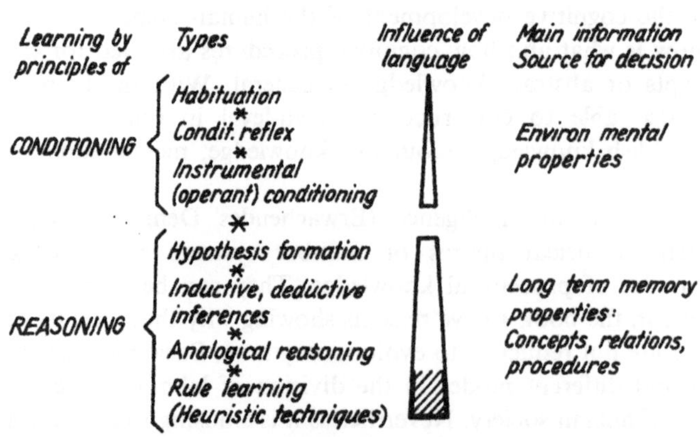

Fig. 5. Main differences of learning capabilities: learning by conditioning and learning by reasoning.

Altogether I have indicated what and how evolutionary factors, detected by Charles DARWIN, can help to understand the growth of learning processes in evolution and the qualitative changes in learning capabilities as well.

References

[1] BITTERMANN, M. E.: The Comparative Analysis of Learning. Science **188** (1977), 699—709.
[2] FISCHEL, W.: Psyche und Leistung der Tiere. Berlin (1938).
[3] GILLAN, D. J.: Ascent of Apes. Human and Animal Mind. 25th Dahlem Workshop (1981).
[4] HARLOW, H. F.: In: KOCH, S., Psychology: A Study of a Science. New York (1959) Vol. **2**, 531—581.
[5] HUNT, J. mc V.: Personality and the Behaviour Disorders, New York (1944).
[6] KENDAL, E. R.: Cellular Basis of Behaviour. An Introduction to Behavioural Neurobiology. New York (1976).
[7] PAVLOV, I. P.: Gesammelte Werke, Bd. 3, Berlin (1953).
[8] SCHNEIRLA, T. C.: Selected writings. Ed. by ARONSON, L. R. et al. New York (1972).
[9] SKINNER, B. F.: In: KOCH, S., Psychology: A Study of a Science, Vol. 2 New York (1959).
[10] THORNDIKE, E. L.: Animal Intelligence. New York (1911).
[11] THORPE, W. H.: Learning and Instinct in Animals. London (1956).
[12] VORONIN, L. G.: Psychol. Bulletin **59** (1962).

Discussion

GHILAROV: Are there any data on the ability to learn in invertebrates other than higher insects and Cephalopods?

KLIX: As far as I know we have all modes of learning until instrumental conditioning in invertebrates (BITTERMANN, 1975). I assume that the share of higher levels depends on the movability of the animals in their life-space (i.e. in Spongiae lower levels should more dominate than in Arthropodae or especially in Cephalopodae). Whether there is trial and error learning in the sense I have defined it here is not definitely clear. I have some doubts but cannot exclude that it exists.

GHILAROV: Is it possible to discriminate conditioned reflexes and learning (of reasoning type) in invertebrates?

KLIX: I think there are some learning capabilities in invertebrates which remind us to reasoning, e.g. the calculation of stimuli probabilities. It might be that there are the precursors of the final (and higher) type. A machinery of calculating different portions is a possible performance of a few thousands of nerve cells. The gap to the inference-like reasoning type is due to a decidability between different modes of calculating proportions. I think such a type of decision in learning is beside the capability of invertebrates (see SCHNEIRIA, 1972).

HÖRZ, H.: In the summary is written "analogues (or homologous?)". Can you give an explanation of the difference between analogous and homologous?

KLIX: These terms are meant in the strict biological sense: We have similarities between pre-human primates and human being with regard to some typs of reasoning. F. I. GILLAN detected some essential properties of analogeous reasoning in chimpanzees. The point is as follows: is this type of analogeous reasoning the precursor of human analogical reasoning or is there only an overt or superficial similarity.

NOVAK: Earlier the conditioned reflexes have been viewed as a higher form of adaptation than the innate reflexes. Now it appears that already in Protozoons the nerve paths are capable of both types of answer principally. The innate (unconditioned) reflexes developed in the groups like insects and most invertebrates, where the small size of the body and limited numbers of nerve cells does not favour a complex conditioning and the short lifespan does not give time for learning. It is also not so much need for it so as the specific micro-environment in the short period of life is rather constant. In vertebrates, on the other hand, both preconditions and needs favour the development of conditioned reflexes and the evolution of learning of the basis of natural selection.

There are cases, however, where also in invertebrates learning is useful for survival, like in the orientation of the honey bee. And here we see that it may reach even a very high level like in the homing bee at their first leaving the hive.

KLIX: Dr. NOVAK, I agree with your statement.

N. JUNG: One remark on the relation between different learning types and the drives or motivations in the sense of ethological definitions. Is it not true that not only environmental complexity causes the development of a new learning type in evolution but that also the main, or dominant, biotope elements determine its occurrence in connection with a drive. Caged quails (gallinaceous birds) for instance, do not learn that the characteristic flight reaction when a flight enemy (Flugfeind) appears, is by starting vertically rapidly because the cage is closed by glass. I.e. adaptation to a certain biotope structure develops only the particular learning mechanism needed to reach the target of a drive. It is necessary to ascertain not only whether a certain learning type exists in a species or not but also in connection with which environment-dependent drives or urges it has evolved. By the way, in oligophrenic patients in psychiatric practice one can see, that cognitive learning also only takes place if a drive is activated or released. If not, no "learning motivation" arises, and there is no learning. Including the ideas of TEMBROCK in the discussion, namely that behavioural mechanisms in direct contact with their targets (= objects) evolved in early evolutionary stages and that in later stages the need for distance-bridging (animal-target) behaviour arose to an increasing extent (but still at the level of instinctive behaviour) this leads me to the suggestion that dependence on intra- and interspecific competition became a very important cause for the evolution of new genetically fixed behavioural mechanisms, i.e. firstly drives, from which there arose the need to evolve further learning mechanisms in order to adapt to short-time changes in the environment. And as for the complexity of the biotope, MCARTHUR has shown that increased complexity may result in greater diversity of species in birds.

Sektion Philosophie, Humboldt-Universität, Zentralinstitut für Philosophie der Akademie der Wissenschaften der DDR, Berlin

Man as a Biosocial Entity From a Philosophical View

H. E. HÖRZ and H. HÖRZ

In our contribution to the discussion we shall formulate a few theses on the extensive debates regarding man as a biosocial entity. We should like to characterise a few standpoints in this respect, to draw attention to two problems and to present a brief summary.

I. Viewpoints

1. The characteristic of man as a biosocial entity is rather a heuristic research program than a complex synthesis of research results obtained by individual scientific disciplines on the natural and social factors governing human behaviour patterns. In this context, there is a real danger of one-sidedly overemphasizing natural factors by taking a biologistic and social factors by taking a vulgar sociologistic, standpoint. These biasses can be precluded by considering the natural and social factors as forming a unity. On the other hand, mere summation of the separately investigated factors can lead to eclecticism. It is therefore necessary to investigate the essence of Man as an ensemble of concrete historical conditions in his unity of the natural and social, emotional and rational, conscious, subconscious and unconscious factors that have influenced his individual development.

2. The notion of man as a social being expresses a new quality in biotic evolution. This new quality has various aspects. It includes the development of productive forces and including the role of man as a creative designer and controller of technological processes. Technologies are means by which man impresses his will upon nature, society and his own behaviour. They are based on his insight into the objective relationships and laws of nature, society and conscience. But man differs from animals not only by his increasing mastery over his natural and social environment, and over his own behaviour, by means of technology, but also by the fact that he is able to comprehend the determiniants and laws governing his own cognition and action. The conscious design and construction of the environment characterizes man's gain in freedom. Therefore his new quality as a social being must be considered in the dialectical unity of the effective results of his actions and humanity. Real humanism is the theoretical and practical fashioning of social relationships by using advanced productive forces to attain the maximum possible gain in freedom for freely associated personalities. Hence, humanism is a target function, an evaluation criterion and a strategy of requirements to be satisfied by our activities.

3. Laws of nature are modified in human behaviour by the new quality of Man as a

social being. Under social conditions, possibilities are realized or put into practice that are indeed contained in the field of possibilities given by the laws of nature, but would otherwise have only a very low probability of being realized. This is shown by the breeding of plants and animals, but also in the role of psychic factors in fashioning individual behaviour. The statistical structure of social laws must be regarded as a unity formed by the tendency in the system's behaviour and by the field of possibilities which are inherent in the patterns of behaviour displayed by groups and individuals but are realized in a conditionally accidental way according to a probability distribution. Social laws have natural conditions of existence which are investigated as laws by the natural sciences themselves. This implies the genetic-biotic genesis of social behaviour, the natural conditions of social existence, such as the natural environment, nutrition energy and raw materials, and the natural foundations of man with genetic and biotic predispositions. These different factors must be analyzed in order to obtain a synthesis.

4. Man is a concrete, historically determined unity of individual productive forces, which comprise the aptitudes and skills he uses to master nature, and personality that employs his rich relationships with nature and society for the active humane practical and theoretical fashioning of reality by acquiring the material and cultural values of Society. Social values are meaningful relationships between entities and mankind and comprise usefulness, morality and beauty. The gain in freedom is at the top of the hierarchy of values. This gives rise to specific humane objectives in the conscious realization of genetic-biotic possibilities which are certainly more comprehensive than those that have been put into practice to date and which are self extending.

II. Problems

1. In the debates with vulgar sociologism, an emphasis is put on the significance of genetic-biotic factors for human behaviour. One even speaks of the rediscovery of the inequality of man [1]. Ample evidence and important detailed considerations have been put forward as proof of the genetic inequality of human beings and of their individuality. These natural differences between human beings are manifested in intelligence, social behaviour and sex, and have, it is argued, hitherto, been concealed by theories of equality and associated actions in history. This allegedly led to demands for general education and to a general striving for affluence. Evolution has given rise to the problem of satisfying the unequal needs of individuals who are unequal in genetic, biotic and social terms. This problem cannot be solved by considering inequality as a fact of nature. The solution must be a social solution based on high levels in the production of material goods, on the conscious and human fashioning of scientific-technological progress and on influences affecting the development of needs. Thus when considering education strategies it is necessary to take into account the existence of two fundamental types of personality: the deductive-logical type who acquires theoretical fundamental knowledge regarding the relationships and laws governing objective reality, and of mathematically formulated theories and how they are illustrated in selected examples, and the inductive-empirical type who utilizes his own experience in the objective struggle with reality to understand theories. Hence, it is necessary to take into account genetic and biotic predisposition, physiological-psychic factors, and the prenatal determinants of individual development, when drawing up education strategies. Furthermore, it is necessary to take into account tendencies inherent

in the development of needs and the possibilities for consciously moulding some of them by propagating outstanding examples that constitute specific group-related or age-related demands on the personality resulting from existing norms and ideals.

2. E. O. Wilson's contention in his book "On human nature" that Marxism is sociobiology without biology and is being vitally threatened by discoveries in human sociobiology is wrong [2]. In the Marxist-Leninist understanding of society, the natural and social differences between individuals play a decisive role in determining the communist principle of distribution according to the different typical needs. Although there were, and are, on-sided attitudes among Marxist theorists, there can be no doubt regarding the justification of the basic idea to investigate the genetic, biotic and physiological foundations of human existence and the natural evolution of social behaviour in biotic populations as preliminary forms for the social behaviour of human beings. Sociology has detected an ecological niche in scientific research, of which many scientists have become increasingly clearly aware in the second half of our century — we need only think of the upsurge in the behavioural sciences. In this context, the exploration of the natural prehistory of human social behaviour, of genetic-biotic predisposition in the individual interpretation of reality by man and of the natural conditions for human existence can easily lead to philosophical reductionism. This danger does not affect from the validity of well-conceived scientific reductions of social behaviour to its natural evolutionary or specific social determinants for research purposes: an overall picture can be obtained only with the help of detailed investigations. In other words, sociology has an important contribution to make to the investigation of social behaviour patterns by studying the evolution of their natural foundations. Philosophical reductionism occurs only if the scientific cognition of the reduced phenomenon, and this of the genetic and biotic predispositions for social behaviour, is used to explain the entirety, namely human behaviour strategies in cultural and personal development. However, today sociobiology is rather a heuristic research program than a formulated theory.

III. Summary

It is possible to summarize the previous remarks to a certain extent. Firstly, to ignore the natural foundations of human existence means to go back to the level of utopian social theories of former times. The insights provided by the biosciences have made it essential to take into account the natural inequality of individuals realized, however, as a phantastic vision of the future, the construction of man from his genetic and biotic foundations is discussed, to take into account the existence of objective chance and the relationships between the behaviour of elements and system laws. In such considerations, the new quality, i.e. the social character, of man ignored. Laws take effect in accidented events. Therefore, humane viewpoints must prevail in experiments with and on humans, especially if they involve intervention in the genetic material of man. In this context the supreme criterion must be the integrity of the personality, i.e. risks must be minimized, the social benefit must be established, the freedom of decision of the people concerned must be ensured, and those performing the experiment must be fully aware of their responsibility.

Secondly, investigating social behaviour patterns within the fabric of their determinants it is possible to ignore the genetic and biotic basis, if this does not substantially influence the variability of the behaviour strategies, but is a relatively constant basis for social behaviour. A gain in freedom requires strategies based on knowledge of the objective field of possibilities. The essential conditions for realising these possibilities are mostly of a social nature. But it would be wrong to neglect the natural foundations of the field of possibilities even if this realisation requires different behaviour strategies. Thus, for instance, the causes of wars cannot be deduced from the biotically conditioned behaviour patterns of the individual. In contrast, education strategies must take into account the natural basis of human behaviour patterns in certain types of personality.

Third, the idea of development must not only be applied to biotic evolution. The development concept is also essential to social theory. The relationship between man and nature develops under concrete historical conditions. Thus, it is much more difficult to grasp the difference between the animal and man in primeval society than in today's highly developed human society with its productive forces based upon automation, data processing and program control. Only the termination of the preliminary history of mankind and the development of human society without class divisions as a human expression of relationships between people provided with a highly developed material and technical basis can make it possible to compensate to a large degree for the typical natural inequalities.

References

[1] C. D. DARLINGTON: Die Wiederentdeckung der Ungleichheit. Frankfurt a. M. (1980).
[2] E. O. WILSON: On human nature. New York (1978).

Medizinische Akademie Magdeburg

Some Aspects of the Mutual Relationship Between Biotic and Social Needs

W. STANGE

Karl MARX described needs as "the offspring of social production and intercourse . . ."

The investigation of needs reveals important relationships between the factors involved in human evolution. Some authors assume that they are the "motor of motivation" (for instance A. N. LEONTJEW [6]) or "fundamental stimulant of activity" (for instance G. DILIGENSKI [2]). It has generally been the social aspects of human needs that were investigated, and the biotic prerequisites of needs have received only scant attention. According to Friedhart KLIX [5] we must find out which factors induce needs. He suggests that a distinction be made between biological and social influences on human perception, recognition, decision and action. These are the factors involved in the origin of needs. As L. SEVE remarked, personality is not merely a social extension of biological individuality. Human needs developed in connection with the evolution of man. They are not fixed and unchangable regulators of our life. It is necessary to consider the relationship between biotic and social needs at two levels:

— firstly, the historical level

and

— secondly, the present level.

The first level starts with the precursors of human needs. DARWIN [1] stated that animals have needs and that they have not only the ability to love but also the need to be loved. It was a long road from the needs of animals to needs of a new quality — human needs — and the way involved many steps. Consequently various hypothesis and theories have been formulated to explain them.

I wish to illustrate this by a brief look at psychology: My first example was taken from H. A. MURRAY [9]:

Needs	
Abasement	Harmavoidance+
Achievement	Infavoidance
Affiliation	Nurturance
Aggression	Order
Autonomy	Play
Counteraction	Rejection
Defendance	Sentience+
Deference	Sex
Dominance	Succorance
Exhibitation	Understanding+

(only three needs (+) do not require immediate social relations).

The second example is from A. MASLOW [8] (who became president of American Psychological Association in 1967):

1. basic needs (hunger, security ...) = deficiency needs
2. meta needs (justice, goodness ...) = growth needs.

These interpretations regard the relationships between biotic and social needs only from a contemplative standpoint. But both underestimate human activity! Speaking in terms of H. HÖRZ' conception of development, we can say that from the standpoint of dialectical and historical materialism we have to take into account the following characteristics of needs:

— objective determination (biotic and social)
— historical character (development)
— reflection of needs in consciousness

A philosophical scrutiny of specific sciences from this standpoint indicates that the results are interesting. For example, G. TEMBROCK [10] developed a matrix of needs of a hypothetical character (in the author's opinion). He gives us an idea of dialectical interrelationships between biotic and social needs. That is an important step in the progress of science and very interesting for the development of historical materialism.

Summary

It is necessary to obtain detailed knowledge of how biotic and social needs are related. This means that we have to delve deeper into the evolution of needs.

References

[1] Ch. DARWIN: Die Abstammung des Menschen und die geschlechtliche Zuchtwahl, Leipzig o. J. I u. II.
[2] G. DILIGENSKI: Probleme der Theorie der menschlichen Bedürfnisse, Sowjetwissenschaft/Gesellschaftswissenschaftliche Beiträge **30** (1977), 256—272.
[3] R. EISLER: Wörterbuch der philosophischen Begriffe, Berlin 1927.
[4] H. HÖRZ and U. RÖSEBERG: Materialistische Dialektik in der physikalischen und biologischen Erkenntnis, Berlin (1981).
[5] F. KLIX: Erwachendes Denken, Berlin (1980).
[6] A. N. LEONTJEW: Tätigkeit, Bewußtsein, Persönlichkeit, Berlin (1979).
[7] K. MARX: Grundrisse der Kritik der Politischen Ökonomie (Rohentwurf), Berlin (1953).
[8] A. MASLOW: Toward a psychology of being, New York (1968).
[9] H. A. MURRAY: Explorations in personality. A clinical and experimental study of 50 men of college age, New York (1963).
[10] G. TEMBROCK: Moderne Verhaltensforschung, In: Wissen aus erster Hand, Berlin (1981), p. 63—79.

Concluding remarks

M. S. GHILAROV

I thank the organizers for the great honour and the privilege to make some concluding remarks.

The present Colloquium was both representative and instructive. Various aspects of biology were discussed here from various view-points showing that the immortal Charles DARWIN's thoughts are nowadays still actual both for the development of natural sciences and for philosophy, for the materialistic „Weltanschauung". This actuality to day, more than a century long, of DARWIN's concepts is connected with the correct philosophical, methodological, dialectic materialistic background of amount of facts. We have met here representatives of various countries, of different specialities, and I dare say of some different ideologies too.

The facts of the wide distribution of the creationism and of so called "scientific creationism" in the U.S.A., so brilliantly demonstrated by Dr. RUSE, show that the struggle for DARWINism has not lost its significance in our time too. During our Colloquium there was shown an old film, a very naive film on some features of the famous Austrian zoologist Paul KAMMERER. KAMMERER's position was wrong, but he was a scholar, one of those, whose mistakes promoted the general development of biological sciences, whereas creationism, especially "scientific" one, is an abdication from science.

But let me return to the development of Darwinism at our time, to "DARWIN — today". A new wave of the vivid interest to the problems of Darwinism began in the thirties-fourties of our century. It was connected with the "new systematics" — with population species concepts versus typological one, with so called synthetic evolution. For a time it seemed that evolutionary biology has solved its problems. But new problems always arise. The problem of criteria of the species remains partly unsolved — the biological concept of species by Ernst MAYR and other representatives of "new systematics" encounters great difficulties with regard to agamospecies, parthenospecies, hybrids etc. The transfer of genes from viruses to the bacterial cells becomes a new complication showing new possibilities of reticular evolution, whereas interspecies hybridization gives its shock to the concepts of strictly monophyletic phylogenesis.

From another view-point Willy HENNIG's cladistic concepts of the monophyletic development of organic world have destroyed the previousely established and accepted understanding and evaluation of higher taxa, based on gradistic premises.

New achievements of molecular biology and biochemistry provide new criteria to evaluate previous phylogenetic conclusions. Experiments with the hybridization of DNA, presented at this Colloquium by Dr. A. S. ANTONOV, prove the perspectiveness of this new approach.

Appreciation of the "selfish gene" as a unit of selection tries to change our modern concepts of the whole organism as such an unit. The fact that natural selection deals with phenotype, being the subject of selection and elimination, whereas as the result genotypes are selected, complicates the scheme of genotype survival.

The dialectic unity of casuality and regularity makes it necessary to take into the account the possibility to find the trends of evolution. Feed-backs realized in correlations (including ecological correlations) allow us to reveal directions of the phylogenesis and to comprehend the facts of the acceleration of evolutionary process.

After DARWIN interest was drawn to the origin of the species, to the splitting of species, to general problems of divergence, of radiation, to phylogenetic trees. Now the interest becomes evident to the regularities of the evolutionary process, solved on the base of studies of convergence. As to phylogenetic trees the attention to a problem rises when new ideas appear. One of stimuli for reviving interest to phylogenetic investigations was W. HENNIG's concept of cladistic strictly dichotomic mode of speciation and of forming higher taxonomic groups. "Gradistic" approaches by E. MAYR and "cladistic" ones by W. HENNIG are now broadly discussed by both taxonomists and evolutionists.

The proposal to use mathematical methods for solving taxonomic and phylogenetic problems made by R. SOKAL and others regarding all characters as equipotential, ("numerical systematics") or by E. S. SMIRNOV asserting the hierarchy of characters ("taxonomic analysis") help to correct already proposed systems but are insufficient for their original construction.

The use of data of branches of biology other than morphology (biochemistry, physiology, ecology, ethology) enlarges the possibilities to correct our knowledge and hypotheses on the phylogenetic relations.

E.g. the "hybridization" of DNA represents such a method which promises to help in solving complicate taxonomic problems, especially those concerning higher taxa.

Both the analysis of many existing species of polyploid fishes, lizards, earthworms, potworms etc. and experimental obtaining of silkworm polyploids, and their crossing prove nowadays the possibility of netlike evolution shaking in some respects the cardinal concept of monophyletic patterns of phylogenesis.

Experimentally it becomes possible to transfer the gene of human interferon into *Escherichia coli* or of human insulin into the bacterial cell what allows us to synthesize organisms with hereditary characters acquired from different phyla of orgnisms. For the practical human activity, especially for the agriculture approaches it is of interest to prognosticate, to forecast, which underivable species (pests, weeds, pathogenes) can adjust themselves to cultivated crops, become of negative economic importance. This is possible on the base of knowledge of those main characters according to which the natural selection propeeds in the course of the formation of secondary biocenoses, of agrocenoses. The comprehension of regularities of natural selection of organisms, of biocenoses in the environment changed by mans' activity becomes to be of the primary importance in the recent period of biosphere transition to noosphere, using V. I. VERNADSKY's terminology. The hereditary variability of organisms is the premise of DARWINian concept of evolution. Recently enormous anthropogenous changes in our environment increase its mutagenicity. Studies of these changes, of the degree of mutagenicity is one of urgent problems of theoretical and practical investigations, DARWINists have to do.

At this Colloquium problems connected with anthropogenesis and biosocial evolution were discussed too, from various viewpoints and using data of various branches of science.

The discussion was vivid and fruitful and proved that further common efforts of sociologists, anthropologists, archeologists, psychologists, primatologists etc. are quite indispensable for the productive work on these problems which are tremendously interesting for the humanity. Memory, collective memory, learning, ethics, many problems are to be elucidating both from biological and sociological points.

At the present colloquium all problems were discussed in light of the dialectic materialism — a methodology spontaneously and fruitfully used by DARWIN and now consciously applied by us. I would point out the significance of DARWIN's pioneer work on biological factors of soil formation, all significance of which has become clear only in the middle of our century. And I am glad to stress that the first international magazine on soil biology — "Pedobiologia" has been organized just in GDR in 1960.

Now some words concerning this Colloquium itself. I am sure that all participants appreciate it as reproachless. We are highly grateful to the main organizers — Prof. E. GEISSLER, Prof. H. HÖRZ, Dr. M. KOCH. Studying soil animals I know what a great work is done by hidden living organisms — so we have to thank all those organizers who remain unknow for us, as well as those whom we do know for their great successive efforts in preparing and realization of this Colloquium.

Summarizing I dare say: well done! And using the title of SHAKESPEARE's play it is possible to conclude that "End well — everything well".

We all wish successes to the next Kühlungsborn colloquium.

Autoren und Diskussionsteilnehmer

ANTONOV, A. S., Prof. Dr., A. N. Belozersky Lab. of Molecular Biology and Bioorg. Chemistry, Moscow State University, Moscow 117234

BRETSCHNEIDER, J., Dr., Friedrich Schiller-Universität, Sektion Biologie, 6900 Jena, Am Steiger 3

BRODA, E., Prof. Dr., Institut für Physikalische Chemie der Universität Wien, IX, Währinger Str. 42, A-1090 Wien

COLP, R. jr., Dr., 993 Park Avenue, New York, New York 10028, USA

DAREVSKI, I. S., Prof. Dr., Zoological Institute, Academy of Science of the USSR, 199164 Leningrad, UdSSR

EBELING, W., Prof. Dr., Sektion Physik, Bereich 04, der Humboldt-Universität, 1086 Berlin, PSF 1297

FÖRSTER, I., Dr., Institut für Marxismus-Leninismus der Medizinischen Akademie Magdeburg, 3010 Magdeburg, Leipziger Str. 44

FROLOV, I. T., Prof. Dr., Scientific Council for Philosophical and Social Problems of Science and Technology, Academy of Sciences of the USSR, 29, Ryleyev street, Moscow 119034

GAUTIER, C., Dr., Equipe Evolution Moleculaire, Laboratoire de Biometrie, Université Lyon 1, F-69622 Villeurbanne Cedex

GEISSLER, E., Prof. Dr., Zentralinstitut für Molekularbiologie, Abt. Virologie, der AdW der DDR, 1115 Berlin-Buch, Robert-Rössle-Str. 10

GHILAROV, M. S., Prof. Dr., Comité National des Biologistes de l'Union Soviétique, 33, Av. Lénine, Moscou V-71, URSS

GOUY, M., Dr., Equipe Evolution Moleculaire, Laboratoire de Biometrie, Université Lyon 1, F-69622 Villeurbanne Cedex

GRANTHAM, R., Prof. Dr., Equipe Evolution Moleculaire, Laboratoire de Biometrie, Université Lyon 1, F-69622 Villeurbanne Cedex

HAILA, Y., Dr., Department of Zoology, Division of Morphology and Ecology, University of Helsinki, P. Rautatiekatu 13, SF-00100 Helsinki 10

HERRMANN., J., Prof. Dr., Zentralinstitut für Alte Geschichte und Archäologie der AdW der DDR, 1080 Berlin, Leipziger Str. 3—4

HOFEMEISTER, J., Dr., Zentralinstitut für Genetik und Kulturpflanzenforschung der AdW 4325 Gatersleben

HOFFMANN, S., Dr., Martin-Luther-Universität, Sektion Chemie, 4020 Halle, Weinbergweg 16

HÖRZ, H., Prof. Dr., Zentralinstitut für Philosophie der AdW 1080 Berlin, Otto-Nuschke-Str. 10/11

HÖRZ, Helga E., Prof. Dr., Sektion Marxistisch-leninistische Philosophie der Humboldt-Universität, 1080 Berlin, Unter den Linden 6

JENSSEN, K. J., Prof. Dr., Institut für Filosofi, Aarhus Universitet, Bygning 328, 8000 Aarhus C, Denmark

JUNG, N., Fachkrankenhaus für Neurologie und Psychiatrie, 1113 Berlin, Herzbergstr. 73

JUNG, F., Prof. Dr., Zentralinstitut für Molekularbiologie, 1115 Berlin-Buch, Robert-Rössle-Str. 10

KEYNES, R. D., Prof. Dr., Physiological Laboratory, Downing Street, Cambridge. CB2 3EG, U.K.

KIRSCHKE, S., Doz. Dr., Sektion Marxistisch-leninistische Philosophie der Martin-Luther-Universität, 4010 Halle (Saale), Große Steinstraße 73

KLIX, F., Prof. Dr., Sektion Psychologie der Humboldt-Universität 1020 Berlin, Oranienburger Str. 18

KRÜGER, D., Dr., Humboldt-Universität, Institut für Virologie, 1040 Berlin, Schumannstr. 20/21

KUHN, H., Prof. Dr., Max-Planck-Institut für Biophysikalische Chemie, Postfach 968, D-3400 Göttingen

LEY, H., Prof. Dr., Humboldt-Universität, Bereich Philosophie, Sektion Marxismus-Leninismus, 1080 Berlin, Universitätsstr. 3b

LÖTHER, R., Prof. Dr., Akademie für ärztliche Fortbildung der DDR, Sektion Marxismus-Leninismus, Lehrstuhl Philosophie, 1134 Berlin, Nöldnerstr. 34—36

MENDE, W., Institut für Geographie und Geoökologie, Bereich Ökonomie und ökologische Systeme, 1199 Berlin, Rudower Chaussee 5

NAUMANN, F., 1600 Königswusterhausen, Wilhelm-Pieck-Str. 35

NOVAK, V., Prof. Dr., Department of Evolutionary Biology, Czechoslovak Academy of Sciences, Na Folimance 5, CS-120 00 Praha 2

PASTERNAK, G. Prof. Dr., Forschungszentrum für Molekularbiologie und Medizin der AdW der DDR, 1115 Berlin-Buch, Robert-Rössle-Str. 10

PETERS, G.; Prof. Dr., Museum für Naturkunde der Humboldt-Universität, 1040 Berlin, Invalidenstr. 43

PFEIFFER, A., Prof. Dr., 8020 Dresden, Tetschener Str. 32

RAPOPORT, S. M., Prof. Dr., Institut für Physiologische und Biologische Chemie der Humboldt-Universität, 1040 Berlin, Hessische Str. 3—4

ROSENTHAL, H. A., Prof. Dr., Humboldt-Universität, Bereich Medizin, 1040 Berlin, Schumannstr. 20/21

ROSENTHAL, S., Prof. Dr., Zentralinstitut für Molekularbiologie der AdW der DDR, Abt. Zelldifferenzierung, 1115 Berlin-Buch, Robert-Rössle-Str. 10

RUSE, M., Prof. Dr., University of Guelph, College of Arts, Department of Philosophy, Guelph, Ontario, Canada, N1G 2W1

SCHOLZ, G., Dr., Zentralinstitut für Genetik und Kulturpflanzenforschung der AdW, 4325 Gatersleben

SCHUSTER, P., Prof. Dr., Institut für Theoretische Chemie und Strahlenchemie der Universität Wien, Währinger Str. 17, A-1090 Wien

SENGLAUB, K., Prof. Dr., Museum für Naturkunde der Humboldt-Universität, 1040 Berlin, Invalidenstr. 43

STANGE, W., Doz. Dr., Institut für Marxismus-Leninismus der Medizinischen Akademie Magdeburg, 3010 Magdeburg, Leipziger Str. 44

STEINMÜLLER, K., Dr., 1058 Berlin, Lychener Str. 73

TEMBROCK, G., Prof. Dr., Sektion Biologie der Humboldt-Universität, Bereich Verhaltenswissenschaften, 1040 Berlin, Invalidenstr. 43

ULLRICH, H., Dr., Zentralinstitut für Alte Geschichte und Archäologie der AdW der DDR, 1080 Berlin, Leipziger Str. 3—4

URBANEK, A., Prof. Dr., Institut für Geologie der Universität Warschau, Warschau, VR Polen

VOGEL, F., Prof. Dr., Institut für Anthropologie und Humangenetik, Im Neuenheimer Feld 328, D-6900 Heidelberg

WENIG, K., Dr., Zentralinstitut für Philosophie der AdW der DDR, 1080 Berlin, Otto-Nuschke-Str. 10/11

YEFIMOV, Y. I., Dr., Institute of Animal Morphology, Academy of Sciences of the USSR, 33, Leninskij prospekt, Moscow V-71, USSR

ZIRNSTEIN, G., Dr., Karl-Sudhoff-Institut, 7010 Leipzig, Palstr. 33

YEFIMOV, Y. I., Dr., Lehrstuhl für Philosophie der Akademie der Wissenschaften der UdSSR, Nabereshnaja Makarowa 4, Leningrad 119164, UdSSR

ZIRNSTEIN. G., Dr., Karl-Sudhoff-Institut, 7010 Leipzig, Palstr. 33

Bei Fragen zur Produktsicherheit wenden Sie sich bitte an:
If you have any questions regarding product safety,
please contact:

Walter de Gruyter GmbH
Genthiner Straße 13
10785 Berlin
productsafety@degruyterbrill.com